W9-BZS-342

Praise for Gordon Hempton and his recordings

"Hempton brings new meaning to the word 'purist.' He meticulously scouts out sites for his soundscapes . . . is totally in touch with his environment as a listener, and . . . committed to bringing what he hears to a wider audience."

—*Smithsonian*

"Gordon Hempton is as rare as the thing he cherishes."

—*Los Angeles Times*

"Accurate and authentic as a tone poem from heaven."

—Vincent C. Siew, former Minister of Economic Affairs, Republic of China

"Instead of using instruments, [Hempton's recordings] open the ears to the sounds of the environment. I think it's beneficial for people to be made aware of this type of sound."

—John Cage, composer, as quoted in *People* magazine

"One of the most highly reputed and best-selling environmental recording artists in the world."

—allmusicguide.com

To Esther ~ to one of the people who have been so important for all of my life! I love audiologists. :) ♡ Rebecca

ONE SQUARE INCH OF SILENCE

One Man's Search

for Natural Silence in a Noisy World

To Esther and the world of listening in so many ways — Gordon

GORDON HEMPTON

and JOHN GROSSMANN

Free Press

New York London Toronto Sydney

FREE PRESS
A Division of Simon & Schuster, Inc.
1230 Avenue of the Americas
New York, NY 10020

First Free Press hardcover edition March 2009

FREE PRESS and colophon are trademarks of Simon & Schuster, Inc.

For information about special discounts for bulk purchases, please contact Simon & Schuster Special Sales at 1-800-456-6798 or business@simonandschuster.com

The Simon & Schuster Speakers Bureau can bring authors to your live event. For more information or to book an event contact the Simon & Schuster Speakers Bureau at 866-248-3049 or visit our website at www.simonspeakers.com

Manufactured in the United States of America

1 3 5 7 9 10 8 6 4 2

Library of Congress Cataloging-in-Publication Data
Hempton, Gordon.
One square inch of silence: one man's search for natural silence in a noisy world / Gordon Hempton and John Grossmann.—1st ed.
p. cm.
1. Hempton, Gordon—Anecdotes. 2. Ecologists—United States—Anecdotes.
3. Nature sounds—Recording and reproducing. I. Grossmann, John. II. Title.
QH31.H358A3 2009
333.78'2160973—dc22 2008036949

ISBN-13: 978-1-4165-5908-5
ISBN-10: 1-4165-5908-6

To every contributor to the Jar of Quiet Thoughts. Your heartfelt writings helped me realize I was not alone in my thirst for natural silence and helped me muster the courage to leave my more preferred reclusive existence and begin two journeys—crossing America and writing this book.

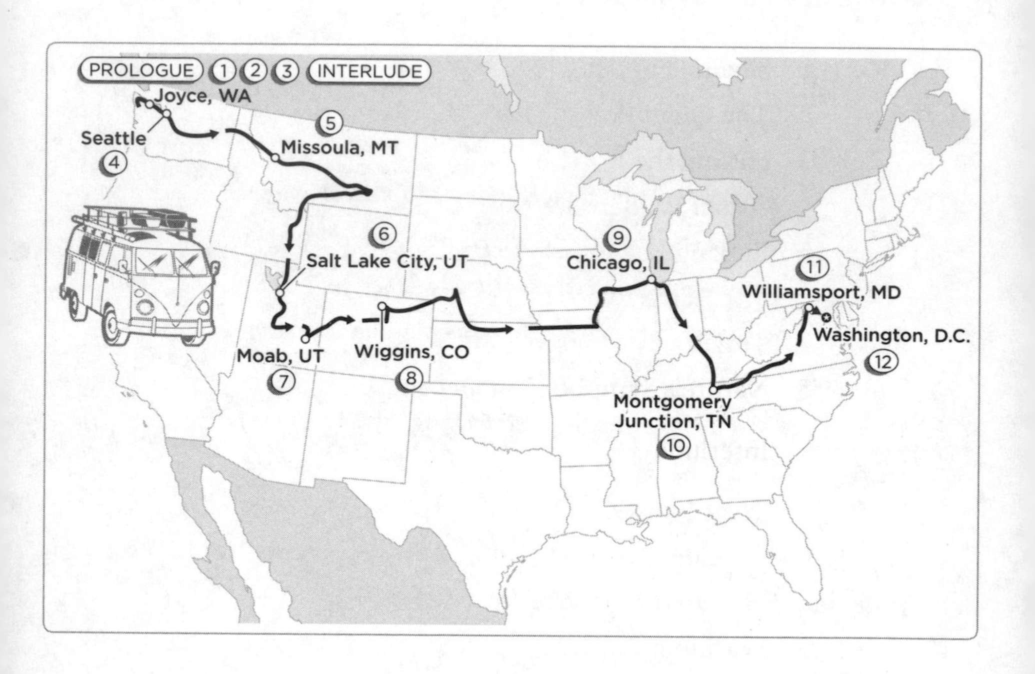
PROLOGUE
1
2
3
INTERLUDE
Joyce, WA
Seattle
4
5
Missoula, MT
6
Salt Lake City, UT
Moab, UT
7
Wiggins, CO
8
9
Chicago, IL
Montgomery
Junction, TN
10
11
Williamsport, MD
Washington, D.C.
12

Contents

ONE SQUARE INCH OF SILENCE

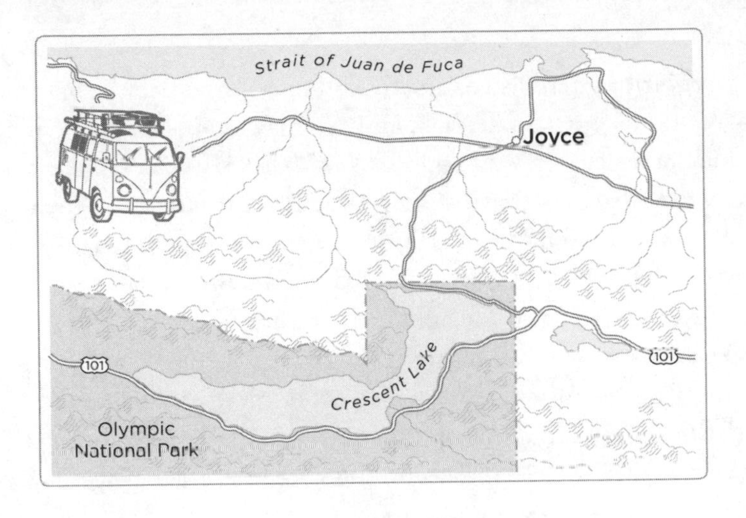

Prologue

Sounds of Silence

"The day will come when man will have to fight noise as inexorably as cholera and the plague." So said the Nobel Prize–winning bacteriologist Robert Koch in 1905. A century later, that day has drawn much nearer. Today silence has become an endangered species. Our cities, our suburbs, our farm communities, even our most expansive and remote national parks are not free from human noise intrusions. Nor is there relief even at the North Pole; continent-hopping jets see to that. Moreover, fighting noise is not the same as preserving silence. Our typical anti-noise strategies—earplugs, noise cancellation headphones, even noise abatement laws—offer no real solution because they do nothing to help us reconnect and listen to the land. And the land is speaking.

We've reached a time in human history when our global environmental crisis requires that we make permanent life-style changes. More than ever before, we need to fall back in love with the land. Silence is our meeting place.

It is our birthright to listen, quietly and undisturbed, to the natural environment and take whatever meanings we may. Long before the noises of mankind, there were only the sounds of the natural world. Our ears evolved perfectly tuned to hear these sounds—sounds that far exceed the range of human speech or even our most ambitious musical performances: a passing breeze that indicates a weather change, the first birdsongs of spring heralding a regreening of the land and a return to growth and prosperity, an approaching storm promising relief from a drought, and the shifting tide reminding us of the celestial ballet. All of these experiences connect us back to the land and to our evolutionary past.

One Square Inch of Silence is more than a book; it is a place in the Hoh Rain Forest, part of Olympic National Park—arguably the quietest place in the United States. But it, too, is endangered, protected only by a policy that is neither practiced by the National Park Service itself nor supported by adequate laws. My hope is that this book will trigger a quiet awakening in all those willing to become true listeners.

Preserving natural silence is as necessary and essential as species preservation, habitat restoration, toxic waste cleanup, and carbon dioxide reduction, to name but a few of the immediate challenges that confront us in this still young century. The good news is that rescuing silence can come much more easily than tackling these other problems. A single law would signal a huge and immediate improvement. That law would prohibit all aircraft from flying over our most pristine national parks.

Silence is not the absence of something but *the presence of everything.* It lives here, profoundly, at One Square Inch in the Hoh Rain Forest. It is the presence of time, undisturbed. It can be felt within the chest. Silence nurtures our nature, our human nature, and lets us know who we are. Left with a more receptive mind and a more attuned ear, we become better listeners not only to nature but to each other. Silence can be carried like embers from a fire. Silence can be found, and silence can find you. Silence can be lost and also recovered. But silence cannot be imagined, although most people think so. To experience the soul-swelling wonder of silence, you must hear it.

Silence is a sound, many, many sounds. I've heard more than I can count. Silence is the moonlit song of the coyote signing the air, and the answer of its mate. It is the falling whisper of snow that will later melt with an astonishing reggae rhythm so crisp that you will want to dance to it. It is the sound of pollinating winged insects vibrating soft tunes as they defensively dart in and out of the pine boughs to temporarily escape the breeze, a mix of insect hum and pine sigh that will stick with you all day. Silence is the passing flock of chestnut-backed chickadees and red-breasted nuthatches, chirping and fluttering, reminding you of your own curiosity.

Have you heard the rain lately? America's great northwest rain forest, no surprise, is an excellent place to listen. Here's what I've heard at One Square Inch of Silence. The first of the rainy season is not wet at all. Initially, countless seeds fall from the towering trees. This is soon followed by the soft applause of fluttering maple leaves, which settle oh so quietly as a winter blanket for the seeds. But this quiet concert is merely a prelude. When the first of many great rainstorms arrives, unleashing its mighty anthem, each species of tree makes its own sound in the wind and rain. Even the largest of the raindrops may never strike the ground. Nearly 300 feet overhead, high in the forest canopy, the leaves and bark absorb much of the moisture . . . until this aerial sponge becomes saturated and drops re-form and descend farther . . . striking lower branches and cascading onto sound-absorbing moss drapes . . . tapping on epiphytic ferns . . . faintly plopping on huckleberry bushes . . . and whacking the hard, firm salal leaves . . . before, finally, the drops inaudibly bend the delicate clover-like leaves of the wood sorrel and drip to leak into the ground. Heard day or night, this liquid ballet will continue for more than an hour after the actual rain ceases.

Recalling the warning of Robert Koch, developer of the scientific method that identifies the causes of disease, I believe the unchecked loss of silence is a canary in a coal mine—a global one. If we cannot make a stand here, if we turn a deaf ear to the issue of vanishing natural quiet, how can we expect to fare better with more complex environmental crises?

Gordon Hempton
—Snowed in at Joyce, Washington

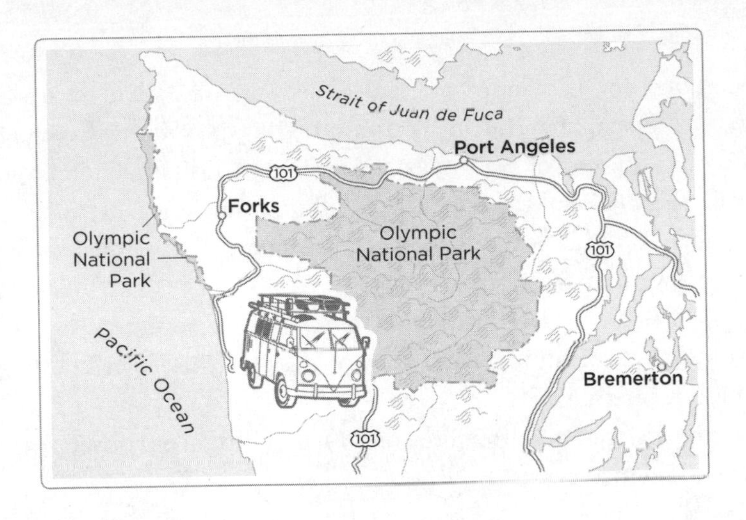

1 Silent Thunder

In this silent place, the noise is deafening.

—Kathleen Dean Moore, director of the Spring Creek Project, Oregon State University

It was a clear autumn night in 2003 when a *thumping* sound woke me from a deep sleep. As always, my bedroom window was wide open to give me the feeling of camping and to enable me to listen. I live in a rural town so quiet that it is possible to hear for miles. Out of the silence, I heard something new.

It was a drumming *thump-thump* sound that I took for pistons churning on a freighter or, perhaps, a new class of supertanker. The sound path had to be a distance of 10 to 15 miles, halfway across the Strait of Juan de Fuca, up Crescent Beach and the hill to my house on Washington's remote Olympic Peninsula. I take pride in living in a place this quiet.

I listen to the world—this is my job and my passion as an acoustic ecologist. I've recorded on every continent except Antarctica. My recordings are used in everything from video games and museum exhibits to nature albums, movie soundtracks, and educational products. More than 25 years of recordings in all manner of natural settings have swelled my sound library to 3,000 gigabytes. I've captured the flutter of butterfly wings, the thunderous booming of waterfalls, the jet-like swoosh of a bullet train, the wisp of a floating leaf, the passionate trill of a birdsong, the soft coo of a coyote pup. I'd rather listen than speak. Listening is a wordless process of receiving honest impressions.

My specialty is quiet, although I record other sounds, too. Quiet is almost too faint for human ears, but not quite, if you learn to listen carefully. I listen carefully.

The thumping of that freighter that October night surprised me. It felt closer than seemed possible.

If asked to choose my favorite sound in the world, I doubt that I could do that easily. If forced, I might say it's the dawn chorus of songbirds, the sound of the rising sun as it circles the globe. But that would disregard the murmur of winged insects as heard over many square miles in the Kalahari Desert, and if that were my favorite sound, that would ignore the hoot of an owl and the way it bounces off the cypress trees in Louisiana, and also ignore the clang of a church bell after it has echoed down the narrow stone streets of an Austrian village. If I had to supply a single answer to that question, my favorite sound in the world would be the sound of anticipation: the silence of a sound about to be heard, the space between the notes.

Thump-thump, thump-thump. Ten to 15 miles is a reasonable distance for deep, low frequencies under perfect atmospheric conditions, but there was something different about this sound, barely perceptible, and this is why I thought it came from a new class of ship. The last thing on my mind was that this thumping noise was the first sign of a hearing loss. Me, of all people, the Sound Tracker. It would have been like a mezzo-soprano wondering if she had nodes on her vocal cords or a painter suspecting he had muscular dystrophy.

But as the days turned to weeks and the weeks to months, there was no escaping the obvious. I could no longer practice my profession. My head was a cauldron of hums and buzzes and distorted sound. I could barely make sense of what people said to me. If more than one person was speaking in a room, the experience was so disorienting that I could only sit and watch others talk. Instead of words, I heard a strange sound similar to a faint AM radio playing country music from the other end of a long hallway;

the words blended together, making the experience unintelligible. I learned to limit my exposure to stressful activities, especially if loud sounds were present; they would only increase my internal raging cacophony and drive me nearly mad. I frequently had to ask my son and daughter to repeat what they'd said and speak slowly. Sentences became shorter, meanings shallower, life duller. I avoided being with people. I went into debt, lost clients, and hovered on the edge of financial and emotional ruin.

I received lots of advice. "You have surfer's ear. It's time for you to grow up and stop bodysurfing. If you stopped bodysurfing," my father said (often), "your problems would go away." The go-away theory: If you stop this or stop that the problem will go away. My hearing loss was being caused by something. Roy, my provider of farm-fresh eggs, advised that if I tipped my head to the side and held a candle below it, my earwax (the presumed problem) would melt out. He offered his wife, who'd bring a funnel, and said I would really be amazed at how much wax will pour out. My cousin suggested, "Sit next to the stove, turn the burner on high. And eat more broccoli." His belief was that the infrared radiation would heat my skull and speed the natural process of healing. And the broccoli? I never bothered to ask. My friend Donna said, "Maybe God is trying to tell you something. Maybe you have spent too much time listening to the outside world and not enough time listening to what is in your heart."

Yes, I was desperate. I removed all my earwax (without the candle and funnel), sat beside the stove, and searched my heart. The only thing that I couldn't do was give up bodysurfing.

My heart told me that I was born to be the Sound Tracker. My earliest memories of aural solitude had come at the bottom of a pool. As a child, I would hold my breath and lie on the bottom of the pool until the world ceased to exist. Even when my lungs began to burn and my body screamed for oxygen, I held on to my solitude. Then suddenly, and often involuntarily, I would point my head upward and with a kick launch myself to the surface, where I would burst into a world of oxygen and sound. "How many seconds was I down there this time?" I would ask myself before checking the pool clock by the lifeguard's chair. I can remember one guard saying, "Nice job. I'm impressed." But that was a kid's game. As I grew up, I sought to do more serious things. I studied botany. I wanted to become a plant pathologist.

In the fall of 1980 I was on that path, driving from Seattle to Madison, Wisconsin, to graduate school, when I pulled off I-90 and slipped down a side road until I found a place to rest for the night, a recently harvested cornfield. Hands behind my head and ready for a deep rest, I lay between two rows of stubby, shorn stalks. I heard a wonderful, layered chorus of chanting crickets and began to smell the dampness of an approaching storm. There, on the prairie, the thunder rolled in from far away, signaling rain long before it arrived. Again and again this thunder boomed and echoed, growing ever louder—magnificent, deep, primordial, soul-shaking sounds. I'd never heard thunder like this before.

Hours later and thoroughly soaked, I thought, "How could I be 27 years old and never truly have listened before?"

My life changed that night in the cornfield, though I didn't fully appreciate it at the time. It took me a few months to realize that graduate study at the University of Wisconsin was not the path I wanted to pursue. I felt a new yearning, one I understood better after reading John Muir describe his life-changing epiphany as "soul hunger." Since then I've been around the globe three times, recording the sounds and silences of nature. My hearing had become my life, my livelihood. My hearing was everything.

Three doctor visits and a CAT scan later, I learned that my hearing loss was due to a problem in my middle ear. But nothing could be done, at least, the doctors said, without the risk of making matters worse. Worse? The best thing to do, I was told, was to be fitted for a hearing aid and hope that the matter cleared up on its own.

To even suggest a hearing aid was an outrage. Nearly all hearing aids are designed primarily to amplify and clarify human speech, to hear what a person has to say. They do not make music more enjoyable or nature sounds more audible.

Back home, in a fit of private anger, I said out loud, "I just want my old life back!" So I examined everything I had done a year before my hearing loss and everything that I'd been doing during my hearing loss, regardless of perceived significance.

I had recently turned 50, and to celebrate this I began taking supplements that were recommended to me by my brother, who is a physician and had been on a rigorous vitamin and hormone regimen himself: high-potency B-complex, potassium, calcium, alpha lipoic acid, to name a few.

And to top off my new look, I also took Rogaine, applying it to my head, to thicken my thinning hair. "A little is good, so a lot must be better," was my philosophy. I poured Rogaine onto my head like hair tonic and sometimes felt it drip down my scalp and around my ears. All of this, my ear doctors reported, had nothing to do with my hearing loss. Nevertheless, out of desperation, I discontinued all supplements and put away the Rogaine.

Then, about two months after discontinuing the supplements, as if God himself had spoken to me, I experienced a sudden onset of completely normal hearing. Sitting in my grandfather's rocking chair next to my woodstove, I realized I could hear the crackle of the fire and the once-familiar gurgling of the refrigerator. Then, as quickly as it had returned, my hearing vanished again.

I continued to abstain from all supplements. Time became my ally, not my enemy. Brief periods of normal hearing came more frequently and lasted longer, then blended together, fashioning an encouraging, nearly normal six months. I shared the good news with my brother, Robert, who suggested I resume, temporarily, the regimen of vitamins and Rogaine, to determine if this was indeed the cause. Right. Though of a scientific mind, that was one experiment I wanted no part of. I never looked back and have counted my blessings ever since. Today my hearing has fully recovered.

We've all heard it said: "There are no accidents. Everything happens for a reason." When I hear this, I think of the great naturalist John Muir, who lost his eyesight in an industrial accident while working as a young man at an Indianapolis carriage factory. Thrust into total darkness, alone, and desperately wishing that he could once again see, to fully enjoy the natural world as God intended, Muir vowed that if his sight should ever return he would devote himself to "the inventions of God" and not to the inventions of man. When his sight did eventually return, he began a 1,000-mile walk to the Gulf of Mexico, "along the leafiest and least trodden path possible," on his way to becoming the man Americans know best as the father of our national parks.

At a time when the world was perhaps most musical, in the mid- to late 1800s, Muir was a dedicated and perceptive nature listener. I've deemed him my mentor over the years; I think of him as a nature sound recordist who used pen and paper. On every page of his journals he delves deeply into the finer points of listening and describes, with an "ear calmly bent," the music of nature, such as this description of Yosemite Falls:

> This noble fall has by far the richest, as well as the most powerful voice of all the falls of the Valley, its tones vary from the sharp hiss and rustle of the wind in the glossy leaves of the live-oaks and the soft, sifting, hushing tones of the pines, to the loudest rush and roar of storm winds and thunder among the crags of the summit peaks. The low bass, booming reverberating, tones, heard under favorable circumstances five or six miles away, are formed by the dashing and exploding of heavy masses mixed with air upon two projecting ledges on the face of the cliff.

In the spring of 2005, my hearing restored, my career as the Sound Tracker back on track, I asked myself, "What good is perfect hearing in a world filled with noise pollution?" After a good bit of thought, I resolved to make good on a quiet conservation project I'd conceived of years earlier.

One Square Inch of Silence was designated on Earth Day 2005 (April 22), when, with an audience of none, I placed a small red stone, a gift from an elder of the Quileute tribe, on a log in the Hoh Rain Forest at Olympic National Park, approximately three miles from the visitors center. With this marker in place, I hoped to protect and manage the natural soundscape in Olympic Park's backcountry wilderness. My logic is simple and not simply symbolic: If a loud noise, such as the passing of an aircraft, can affect many square miles, then a natural place, if maintained in a 100 percent noise-free condition, will likewise affect many square miles around it. Protect that single square inch of land from noise pollution, and quiet will prevail over a much larger area of the park.

My hope is that this simple and, I believe, inexpensive method of soundscape natural resource management will prove both an inspiration and a helpful mechanism for the National Park Service to meet several underattended, codified goals, namely, preserving and protecting the natural soundscapes of its parklands and restoring those soundscapes degraded by human noise.

Begun solely by my initiative, One Square Inch of Silence remains an independent research project. It has the qualified support of Olympic National Park officials. Park Superintendent Bill Laitner hiked with me to

OSI on Easter Sunday 2006 and knows the value of preserving natural quiet through a simple method that noisemakers can understand.

When I periodically visit OSI to monitor for possible noise intrusions, I note the time and, whenever possible, the noise level of the man-made intrusion. Then I attempt to identify and contact the responsible party by mail, explain the importance of preserving our last vestiges of natural quiet, especially in an environmentally protected national park, and ask them to voluntarily avoid such future noise intrusion. With my letter I include an audio CD that gives examples of the listening experiences that their corrective actions will help save. The audio CD ends with a noise intrusion, making it easy to understand that noise causes real destruction of the park experience. I post these intrusions and contacts in the News section of my website, www.onesquareinch.org, so the public can learn who is responsible for breaking the peace.

Olympic National Park was chosen for One Square Inch because it has a diverse natural soundscape combined with substantial periods of natural quiet. Unlike other national parks, such as Yellowstone, Grand Canyon, or Hawaii Volcanoes, where noise debates are long standing, air tourism at Olympic is still in its infancy. It has no through roads, no scenic drive to its highest peak. To reach its backcountry, you must go on foot. Because there are few noise intrusions in this wilderness, noise sources are easier to identify than at other parks. Each habitat type (alpine glaciers, rain forests, lakes and streams, and wilderness beaches) provides meaningful examples of soundscape beauty that listeners can easily identify and enjoy. But all such pristine experiences remain endangered.

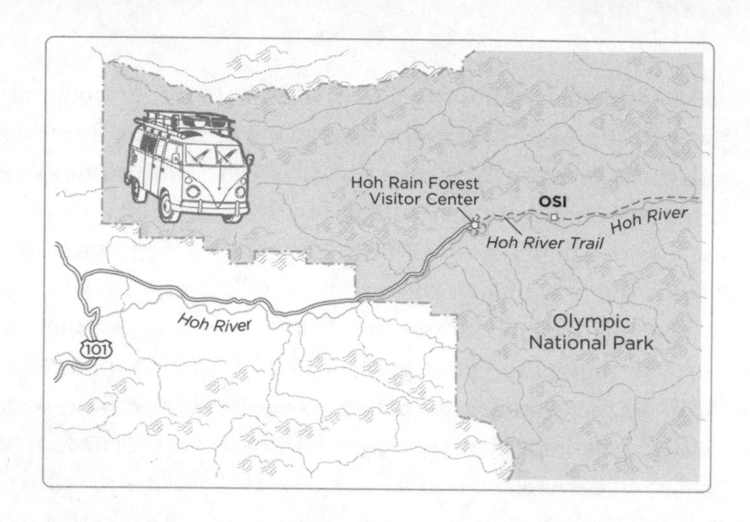

2 The Quiet Path

See how nature—trees, flowers, grass—grows in silence; see the stars, the moon and the sun—how they move in silence. . . . We need silence to be able to touch souls.

—Mother Teresa

Good things come from a quiet place: study, prayer, music, transformation, worship, communion. The words *peace* and *quiet* are all but synonymous, and are often spoken in the same breath. A quiet place is the think tank of the soul, the spawning ground of truth and beauty.

A quiet place outdoors has no physical borders or limits to perception. One can commonly hear for miles and listen even farther. A quiet place affords a sanctuary for the soul, where the difference between right and wrong becomes more readily apparent. It is a place to feel the love that connects all things, large and small, human and not; a place where the presence of a tree can be heard. A quiet place is a place to open up all your senses and come alive.

Sadly, though, as big as it is, our planet offers fewer and fewer quiet havens. This is especially true in developed nations, where the high consumption of fossil fuels translates into noise pollution. It's come to this: there is likely no place on earth untouched by modern noise. Even far from paved roads in the Amazon rain forest you can still hear the drone of distant outboard motors on dugout canoes and from the wrist of a native guide the hourly beep of a digital watch. The question is no longer whether noise will be present, but how often it will intrude and for how long. The interval between noise encroachments (measured in minutes) is the measure of quiet these days. In my experience, a silence longer than 15 minutes is now extremely rare in the United States and long gone in Europe. Most places do not have quiet at all; instead, one or more noise sources prevail around the clock. Even in wilderness areas and our national parks, the average noise-free interval has shrunk to less than five minutes during daylight hours. By my reckoning, the rate of quiet places extinction vastly exceeds the rate of species extinction. Today there are fewer than a dozen quiet places left in the United States. I repeat: fewer than a dozen quiet places and by that I mean places where natural silence reigns over many square miles.

In 1984, early in my career recording nature sounds, I identified 21 places in Washington State (an area of 71,302 square miles) with noise-free intervals of 15 minutes or longer. In 2007 only three of these places remain on my list. Two are protected only by their anonymity; the third lies deep within Olympic National Park: the Hoh Rain Forest in the far northwest corner of the continental United States. I moved near the Hoh in the mid-1990s just to be closer to its silences. In the Hoh River Valley, nature discovery occurs without words or even thoughts—it simply happens. Wondrously. But you have to listen.

And to do that, you first have to silence the mind. On the drive to the Hoh, I begin to shed all pressing thoughts—of work and family and the woes of the world. I generally stop for several hours at the Quileute Indian village of La Push and wash my mind clean, purify myself, in the Pacific Ocean. Summer and winter, the water temperature is always within a few degrees of 50 Fahrenheit. Five millimeters of neoprene closely wrapped around my entire body except for my face protects me from the cold. The rest is up to me. The ocean doesn't know me from a piece of driftwood. And it shows me a different look every time out. On this October morning, the ocean swells are six feet at 10 seconds—six feet high and 10 seconds apart—and rising nicely about 100 yards offshore opposite Lonesome Creek, where the rip to James Island digs a channel in the shore bottom that attracts visitors such as whales, Steller sea lions, harbor seals, otters, and porpoises. I swim out through the surf in the manner of a needle through cloth, diving in front of a wave and holding my breath while the pressure and roar of a six-foot wave pass overhead. Then I skim across the sandy bottom and pop to the surface for a gulp of air before descending again. In time, I make it beyond the farthest breaking wave and rest.

The ocean is a drum. It beats the music of the global weather systems. In ancient times during days of sail, an experienced mariner could tell the weather far out at sea by interpreting the look and feel of the ocean's waves. The Pacific is my cradle, gently rocking my gaze, rising and lowering the offshore fog-framed islands. I look for my "friend," the harbor seal who often greets me with a flipper kick to the ribs to inform me I'm scaring his fish, but I don't see him and turn my attention to the approaching swells.

Unlike a surfer who catches a wave atop his board, I'm a bodysurfer. I employ no board, preferring no layer of separation between myself and the surging, glorious, thrusting wash of the surf. I wait until a wave crests, then with one quick kick I launch myself and join the wave. Changing my shape with changes in water-body pressure, I am able to travel with the wave toward shore until either the wave spills all of its energy or I separate myself voluntarily from the wave—or I wipe out. I'm at home here, where my body is the only brain that I need. All I have to do is choose the right wave, kick-start and drop in on the wave, then let my body respond.

After two hours my breath is peaceful, my body exercised, my mind blissful, and my thoughts clean. I am "innocent" again, as I like to think of it, and ready to listen. I take a quick freshwater rinse in Lonesome Creek, shed my wetsuit for my dry land clothes, and start the 90-minute drive to the Hoh Rain Forest with the car heater on high, or what passes for high in my '64 VW bus.

I pass the Hard Rain Café, a small eatery, but few other signs of civilization on the drive, if you don't count the scarred hillsides, clear-cut forests pushing up new growth. The 18-mile drive on the Upper Hoh Road off Highway 101 is lined on both sides by the tallest living things on earth: Sitka spruce and Douglas fir that stretch nearly 300 feet high, giant western hemlocks, and western red cedars, some as much as 1,000 years old. Pulling into Olympic National Park, I begin to feel the magnificence of this place, the largest, most pristine stretch of temperate rain forest in the Western Hemisphere. Stretching over some 1,400 square miles of the mountainous interior of the Olympic Peninsula, the park is home to bald eagles and northern spotted owls and more than 300 other species of birds. Salmon migrate up many of the park's 12 rivers. Cougars, bears, and Roosevelt elk roam its forests. At least eight species of plants and 18 animal species can be found here and nowhere else in the world, among them, the Olympic marmot, Olympic snow mole, and Olympic torrent salamander. Olympic National Park is nothing short of a national treasure, recognized around the globe as a World Heritage Park and a designated World Biosphere Reserve.

But few value Olympic National Park as I do: with my ears. I believe that Olympic Park is quieter than any other national park or wilderness area in the 390 units of the 84 million acres managed by the National Park Service—including the largest, Wrangle–St. Elias in Alaska, where squadrons of vacationing flightseers break the silence on a clear day. Olympic National Park retains much of its natural quiet for one reason: overcast skies. The region has more than 200 completely cloudy days each year, and even more partially cloudy days that go uncounted. Many of those cloudy days are also rainy days. Rain, and lots of it, is a big deterrent to selling tickets for scenic tours.

Not only is Olympic Park the least intruded upon by human noise, but it also has the greatest diversity of natural soundscapes of any national park that I have been to. Often referred to as three parks in one, Olympic Park has a rugged mountainous interior with glacier-capped peaks, lush forested valleys with the world's tallest trees, and the longest wilderness seashore in the lower 48 states.

You would think that Olympic Park's premier status in both natural quiet and natural soundscapes would earn it special recognition by the National Park Service. Not so. Its unequaled acoustic environment has no special protection, no special management, and not a single person on staff specifically trained in acoustic ecology. It is clearly as vulnerable to the same administrative bungling that has destroyed the natural soundscapes of Grand Canyon and Hawaii Volcanoes National Parks. In the wake of the National Parks Air Tour Management Act, which Congress passed in 2000, requiring the Federal Aviation Administration (FAA) and the National

Park Service (NPS) to plan for air tours over national parks, Olympic Park has started to attract businesses catering to flightseers. One such business, Vashon Island Air, now offers tours on demand and advertises "The Grand Tamale . . . We fly past Mt. Olympus, then down the valley of the Hoh River, the only non tropical rainforest in the world."

Well, the Hoh Rain Forest is my Grand Tamale, too. America's quietest spot. It's the place I have chosen to defend from all human-caused noise intrusions, which, in effect, translates into keeping out all air traffic, commercial flights as well as air tours, because air traffic destroys a hiker's opportunity to listen to nature undisturbed, unimpaired. Those seeking solace away from a noisy world who instead hear an airplane roaring overhead return home unfulfilled, unbathed by the spiritually cleansing power of quiet. By the time the noise has traveled far enough to dissipate below audible levels, many square miles have been consumed by a single prop plane or passing jet.

But just as noise can affect quiet, quiet can affect noise. By keeping even one square inch 100 percent free of noise, or at least attempting to do so, I am able to push back aircraft for many miles and help to preserve natural quiet over much of the entire park. Natural quiet, like clean air and clean water, is part of a delicate ecosystem. So when man-made noise intrudes in the wilderness, the equivalent of crackling static on the phone line of all creatures but man, it impairs the ability of animals to communicate. And wildlife are just as busy communicating as we are.

My initiative requires frequent visits to my little One Square Inch of Silence. I go as often as I can, thank God, and joyfully refresh my spirits. On most visits I observe no noise intrusions, only blissful solitude. Often I pack my high-tech recording gear, equipment capable of recording sounds fainter than the human ear can hear, such as the flapping of butterfly wings. But this time of year, autumn, it is dry and warm, a perfect time to travel light and observe rutting Roosevelt elk.

I reach the Hoh visitors center parking lot just before 3 p.m., eager to don my backpack, which I've stuffed with enough gear and provisions for three days. I kill the engine and swing open the car door. My welcome to the Hoh Valley is most unwelcome: noise coming from two directions, a nearby trail leading to the Hall of Mosses and the ranger station. Reaching in my pack for my sound-level meter, I head off toward the louder of the two noises, the *Brrrrrrrrr* coming from the trail. After 200 yards I come upon a trail crew installing a guardrail as a safety precaution for wheelchairs on this handicapped-accessible trail. *Burrrrrrrrrrrr.*

The sound morphs to *Briiiiiiiiiiiiiiiiiii* as the chainsaw burns more gas to chew deeper, faster. I look at my sound-level meter and observe a reading of 75 dBA from about 35 feet away.

Sound is a slippery reality. Scientists who study the physics of sound commonly measure sound using decibels (dB) on a logarithmic scale named in honor of Alexander Graham Bell. Zero decibels (0 dB) is the threshold of human hearing, the faintest sound the human ear can hear. Ten decibels (10 dB) is 10 times the power of that faintest detectable sound. Twenty decibels (20 dB) is 100 times the acoustic event of that barely audible sound. But because humans hear some sounds more easily than others (the human ear is more sensitive to midfrequencies than low or high frequencies), dB measurements are misleading. So my sound-level meter (Brüel and Kjær SLM 2225) measures more accurately what the human ear hears by using a formula that takes this into account and calculates A-weighted values. What you hear is what you get: a dBA value. Most noise ordinances use dBA, not dB.

"We'll have to trim off the top of the post," says the oldest of the three workers. The chainsaw again roars into action, this time reaching 85 dBA. As each 10-decibel increase above the 30-decibel normal sound level results in a 10-fold increase in power, this is actually 100,000 times the normal ambient sound power or energy level for this time of year at the Hoh. As similar as a swimming pool is to a glass of water.

I approach the crew to ask them if they mind if I observe their activity, holding out my sound-level meter with its blinking light and dBA scale.

"Whatever, so long as that isn't some radioactive thing. We'll be done soon, I promise," says the trail boss, Ben.

"Are you park employees or contractors?" I ask.

"I am permanent trail crew and these guys are seasonal workers. You must be the sound guy."

I lost count long ago, well past 100 visits, of how many times I've hiked in the Hoh Valley, often with my recording equipment. I'm known by park officials, as is my One Square Inch of Silence campaign for soundscape preservation in Olympic National Park.

Eyeing their railing work, I see the cuts don't exceed a few inches, four at the most, in diameter.

"Any reason why you're using power tools rather than hand tools?"

"Right here you could go either way," Ben answers, explaining that personally he likes hand tools, but "obviously power tools are a lot quicker."

I explain my intention of hiking up the Hoh trail to One Square Inch to monitor for noise intrusions and share my hope of finding a herd of Roosevelt elk.

"Heard any elk recently?" I ask.

"Not down here."

No kidding, I think. Elk do venture as far down the valley as the trailheads near the visitors center. I've seen them around here on many occasions. But with a chainsaw roaring in place of a handsaw, elk will surely keep their distance, meaning that anyone in a wheelchair unable to venture more than 100 yards from the parking lot will have no chance to hear or possibly see these magnificent creatures.

On the way back to the parking lot to grab my backpack a passenger car drives by me at a distance of 20 feet (70 dBA); a John Deere tractor passes at the same distance (88 dBA); and in the parking lot, a car about 60 feet away honks its horn, reassuring its owner that it's made good on the push of a remote locking button (90 dBA).

Noise is still coming from the ranger station, so I head there instead of returning to my VW right away. A sign on the powerhouse reads "Ear protection required when equipment operating in area." But the powerhouse itself is silent. I proceed on, bushwhacking through some salmonberry bushes until I arrive at campsite 84 of Loop A. I recognize the ranger immediately. His uniform name badge says D. Ellison, but he goes by the nickname Smokey. Last year I found him on top of the Hoh Rain Forest visitors center with a gas-powered leaf blower doing fall cleanup, which registered 110 dBA on my sound meter.

"Hi, Smokey. You remember me? I'm Gordon."

"Yeah, I remember you."

"I'm headed to One Square Inch and measuring noise levels in the park."

"This is a different leaf blower than before," offers Smokey. "This is a four-valve. We got rid of the two-valve. And we're now using a truck, but in the summertime we use an electric cart."

"We can measure the noise level of your new leaf blower, if you're interested."

"Yeah, I'd be interested."

Smokey explains that after I'd pressed the point about worker noise in the park, there'd been a meeting. "We had a big talk here at the park about noise and went back through the catalogue and got these."

"There should be a sticker on the side that gives it a rating. Can you see one?"

"Category Three, seventy-five dB."

No way, my ear tells me. It's a lot louder than 75 dB.

A pull of the rope starts the motor. Smokey lets it idle, then brings it up to normal operating speed.

"It's ninety-three dBA at a distance of three feet."

This is 60 dBA above the normal baseline ambience for this area, and by far the loudest sound of the day. Good intentions or not, Smokey is still wielding a sonic hurricane in place of a rake.

I double back for my backpack. Its familiar 50 pounds swing easily over my shoulders and I set off. A short way from the visitors center I reach a small wooden bridge that spans a wonderful babbling brook. No salmon here yet; they must be waiting offshore at Hoh Head for the first big rain (42 dBA).

I head up the ancient riverbank, where a sign at the main trailhead reads "Elevation 773." I'm craving quiet, yet from several hundred yards away I hear the work of Ben's trail crew. *Aaaaaaaaaaaaaeeeeeeeek.* Somebody's driving lag screws with a power tool.

Ten minutes later, about a half-mile up the Hoh River Trail, at 3:40 p.m., my sound meter finally becomes a quiet-level meter (22 dBA), just 2 dBA short of the meter's lowest possible accurate measurement. I hear only the distant hush of the Hoh River, low-running and filtered by more than 300 yards of ancient forest. The air is absolutely still. So still, in fact, that autumn stands frozen, as if in a photograph. A number of detached leaves rest on edge, house-of-cards style, atop other leaves and spruce boughs, silently awaiting the next breeze to set them sailing on their way to the forest floor amid the deciduous applause of other leaves flapping loose and taking flight.

My senses are beginning to sharpen. As I proceed along the path, step after step, my body and mind make the transition from the rhythm of the sea to the rhythm of the rain forest.

At the 0.9 mile campsite marker, just short of a mile from the trailhead, I measure 36 dBA. I hear only the Hoh River beyond the trees.

Near the 1.4 mile campsite marker, with the river much closer, only 40 yards from where I stand and partially visible, the sound meter reads 46 dBA. Although I know how others might be seduced by the proximity of the river, I would never set up camp by the rushing water. First, the ambient river noise is too loud to talk comfortably, even at close range—not that that need be accounted for when hiking alone or that conversation is necessarily paramount when hiking with someone else. But the need to raise one's voice and strain to hear suggests a more important difficulty: the difficulty of hearing other sounds, informative sounds, such as the snap of a twig announcing a hungry raccoon, or one of many voices of the raven, or the alarm call of the Douglas squirrel.

Wildlife depend on their sense of hearing to detect the approach of predators and will not remain very long in places where it is difficult to hear, so the chances of wildlife observation are poorer in louder than in quieter locales. White-tailed deer drink here, at river's edge, but not for long. You can watch them listening and see their anxiety. They can move their long, funnel-shaped ears independently, first one, then the other, to determine the exact position of whatever produced the sound they detected. White-tailed deer are among my favorite advisers. Being an important food source for the many cougars that inhabit the park, they seldom remain close to a noisy river or stream for longer than the time required to drink, and while they do, they pause often and look in different directions to compensate for the temporary inability to use sound for their security surveillance.

Besides missing out on wildlife observation, there's another reason not to camp here, so close to the river. The riverbed affords an avalanche chute for frigid air from the mountainsides far upstream. Late in the day this phenomenon often goes unnoticed because the wind is from the west and warm. But in the early morning, when air layers find their own places by temperature alone, warm air rising and cold air falling, the riverbed offers a natural drainage channel for upriver cold air that arrives with its own wind chill. Maybe 50 feet away and 10 feet higher than the river the temperature will be 10 to 15 degrees warmer.

Beyond the 1.4 mile campsite, just off the trail, there's a spectacular view of Mt. Tom from the top of what was once the ancient riverbank, when glaciers were larger and river flow was significantly greater. The trail then passes through a fallen spruce log six feet in diameter, a four-foot column sawn out to reclaim the footpath. Here a chainsaw *was* needed to breach the recently fallen behemoth. The resulting sawdust smells sweet, even delicious, and I detect the scent of dried leaves and nearby mushrooms, the first smells of the day strong enough to hold my attention and cause me to stop.

At the 2.0 mile campsite signpost I drop my backpack and listen: 40 dBA. Distant river sounds. I walk into the campsite itself and measure again: 43 dBA. Finding a clear spot among the vine maples, I measure yet again. The 45 dBA is far higher than I would like, but just then I hear a water ouzel make the sounds *Kerr, kerr, cheep* before breaking into its beautiful song. Writing in the late 1800s, John Muir, whose steps I've carefully retraced through Yosemite Valley, called the water ouzel "the mountain streams' own darling, the hummingbird of blooming waters, loving rocky ripple-slopes and sheets of foam as a bee loves flowers, as a lark loves sunshine and meadows." He described the ouzel's separate songs as

> perfect arabesques of melody, composed of a few full, round notes, embroidered with delicate trills which fade and melt in long slender cadences. In a general way his music is that of the streams, refined and spiritualized. The deep booming notes of the falls are in it, the trills of rapids, the gurgling of margin eddies, the low whispering of level reaches, and the sweet tinkle of separate drops oozing from the ends of mosses and falling into tranquil pools.

How could I make camp anywhere else?

With daylight fading, I tug my tarp from my backpack. The 9-by-12-foot tarp is an old friend, purchased in Seattle in the 1970s at what was then a fledgling, single-store purveyor of outdoors gear called REI, long before it and another Seattle company became national brands. Worn thin like a favorite pair of corduroys, and nowadays lacking most of its waterproof coating, it borders on translucent but remains serviceable. I enjoy setting it up, first sizing up potential sleeping spots, then stringing it from suitable branches and bushes, tying off the light but strong, Boeing surplus, braided, waxed line through each corner eyelet, making sure to tightly tilt the tarp into the prevailing weather, fashioning not so much a roof over my head as an umbrella. Today's blue sky offers no guarantee of a dry night. This is, after all, a rain forest. When this funnel-shaped valley captures the moist Pacific air as it rises to the nearby mountains, the rain falls with unimaginable force and volume: 13 feet in an average year, real snorkel weather. As much as I like to sleep completely out in the open, even on a clear night in the early fall before the rainy season, I remind myself that the Hoh could also be spelled H_2O.

I'm having dinner—some hunks of sourdough bread with cheddar cheese—when a jetliner intrudes overhead. I note the time, 5:25 p.m., but not the decibel count, because I'm eating. Before setting off on an afterdinner walk, I hang my food in a red alder tree about 30 feet up and out on a limb 10 feet from the trunk to protect it from raccoons and black bears. I fill my smaller shoulder bag with a headlamp, a camera, and the sound-level meter and head off farther up the valley. It's just past 6 p.m., close to sunset. Moonrise will be delayed at least two hours by the steep mountainside, but when it appears, the waning moon, three days past full, will shine brightly. I expect the elk will be active.

At the 2.3 milepost and campsite marker I measure an ambient sound-pressure level of 39.5 dBA. This is just 5 dBA less than at campsite 2.0, but it feels *much* quieter.

Decibels take some getting used to for the uninitiated, who are trained to think linearly. If the voice of one person registers 60 dBA, we'd expect two people speaking simultaneously, and hence making twice as much noise, to register 120 dBA. But the correct answer is 63 dBA because decibels are measured on a logarithmic scale. As noise levels decrease, their measurements are also surprising. For example, here in the Hoh Valley, away from the water sounds, the natural quiet is typically around 25 to 35 dBA. On paper this hardly looks quiet, but to most ears it will sound stone silent at first. Only after a period of minutes will small textures appear, typically the subtle sound of distant wind playing high in the forest canopy.

"So what's the big deal?" some might ask, of a jetliner cruising at 36,000 feet over the Hoh Valley and registering 45 to 55 dBA on the ground.

"That's quieter than a conversation." The problem is that the jet noise is so much louder than the quiet ambience: for every 3-dBA increase there is twice as much energy; for every 10-dBA increase an event will sound twice as loud. A noise intrusion of 20 dBA above the quiet of the Hoh Rain Forest is 100 times the natural sound power level! Here in the quiet wilderness, we experience this noise intrusion as a dynamite blast, except a dynamite blast would have *less* impact because it would be shorter in duration and limited to one area rather than a continuous roar that cuts through the silence from one end of the park to the other.

I make it to the next creek, Mineral Creek, and view its lush waterfall in the fading light. I often think of this spot, about a mile short of One Square Inch, as my gateway to quiet because of the beauty of the falls. At nearly 70 dBA, measured from the footbridge over the creek and approximately 75 yards from the falls, the only sound is water in all its guises: its thundering cascade, gurgles in rock enclosures, and distant sprays.

From the sound of the water alone I've learned to distinguish the age of a tumbling stream. Older flows, such as those in Appalachia that escaped the last glaciation, have been tuning themselves for many thousands of years. Their watercourses and stony beds, smoothed to paths of least resistance by the ageless cycles of torrents and floods, sing differently. To my ears, they're quieter, more musical, more eloquent. Youthful streams, with their newly exposed and angular, unsmoothed rocks, push the water aside brashly, with a resulting clatter. In all cases, the rocks are the notes. I sometimes attempt to tune a stream by repositioning a few prominent rocks, listening for the subtle changes in sound.

The more you listen, the more you hear. At Mt. Tom Creek Meadows, where the footing is often soggy, a series of boardwalks overlie the trail. Here, as if walking on top of a long, wooden xylophone, it's possible to discern the condition of each slat by the sound it makes. Decaying boards produce a dull *thud*. Newly replaced boards resound clearly with a bright *bong*. Most boards are somewhere in between.

A bit later, above the distant hush of the Hoh, I hear my first elk and stop to enjoy its high, flute-like bugling. For a few minutes the sky shows through the trees as an incredible deep pastel pink and baby blue, and the turning leaves of the vine maples glow scarlet. Then the forest slips into black. I reach for my flashlight. It refuses to light. Apparently it got switched on accidentally and the batteries are dead. I've got a backup flashlight but want to conserve its juice, so I hold out my pocket-size sound recorder, whose LCD display has an ambient glow that is just bright enough for slow, feeling footsteps on the familiar path.

I make it back in a half-hour. The ouzel is no longer singing. I hear no elk. I kneel to lay my sleeping bag beneath the tarp and slide in. In my heart I have the same feelings that I have late at night gazing at wood embers: reverence, loyalty, devotion, gratitude. I fall asleep to the distant hush of the Hoh.

A couple of hours later I stir. With the moon overhead, I stare dreamily at the overlapping shadows of the vine maple leaves, visible in silhouette through my aged tarp, before drifting back to sleep.

At 2:55 a.m. the first jet intrusion of the day is loud enough to wake me from a dead sleep. I'm too tired to fumble for the sound-level meter to take a reading, but I note the time so that when I'm back home I can log on to the Seattle-Tacoma International Airport WebTrak website to try to identify the plane and the airline responsible for the overflight. Still awake at 3:15 a.m., I hear a second jet.

As I sink deeper into my down sleeping bag, a bag of my own design I call the Worm, my toes rub up against my stove canister. On cold nights I tuck it inside the Worm with me to keep it warm and ready to light, much preferring a banged toe to any delay in the brewing of my morning coffee or tea. I can easily reach the stove because my sleeping bag has a drawstring at each end, permitting flow-through ventilation on warm nights and, when the mercury falls, a snug cocoon when drawn tight, there being no zipper to interrupt the enclosure of down. The Worm also serves me well at daybreak on frosty mornings. I often undo the bottom, freeing my feet, stand, tighten the lower drawstring around my waist, and angle the other opening from atop one shoulder and under the opposite arm, toga style; then I tighten the drawstring, leaving one arm free. I'll fire up the stove, brew some coffee, and after draining my cup, and though long since up and about, finally get out of bed. Soon after inventing the Worm and arranging to have it made for me I tried to patent my design. I didn't mind when I learned that I wasn't the first to dream up a double-drawstring tubular design. What irks me to this day is that the patent is for a bowling pin cover.

The night silence returns. I lie snug in the Worm, each breath clearly visible in the moonlight.

Hoo. Hoo. Hoo-hoo.

After a pause, the call repeats over and over for the next four minutes. The great horned owl is right above me, in the giant Sitka spruce tree that shelters the campsite. To my ears this territorial proclamation goes unchallenged, but the owl's hearing is better attuned to an owl's voice and his position different from mine, so perhaps I'm treated to only one end of a dialogue.

At 3:35 a.m. I believe I'm hearing the beginnings of a third jet intrusion, when the sound takes on more subtle qualities and approaches me as if a breath with thoughts. John Muir describes this diaphanous sonic phenomenon in this fashion: "The substance of the winds is too thin for human eyes, their written language too difficult for human minds, and their spoken language mostly too faint for the ears."

Through my tarp I can see the vine maple leaf reflections dance slightly and a few of the boughs high overhead wave in the moonlight.

Quiet is quieting.

My eyes reopen just after 7 a.m. Snug in my sleeping bag, I'm awaiting the dawn chorus, the onset of birdsong triggered by the gradual increase in ambient light, as one by one different species chime in. I hear an *Ut. Ut. Ut.* What is that? As I listen, man intrudes once more. I hear a low-flying prop plane, probably looking for elk herds near the park's perimeter. It's hunting season, and the Roosevelt elk that wander freely in and out of the park make prized trophies when bagged outside of park borders.

Breakfast is simple: a Balance bar and Red Rose tea. Back home I live to eat, but in the wilderness I prefer to keep my diet simple, browsing like an animal, eating a little here and there, grazing on any huckleberries that the bear and elk leave behind. Real food is too much of a distraction and dulls my sensory edge.

Finally, sipping my tea, I hear the first notes from a western winter wren, a high-pitched twittering that goes on continuously, or so it seems, for nearly a minute from a concealed position halfway up a towering western hemlock tree. Though similar in appearance to the eastern winter wren, its song is completely different. The song of the eastern winter wren is operatic and full; the song of the western winter wren is sharp and narrow. Many songbird species have these kinds of variations and even local "dialects," as ornithologists call them. Clearly, wild creatures use languages that we're just now beginning to decode.

One day, back home in my studio, I decided to do a little experiment with the song of a western winter wren I'd recorded in the Hoh Valley. To my human ears it was a long continuous stream of very fast modulations of amplitude and high frequencies. Though cheery and one of the few birdsongs heard even on the most dismal days of winter, it wasn't exactly sing-along material. I hypothesized that because I speak in one-breath sentences, the winter wren might, too. So I converted the length of the wren's "sentences" from its breath to my breath. Granted, I did this with great speculation, making the assumption that breath length is a function of animal size. Since this experiment was just for fun, I took a simple one-second song sample and expanded it to 12 seconds. The results astounded me. My studio wren sang a song as elaborate as any humpback whale. Since then, each time I hear the western winter wren's twittering, I am reminded of those intricate bends and twists that another winter wren might hear.

Finishing my tea, preparing to set off for One Square Inch, I realize that I can finally hear the river singing. Actually, the entire valley is singing. This phenomenon is so subtle that I have yet to record it successfully, but during optimal listening conditions I have heard it in nearly all of the river valleys I have visited worldwide. The valley must be forested, the river actively flowing and producing a broad-spectrum sound source, the air absolutely still, preferably in the morning, when it has been calm for several hours. Finally, and most important, my ears must be completely relaxed and my mind clear.

This river singing varies in pitch and timbre from river valley to river valley, not so much affected by the river as by the type of vegetation and the size and shape of the valley. It is so distinct and characteristic that I can hum it, though I am always off-key because there are so many layers to it. I do better at imagining it and committing it to memory like a pop tune or advertising ditty and often carry it with me as a personal mantra for days after I return from the trail. Eventually, intruded upon by more recent sounds, it fades from memory, encouraging me to go back for a refill.

I can only speculate about what precisely produces these valley sound signatures, but I imagine it goes something like this: Any broad-spectrum sound source such as a rushing river, waterfall, ocean surf, or even traffic noise sends out sound waves that travel in all directions, colliding with surfaces, penetrating objects, and otherwise becoming changed by the local environment. In environments that contain repetitious structures of similar size and shape, sound waves are modified as they travel through the environment, absorbing, refracting, and reflecting, some frequencies more than others. The result is that what started out as more or less static noise becomes a tune, one that varies with the environmental topography and atmospheric conditions.

I have listened to this environmental music while exploring coniferous forests, pebbled beaches, and canyons. Although it should be possible to hear in any place where conditions are right, I have not heard it in urban areas, perhaps because the patterns are all too big, the listening area too small, and the overall ambience too loud. This landscape music is best heard from a distance of a mile or more from the sound source, far enough so that the harmonics are clear and the local ambience quiet.

I particularly enjoy this kind of music, as compared to, say, the song of a bird, admirable and even inspirational as birdsong is. A whole-valley listening experience is the result of *place,* not an individual performer. I can feel the importance of the living community, how one thing is not more important than the other. It's *everything* that matters. When listening to this music of place, whether here in the Hoh or in the backcountry of Yosemite, I am inspired to be a better neighbor, a better parent, a better child because I feel part of something much bigger: a collective place that makes music and sings to me.

Not a leaf is turning; there is no wind or even a passing breeze. The rocks down by the river show a wet line a foot above the present river line, indicating that the water level fell during the night. The exposed sand pockets among the dry gravel bars have collected moisture for some reason and are cold, much colder than the surrounding areas.

My goal today is to pay another visit to OSI and continue to monitor for noise intrusions. At 8:20 a.m., passing the 2.3 mile campsite, I measure a base ambience of 41dBA. I am surprised because it feels quieter, possibly because the surrounding trees create a nice pocket for warmer, lower tones. The first rays of sunlight have reached the mountaintops, but it will still be several hours before I remove my down jacket. There is no other sound except the *rush-hush* of the river—not a bird, not a squirrel, not an elk. I can hum the quiet music of the valley, and do. It is a sound full of love.

The sound of a jet roar just after 8:20 a.m. yanks my glance skyward, and because I am near the river, one of the few places to see large areas of the sky, I see the jet trail cutting southwest over the park. During the duration of the jet's overflight, I can no longer hear the river sing. Rather than grab my sound-level meter I listen more closely, trying, unsuccessfully, to hang on to the music of the valley.

The Hoh River is at its lowest flow rate of the year. Except for a few early arrivals in the deep pools, the river bears few salmon. The autumn rains will soon flood the river, sending a huge flush of fresh water into the ocean, signaling to the salmon that it is time to complete their life cycle. This low river condition is perfect for seeing the sound that is soon to arrive.

I see sound by studying the stones in the river, which are arranged, not at all randomly, but in a musical score. The largest stones, about the size of basketballs, make resounding thuds whenever they roll along, pushed by the strongest currents. These are in the main channels, and some lie partially buried. The smaller stones that produce the midtones and high pitches have arranged themselves in conspicuous bands corresponding to currents of different strengths of water flow. All are now silent, but when the heavy autumn rains return, the river's song will play loudly, so loud that you can hear the underwater concert from the trail.

During these autumn floods, I have dropped a hydrophone into the river to listen more carefully to the deluge concerto. At first, the music is only noise, loud and various, very similar to concrete going down a metal shoot, but within seconds the raucousness subsides or bursts into another clamor, as a boulder smashes through, crushing other stones into smaller sizes. Occasionally the ear can detect the riverbank eroding, spilling new stones on a seaward journey that may last centuries. The roots of even the largest trees are sometimes exposed in the torrent; I have heard them twist and splinter, like large bones slowly breaking. This is not relaxing to hear, but it is educational, for the sheer force of all this water helps produce another concert.

Stones of all sizes eventually make their way to the park's wilderness beach, where incredible music can also be heard. The stones arrange themselves in tonal bands, the result of having been swept and stroked by countless fingers of winter's storm waves. Many huge driftwood logs still bear enormous root cavities, big enough to walk into, like caves. Sitka spruce is the wood of choice for many of the finest guitars and violins and the soundboards of Steinway pianos because of its anisotropy, or elasticity. Compared to other woods, its uniform fibers vibrate easily. I have often recorded inside what I call "ears of wood," old-growth Sitka spruce logs, uncarved violins, if you will, that vibrate not with the touch of a bow, but with the crash of each ocean wave, and then by its more nuanced backflow as it retreats across the surf smoothed stones.

Whenever I am asked to name one of my favorite sounds, this sound from these ears of wood comes readily to mind. I shared these ears of wood with my students at Olympic Park Institute, where I taught nature sound portraiture in the mid-1990s, but I would guess that fewer than 100 people have heard this incredible surf symphony in the wild. You have to poke your head inside the driftwood log to hear it. To my knowledge, not a single park ranger working at Olympic Park has heard it, which may explain why many of the finest musical logs were moved from Rialto Beach during the repair of a rock jetty in the late 1990s.

At 9:55 a.m. a propeller-driven airplane makes a wide circle over the Hoh Valley with a noise impact of 63 dBA. Because it is such a clear day (not a cloud in the sky) this activity is likely flightseeing, a scenic drive in the sky. In addition to the company offering the Grand Tamale tours, an outfit called Rite Brothers out of Port Angeles, Washington, offers similar flights, as do other businesses as far away as Victoria, British Columbia. These superficial "park visits" are at best gee-whiz aerial views that titillate—and tap only one sense. I wonder whether these operators and flightseers worry at all that their flight impairs the park experiences of others below, visitors who have a birthright to enjoy wilderness solitude?

Shortly after 10 a.m. I reach my One Square Inch turnoff landmark on the left side of the trail, a stilted Sitka spruce that offers an opening large enough to go inside for shelter during a heavy rainstorm. A short walk brings me to my sacred spot within this special forest, a chest-high, moss-covered log that provides a noble, if oversized, pedestal for my tiny OSI marker, a simple red stone. This red stone was given to me by David Four Lines, the former cultural elder of the Quileute tribe, whose reservation lies at the mouth of the Quileute River, a sacred place where other rivers—the Bogachiel, the Calawah, the Sol Duc, and the Dickey—come together. In the right light, the stone, which David Four Lines used to smooth ceremonial wood carvings, seems to transcend mere rock, resembling living flesh or a piece of sushi-grade tuna.

As always at OSI, I listen for noise intrusions, noting now that the base ambience measures 28 dBA and consists mostly of the sounds of the river arriving through several hundred yards of forest. A Douglas squirrel chatters away from a hemlock branch 50 feet up (50 dBA). In the animal world of listening, the loudest sound in an ambience is an important one. Normally, the loudest sound is often made by the creature at the top of the food chain, who feels most secure and least at risk of predation—this morning's flightseers, for instance. And this squirrel, too—although perhaps foolishly, should there happen to be a hungry owl nearby.

At 10:10 a.m. I record another high-altitude noise intrusion (40 dBA): jet aircraft. Based on the sound alone I would normally picture the jet traveling south to north, but I have learned that the mountainsides bounce the sound around so well that the jet might be traveling west to east or even north to south. Nevertheless, I often mark the apparent direction, even though, without visual contact, this information remains uncertain, for it may aid me in identifying the aircraft when I'm back home, sitting at my computer. Just as this jet noise fades out at 10:13 a.m., I hear a prop plane over the north ridge of the Hoh Valley toward the city of Forks (39 dBA).

At 10:21 a.m. a plane flying west to east produces a noise impact of 68 dBA. This is very loud, especially when compared to the 28 dBA base ambience. Recalling that each 3-dBA jump on the meter signals roughly a doubling of the sound wave power of the noise intrusion, that means the 40-dBA increase represents more than a doubling of the audible acoustic energy *13 times over.* If you had a dollar and doubled it 13 times it would grow to $8,192.

I've come to think of silence in two ways.

Inner silence is that feeling of reverence for life. It is a feeling we can carry with us no matter where we go, a sacred silence that can remind us of the difference between right and wrong, even on a noisy city street. It resides at a soul level.

Outer silence is different. It is what we experience when we are in a naturally quiet place without the modern noise intrusions that can remind us of modern issues beyond our control, such as economic aggression and the violation of human rights. Outer silence invites us to open up our senses and get connected, once again, to *everything* around us. No matter in what direction you look it is all the same connection. Outer silence can recharge my inner silence. It fills me with gratitude and patience. I don't think I have been either tired or hungry while in a place of outer silence. The experience of *being there* is so complete. And then, after I return home, I generally sleep long and hard.

Fountains of Youth, that's what John Muir called our national parks. Already, less than 24 hours since my arrival in Olympic National Park, I can feel my senses heightening, including my sense of smell, along with my sense of hearing. In the morning's still, moist air I detect undisturbed pockets of scent—sweet, sometimes musky, occasionally herbal.

At 10.34 a.m. another prop plane travels east to west at 59 dBA. More flightseers on a clear sunny day? Or have hunters taken to the air to pinpoint one of the elk herds that wander in and out of the park boundaries? Four noise intrusions in just over a half-hour. This is the highest rate I have observed in the 18 months I have been logging noise intrusions at OSI.

I have never met a person who thinks that aircraft noise belongs in a wilderness. In fact, I have played my recordings of nature to children who, hearing an aircraft intrude, ask in disbelief, "What is that?" When I tell them, they ask, "Why is that allowed?"

Fact: In 1992 aircraft noise was audible in Yosemite Valley more than 50 percent of the time, according to a park ranger who shared with me the results of a noise study he undertook because nobody else wanted the job.

I reach beneath the One Square Inch log and pull out my Jar of Quiet Thoughts. Actually, it's not a jar at all. I did begin with a jar some eight months after establishing One Square Inch of Silence, leaving a pencil and some paper inside and inviting the thoughts and impressions of pilgrims to my designated quiet sanctuary. But the jar's screw-top lid proved no match for the H_2O of the Hoh. When I returned a few weeks later, the jar looked like an aquarium, with paper and pencil swimming inside. Now I use an antique metal ice cream container, a quart-size cylinder from the early 1900s that customers would bring to the dairy and get filled with hand-packed ice cream. It's about five inches in diameter and seven inches high; most important, its metal lid overlaps the cylinder by about two inches and fits snugly because I've added a layer of foam and a piece of rubber inner tube to tighten the seal. Inside I've also slipped in a few silica gel packets to serve as desiccants.

Thoughts left in the jar are private to this place. They can be read only by those who visit One Square Inch. Today I find a ten-dollar donation to One Square Inch of Silence among the 50 or so notes. One note tells of a marriage proposal, made right here. A silent proposal, of course. I am encouraged. This is the way things happen, one step at a time, same as any other trail. Could it be any other way?

10:46 a.m. A jetliner intrudes. 36 dBA.

Most visits to OSI produce no observations of noise intrusions, but already, in less than an hour, there have been seven. I wonder if this is because the day is so clear; there is no layer of clouds to reflect the noise. I can't remember another day in the Hoh as clear or as calm. The sonic insults continue apace: four more jetliner intrusions in the next half-hour. This makes 11 noise intrusions in the past hour and six minutes, all of them from aircraft. I remind myself not to be led by anger.

As I walk back to the trail there's one last jet intrusion at 11:20 a.m. But I don't reach for my sound meter. I need lunch. I need direction. I need some answers. Where did all these needs come from? I need to make sense.

As I pass the waterfall at Mineral Creek near Mt. Tom Creek Meadows a jet intrusion is so loud that it is clearly audible over the roar of the waterfall. I look at neither my watch nor the sound-level meter, but stare only at the cascading water, which takes its own form, a fluid without a vessel. Why can't Olympic National Park become an FAA-assigned no-flight zone? I'd like to meet the person or persons who believe that there is any reason good enough to have this kind of noise in the Hoh Rain Forest.

Walking back to camp, I am thankful that my frightening brush with hearing loss proved fleeting, thankful I can hear the river singing. The air is sweet with dried alder leaves and grasses, sorrels and mushrooms. I plan to bathe in the Hoh near my campsite and dry in the sunshine, a rare treat at any time of year.

After washing up and eating, I feel ready for another walk, this time in search of elk. I head a bit up the trail, freezing when I hear a new sound, a faint, dry, tinkling sound coming from a salal bush low to the ground. I look closer and see that a clump of hemlock needles decorates the bush; the needles have fallen more than 100 feet! This sound reminds me of a late-night campfire after the fire has finished its rage, when the heartwood is solid ember and the wood tissue has been hollowed and begins to collapse like a glass ornament slowly breaking.

To the ear of the animal, a sound as simple and soft as dried hemlock needles falling onto salal leaves speaks of

> *Security*: This is a quiet environment where it is possible to detect very delicate sounds, such as the approaching footsteps of a predator. This is an unlikely spot to be at risk and a good place to bed down if you are tired.
>
> *Remote location:* It takes many square miles of isolation to produce an ambience of such simple character. There are few places left in the United States, let alone the world, where such a simple sound can be heard unadulterated by noise pollution.
>
> *Vegetation:* This is a tall coniferous forest. Besides the fact that the sound of falling needles is different from the sound of falling broad leaves, the wind that caused these needles to be released is not evident in the sound of their impact on salal. This indicates that the forest canopy is high overhead, very high, in fact, or there would have been at least a stir of other leaves within the forest understory.

This tall coniferous forest, in fact, offers cathedral-like acoustics with a reverberation time lasting about two seconds, which adds time to the interpretation of sound events. There is also its distinctive microclimate: the forested space is walled-in and thus more sheltered from temperature and weather extremes experienced in open spaces and high above the forest floor in the treetops. Moderate microclimates are less demanding on warm-blooded animals to regulate temperature than open, exposed areas, and therefore allow more leisure activities, such as resting and socializing.

Western winter wren at 50 feet. 40 dBA.
Red-breasted nuthatches and chestnut-backed chickadees at 30 feet. 45 dBA.
1:45 p.m. A helicopter passes over and along the ridge north of the Hoh Valley. 50 dBA.

Unlike the previous overhead noise intrusions, the National Park Service itself may be responsible for this one. The NPS uses helicopters for various park jobs, including counting elk in the Hoh Valley. I confirmed this in an e-mail correspondence with an NPS public information officer named Barb Maynes, who wrote that helicopter flights for counting elk remained "high enough above the canopy that the downdraft from the helicopter did not cause upper tree limbs to move, or epiphytic plants or duff to fall from the upper canopy." In other words, she addressed visual impacts of the helicopter flights but said nothing about the potential noise impact on the elk herds or the rest of the wildlife community or the degradation of the natural soundscape, which the Park Service's own management plan specifies it must preserve to the greatest extent possible. Are they doing so?

At 2:25 p.m. I find myself at a big leaf maple grove just as the day's first strong breeze rolls up the valley. I hear a little rustle and the loosening of the first leaves and watch these brittle gliders rock and swirl before coming to a temporary rest on the thick ground layer of ferns. Individual leaf impacts on fern fronds average 30 dBA at six feet. The entire event peaks at only 40 dBA, like a breath in the silence. Walking through the dry leaves creates a sound of 45 dBA; when I switch to a kid's foot shuffle, I manufacture one of the loudest events in the forest: 65 dBA. A solitary bumblebee whizzing by might register 34 to 44 dBA.

Later I lie down in one of the giant maple groves to take a nap. I drift off watching the colorful patterns of maple leaves, both the big leaf maple, which is yellow with brown spots, and the vine maple, which turns to bright reds, oranges, and yellows.

I wake to a jet intrusion at 6:04 p.m. (44 dBA). I return to camp and pre pare dinner near the river's edge and away from camp, where food smells might attract unwanted dinner guests, such as a black bear or a raccoon. Even with the relatively high dBAs produced by the river, I can hear part of the evening rush hour at Seattle-Tacoma International Airport, about 75 miles away as the crow flies. I add to my time sheet of jet intrusions: 7:55 p.m., 8:00 p.m., 8:15 p.m., 8:20 p.m., and 8:30 p.m.

I wake during the night to see the riverbed brightly illuminated by the moonlight, the trees framing the view from my sleeping bag clearly silhouetted. I can hear the distant bugles of elk coming from more than a mile farther to the west. I drift back to sleep vowing to find a herd tomorrow.

At 7:30 a.m. I stir and gaze out at the first morning light. I spend the morning in quiet observation of the natural wonders around me. A solitary tree frog at 30 feet (55 dBA). Its voice is nearly as loud as a casual human conversation and perfectly suited to the human ear—slow, deliberate, and clear, similar to a dry rubber hinge.

The hiking trail shows elk hoofprints without dry or sagging edges. Recently made prints. The vegetation bears many browse scars, so a large herd must be nearby, but not too close, as I cannot smell their sweet musk odor. I look to the riverbed and then to the forest; they could be anywhere, I think. Their usual behavior is to seek the solitude of the forest during the day and the openness of the riverbed at night, particularly in bright moon light. I decide to wait here.

4:12 p.m. A small prop plane rides along the north ridge. 44 dBA.

At 4:50 p.m. and still in the same spot, I hear an elk bugle coming from the direction of the south-facing slope of the north ridge, loud and clear. I decide that I have waited long enough and head off-trail and into the rain forest on the preferred path used by the rest of the wildlife—white-tailed deer, black bear, and cougar, to name a few. I carefully pick my way through a dense patch of young Sitka spruce, pricklier than salmonberry bushes, trying not to become a blip on the auditory radar of the Roosevelt elk. They do not seem to see very well, but they are good listeners. My steps are slow and rounding. I lay each foot down at the edges and then curl my step inward and forward to spread my energy and quiet my footfalls as much as possible.

I see one: a large bull elk standing in a moss-covered opening between two large hemlock trees. His rack is magnificent. Then I hear *Ew* coming faintly from deeper in the forest and look in that direction. I spot more than eight females. Behind them stands another bull, even larger than the first. Before I can count his points he moves behind the wildwood, a twisted mass of moss and vine maple. Another *Ew* sound, and I look to see more females. The herd numbers more than 30 elk: four adult males, more than 20 adult females, and several immature elk.

We can thank the Roosevelt elk, perhaps the ancestors of this particular herd, for the very existence of this wilderness preserve. Olympic National Park was established in 1938 from a Federal Reserve that was established to protect the Roosevelt elk. Understandably. The sound of the mature male is one of the most beautiful natural sounds that I have heard. From a distance of a quarter-mile or more, the preferred listening position, the sound is flute-like: a long, drawn-out, whimsical note that lifts slightly in pitch before the echo travels great distances. At closer distances the same sound is different: it is brassy, but still long and smooth, and often ends with a series of three or more stiff grunts. This sound is not nearly as musical to my ears because the acoustics of the towering rain forest, its habitat, have not *sweetened* it. At close range, the same sound is an aggressive, adrenaline-filled bellow with deep gurgles and fear-inspiring grunts intended to dissuade all serious contenders and convince them to back off—myself included. But I remain.

The *Ew* of the females and young appears to be a contact call that allows members of the herd to monitor their relative positions. It is important to spread out far enough to allow efficient feeding, but not so far as to lose contact with one another. This call also appears to convey several emotional states, such as distress, loss, and even sexual readiness.

I have also heard male elk produce a sound like a bark. This impulse sound is loud and distinct, an alert to all those within hearing range to be on guard because something unusual is in the air. This bark may precede a stampede, and since these animals are as large as horses, the bark of an elk deserves attention and respect.

I work my way closer, to about 100 feet, and aim my camera, careful to use the telephoto lens in manual mode to disengage the motor of the automatic focus. The vegetation is thick, the light is weak, so I continue to move forward in stealth mode, looking for emergency exits along the way. At this time of year the testosterone-primed males are unpredictable, and I would never want to accidentally come upon one and make it feel challenged. I see several potential shelters, root cavities at the base of giant trees large enough for me to pop inside but too small for a rack exceeding four feet.

The elk herd has picked a beautiful forest amphitheater in which to leisurely spend the day, a big open space draped with moss so thick it looks sprayed on with a giant flocking gun. I make a mental note to return here in springtime with my sound-recording equipment to make a portrait; the trees will host many birds and provide wonderful acoustics.

The elk are browsing at a very leisurely pace on vine maple leaves, huckleberry, and salmonberry. The sound of their activity, mostly soft twig snaps, without grunts or vocalizations, measures 32 dBA at a distance of 75 feet.

A male emits a loud bark. I've been busted. Must have been the sound of the camera shutter that gave me away. The herd heads away from me toward the river. I rattle off a few more shots but don't follow. I am not interested in changing their behavior. After they leave, I hightail it back to the main trail and go east, up the valley to where I know I can intercept the herd if they continue toward the river.

Yep, good call. Just as I get to the Mt. Tom overlook location, a large male with a respectable rack steps into the trail about 50 feet in front of me and faces off. The rest of the herd passes behind him, and eventually the Grand Bull himself crosses, and all proceed down the riverbank to the flood plain. I take a position on the bank and continue to take pictures and observe their habits.

The Grand Bull is easily the largest elk that I have ever seen in my 25 years of visiting the Hoh. He goes unchallenged, ruling the herd, if not the valley. I dub the other three males the Executive (the one who blocked the trail) and the Two Cowboys (who are always jousting with each other). The Executive apparently earns his time with the females in exchange for taking care of a lot of the Grand Bull's business. The Two Cowboys have figured out that it is going to be a while before they have a chance to mate and spend most of their time heads down, antlers locked, grunting and pushing and generally venting their frustrations on each other while gaining strength and skill at fighting.

At 5:35 p.m. a military jet booms up the valley, but at a high altitude, keeping the noise impact down to 50 dBA. The event is very quick, with a fast onset and somewhat longer decay—less than a minute in all. Nonetheless, a distraction. I look back at the tussling Cowboys and hear them kicking up the cobblestones in the riverbed as they try for better footing. Their antlers make an unusual sound, similar to wood but denser, and their grunts at this distance of more than 100 feet sound more like complaints of exhaustion than bursts of bravado.

At 5:55 p.m. another jet intrusion breaks the spell, and I head back to camp, fix dinner, and go to bed early.

I wake at 2:45 a.m. and listen to the river sing through the valley. It is a strumming-humming sound with surprising changes. It would surely rank high on my life list of favorite sounds. What other sound comes from an entire place, and due principally to the plant life, no less! Surely the dawn chorus would be among the finalists, too, beautiful and inspiring, distinct and expressive. But this sound, the Hoh-hum, is sacred, made all the more so by its simplicity. I am going to try to hum some of it here on the page.

First of all, there is a low, river-rush layer of sound. A *Psssssssssssssssssssssss*. Kind of like that. Then there is the very faint echo of the actual gurgles—*Lelelelgurlgle*—which is bouncing off nearby surfaces. Then there's an *Aaaaaaaaahhhhhhh* layer of sound. Well, it is actually a little higher than that, more like *Heeeeeeeeeeee*. Together they blend finely, ebb in and out of each other, and twist-tie into an almost inseparable whole.

This humming sound is so delicate that it cannot stand up to the slightest noise intrusion. Even the listener must remain absolutely quiet—no talking, no foot or arm movements, and quiet, slow breaths with open mouth. I believe an open mouth improves hearing for two reasons: it straightens the auditory canal and it enables the mouth to serve as a resonant chamber, amplifying faint sounds to more audible levels. This is exactly what children do instinctively when the lights go out: the jaw drops and faint sounds are more easily heard.

Does a place have a soul? Yes, I think so. And a place has intelligence, too. Just look at any clear-cut hillside. What does that place want to do? Heal itself.

This Hoh-hum sound of the entire valley is so reassuring, all-pervasive, and satisfying that I cannot imagine any man-made product from this forest timber producing anything as rewarding, not even a violin. I feel so light and free from the burden of possessions when I am backpacking. I often think about how I will simplify my life even more when I return home.

4:05 a.m. Jet intrusion. 44 dBA.

I climb out of my sleeping bag at 7:00 a.m. under foggy conditions, fix tea, pack quickly, and head out. Within the hour I intercept the same elk herd I saw yesterday, this time reverse-commuting from the riverbed to the forest. The fog has formed a layer 100 feet above the forest floor, allowing good ground visibility. One male shows 10 points and another far more. I am able to take up a position within 30 feet of a browsing male and begin to snap frames. I hear hoof clomps resonate in the wood-laden soil and look to see a buck coming up the riverbank unaware of me. I remain motionless and watch him pass close enough that I can smell the mustiness of his moist coat.

When the elk move on I continue my hike back down the valley, coming next upon a pair of white-tailed deer. I watch, listen, and snap away. Not only have the white-tailed deer advised me on where not to set up camp for the night, but by following their tracks I've discovered some spectacular listening spots. Invariably, again for reasons of safety, they bed down in places where sounds naturally collect. I never pass up an opportunity to sit and stay a while when I come upon a matted and sometimes still warm place on the forest floor where a white-tailed deer has slept.

A few moments later I come to a maple grove. The ambience is profoundly quiet: 20 dBA, the lowest reading of my sound-level meter. Still, it is not silent; there's a changing sense of space as I move. This is the presence of life.

Phip. Something falling from the forest canopy. 39 dBA.
Dee, dee, dee. Chickadee. 31 dBA.
Thump. Thump. Woodpecker tapping. 25 dBA.

All these tiny sounds punctuate the silence of the fog-drenched forest. But walking back down the trail, my stay in the wilderness coming to an end, I hear the hollow sound of internal doubt. I've never experienced so much aircraft noise over the Hoh Valley: jets, prop planes, even a helicopter. Is One Square Inch of Silence enough? Something deep within me is shifting. That much I know.

I eventually arrive back at the Hoh Visitor Center. A sign announces its autumn hours: "Open Friday, Saturday, Sunday 10–4." How can I expect the Park Service to budget for natural quiet management when they can't even afford to keep the Visitor Center open more than a few days a week?

Crossing the parking lot to my VW, I pass a visitor who nods a silent "Hello" and then points his key remote to lock his car. I feel my body tense. But, hallelujah. Instead of a confirming car horn, thanks to some bright (and quiet-minded) automotive engineer, the headlights flash instead. Yes, I think, there's one more vote for quiet.

But then there's a thunderous boom. A moment later, Murray's Olympic Disposal rumbles into sight on its way to the ranger station. It's time to take out the trash.

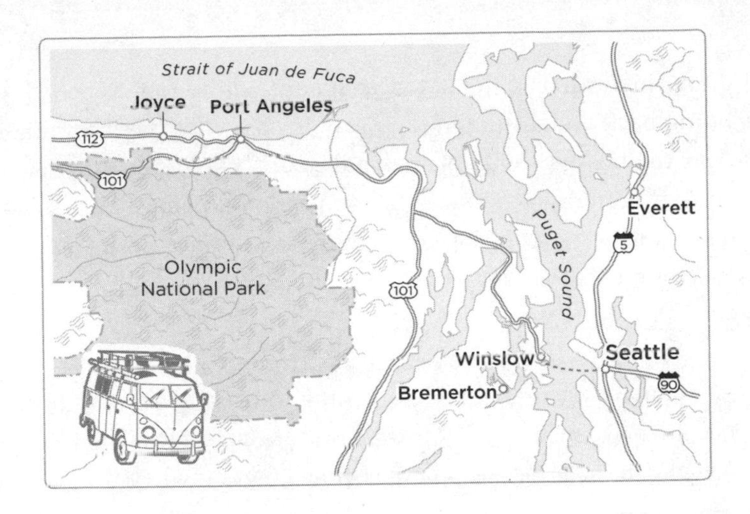

3 Hitting the Road

Go placidly amid the noise and haste, and remember what peace there may be in silence.

—Max Erhmann, American poet

If Joyce, Washington, is not the quietest town in America, it is certainly one of them. My closest and loudest neighbor is a cow, often buried in fog about 500 yards down Thors Road, a county road that most weeks sees three cars other than my own. All of them lost. The house I rent lies at the end of this road. Out my front window I see a few rolling hills that end at Crescent Beach and the Strait of Juan de Fuca. On a clear day the view extends to Vancouver Island, Canada. Behind me lies Olympic National Park.

The loudest sound inside my home is the phone, one short electronic ringtone from the bedroom that signals a call forward to my cell phone, which I normally keep on vibrate. The second loudest sound is my antiquated refrigerator that hums, gurgles, and slurps to do its job—mainly, keep my beer and steaks cold and the Yoplait yogurt fresh for when my daughter visits. The third loudest sound is the hum of my computer workstation, which empowers the business side of my fieldwork as the Sound Tracker and links me through the Internet to paying clients and individual buyers of my recordings, enabling me to live where I please, in a town that numbers about 100. On schooldays I can hear windblown laughter from the playground of Crescent School, a K–12 facility that's about a mile away. Often this childhood glee reminds me to take my own recess. On comes the screensaver, and off I go to bodysurf.

At night I can use my backyard for a recording studio (and do), except in spring, when the courting frogs inhabiting the trout pond past the barn and fruit trees upstage everything else with their all-night hootenanny, a concerto already well represented in my sound library. The local wildlife runs much bigger. Elk, bear, cougars, and coyotes all wander by. My teen-age daughter has grown up looking at wildlife looking at her. She knows the color of a cougar's eyes in a flashlight beam (amber) and what not to do (run) in such a situation.

Joyce is as charming to look at as it is to hear. It has road signs that read "Bytha Way" and "Uptha Creek." Its hub, the Joyce General Store, is the only grocery, gas station, and post office for nearly 20 miles in any direction. Stop in more than once and Leonard and Mary, the proprietors, will thereafter greet you by name. Leonard never tires of retelling the story of his Grand Ole Opry performance, usually banjo in hand, either behind the till or on the outside bench waiting for the next customer to arrive. Only after he describes his musical tour de force does he let on that he played to an empty house after lagging behind on a public tour of the music hall. Leonard's wife, Mary, is a practicing licensed attorney. She makes change at the register and stacks the finest collection of bar candy in Clallam County. The bounty runs more than eight feet down the counter, and she can tell you about each kind, too, should you appear indecisive. My favorite is a chocolate-covered malted foam bar called Violet Crumble. Mary's brother Jim Pfaff is my landlord. Each month for the past seven years my rent check has landed right here, in P.O. Box 1. Jim drives a gravel truck but is an old softy. He once spent a half-day trying to fix the light in my refrigerator, included in the rental, then offered to replace the refrigerator when the light stayed dark. I refused his offer. Lyda, a native Makah, works nine paces to the left of the till behind the counter at the U.S. Post Office. She entices folks to linger a bit longer with a jar of hard candy so she can better perform her other, unpaid job: gathering the news.

She once yelled out to me standing at the grocery counter, "Gordon, do you know Sue's phone number?"

Without thinking, I answered.

She raised her eyebrows and smiled, then said, "Sue forgot to put a stamp on her letter." She held out the envelope with feigned innocence.

"There's no return address, Lyda. What makes you think it's Sue's?" I asked.

"I recognize her handwriting," she said, dialing the number and telling Sue that she just applied a stamp and that Sue could pay the postage on her next visit.

Of course, all this was overheard by several other people, and soon the whole town knew I was seeing Sue. Sue bought her house off the auction block for the price of taxes and then built herself a fence around it out of scrap wood. The fence is pinkish, with boards of different lengths and widths. That's how we met. I stopped the car to tell her that I liked her fence.

When they run from someone or something, many folks head west and north, seeking unsettled country. Here in Joyce, in the mostly forgotten northwest corner of the contiguous United States, I, too, am a fugitive—a fugitive from noise. Most people around here earn their livelihood with a chainsaw. I live by my ears. I listen and record and consult. Businesses from around the world seek my recordings and advice, which I offer freely, up to a point, then charge accordingly. I've also published more than 60 albums of my environmental soundscapes, and thanks to iTunes I no longer need to keep a CD inventory of any consequence. Nowadays it's commerce by mouse-clicks—uploads, downloads, PayPal, and wire transfer—saving me countless hours and freeing me to devote more time to my goal of preserving silence at One Square Inch.

After checking e-mail over my Clallam Broadband connection (run by Leonard), I open an Internet site I've bookmarked called WebTrak that enables me to turn back the clock, pick a date and enter a time, and then ID most planes flying into and out of Sea-Tac Airport, the primary takeoff or destination of flights passing over Olympic National Park, on a tracker resembling a radar screen. My recent trip to One Square Inch left me especially eager to identify the noisemakers, for the list stretched much longer than on past monitoring visits.

One jet overflight that I identify proves particularly disturbing. I see on WebTrak that the noise intrusion of 44 dBA, which woke me at 4:05 a.m., came from a Boeing 777-200 with tail number N787AL. This plane belongs to American Airlines, the first airline to support One Square Inch of Silence by agreeing in 2001 not to fly over Olympic National Park. Yet here was an Asian in-bound American flight to Dallas–Ft. Worth, company headquarters no less, that flew right over the Hoh Valley, breaking that promise and shattering what little confidence I've gained over the years that One Square Inch might be a viable means for preserving natural quiet.

I was angry last week in the Hoh. But now I'm stunned and filled with doubt. Am I living some grand delusion? Tilting at windmills? Do people think I've gone a little wacky from too much time alone in the wilderness? I begin to feel a familiar sensation growing inside me, a ball of quiet heat that I've experienced during other times of self-doubt.

During the winter of 1989 I found myself out of work as a bike messenger when I came down with pneumonia. Weeks went by, and without money or work, we ran out of heating oil and our pipes froze. Back then I was married to Julie. We had a four-year-old son and mounting debts. Nearing rock bottom, I was inspired one day by a beautiful sunrise and the first scattered calls of birds. Then, the sun rising higher into the sky, I heard the chorus of birdsong building and imagined the sunrise circling the globe, propagating this endless wave of birdsong, as it has since the dawn of time. My hopelessness disappeared as, still bedridden, I conceived of The Dawn Chorus Project. One year later, I circled the globe and recorded sunrise on every continent except Antarctica. Two years after I returned, I received the gold statuette that now stands on a shelf above my computer, an Emmy award for Individual Achievement in a Craft: Sound/Audio.

Despair, I can't help but realize, seems to be what puts me in motion. I hiked up the Hoh Valley and placed my One Square Inch stone as I was just emerging from 18 months of despair when my hearing failed me. And now, fearing that my efforts are falling on deaf ears, I'm feeling another project coming on.

The Hoh Valley, even all of Olympic Park, has become too small for me.

I need more. I feel an overwhelming need to connect with people, fellow Americans, and listen to their thoughts about quiet. Do they even think about quiet? I remember that I recently received notice that the secretary of the interior, Dirk Kempthorne, would be in Seattle for a "listening session" about the national parks. I dig up the release:

> On Monday, March 26, 2007, Secretary of the Interior Dirk Kempthorne will host a listening session in Seattle, Wash., to seek suggestions and ideas on President Bush's National Park Centennial Initiative.
>
> The session at Seattle's Town Hall is part of a series of meetings being held around the country. The President's proposal would provide up to $3 billion in new public and private investment during the next ten years to reinvigorate and strengthen national parks by the National Park Service's 100th birthday in 2016.
>
> Participants are being asked to focus their comments on three vital questions:
>
> - Imagine you, your children, or future generations enjoying national parks in 2016 and beyond. What are your hopes and expectations?
> - What role do you think national parks should play in the lives of Americans and visitors from around the world?
> - What are the signature projects and programs that you think should be highlighted for completion over the next 10 years?
>
> "These sessions are a great opportunity to think big and act boldly to develop a plan to prepare national parks for the future," said Kempthorne.

Think big. Act boldly. Indeed. I need my own listening sessions. I need to listen to the land, both in pristine places and workaday, heavily populated places, because the land does have a voice. It's been 17 years since my last coast-to-coast sound safari, when I drove on back roads from the Carolinas to California. What does the country sound like today? Are people inured to the noise around them? I need to take the sonic pulse of America, in words and decibels, and capture a kind of acoustical, cross-country EKG. This time I'll go west to east, from Washington State to that other Washington, and do what I do: listen.

To be a listener requires a certain willingness to become changed by what you have not yet heard. Who knows what I'll hear, or how my itinerary will change along the way? But here's my plan: If, by the time I walk into our nation's capital, emerging from the national park that runs along the C&O Canal bordering the Potomac River, I still believe in the soundness of my quest for quiet, I'll seek meetings with government officials to try to get them to listen to me. First, though, I need to dust off my wheels of choice, my old '64 VW bus.

This is a car with some miles on it, literally and figuratively, and a fitting, though imperfect, traveling companion. Opening the door releases a whiff of past trips, a heady mix of plant fiber, animal hair, cedar kindling, and fast-food flotsam, such as dropped french fries and spilt coffee. The previous owner was a sculptor who used it to haul his sculptures around, but he maintained the Vee-Dub like a work of art. I've since repainted it teal green and customized the interior for my travels. Before selling it to me in Seattle some 15 years ago, the sculptor imparted three rules of the road that go with this iconic van. First, he told me, you have to be living life right to drive an old VW. You can't be in a hurry, because 50 mph is about tops. Second, you can't even be sure that you will get there, so you might as well start to enjoy the journey from mile one. And third, you have to have lots of friends just to find the parts to make repairs. When I asked, "Why are you selling it?," he simply shrugged and said, "Please, don't ask."

I flick the ignition switch underneath the dashboard, then push the starter button. Nothing. Not even a frog croak. Granted, an old six-volt battery doesn't have the oomph of today's 12-volt batteries, but dead is dead. Nonetheless, like all Flower Power Vee-Dub van owners, I'm prepared. I've parked facing out of my carport at the top of my driveway, poised for the inevitable running jump start.

With a release of the emergency brake, I push from the side until the hill starts to take over, then hop inside and pop the clutch. The carburetor is stuck in full throttle, causing the air-cooled engine to scream. I immediately shut off the engine, apply the brake, get out, then lift the engine compartment door and poke at the carburetor to flip over the throttle. Then I push again. My county road, which descends straight and unbroken for a quarter-mile, serves as my AAA. The engine catches, and soon I'm pulling into the parking lot at the Joyce General Store. I let the engine run while I fill the flat spare and buy some Girl Scout cookies from two smiling sisters,

Nora's two girls, who've commandeered the best selling spot in town. Thin Mints, Samoans, and Shortbreads. Then I drive to Happy Motors.

Two brothers, Dave and Moose, our western versions of Click and Clack, the NPR car mavens, run Happy Motors. In their late teens Dave and Moose started bending over '64 VW buses like mine—when those cars were brand new! Their yard, where they moved the business in 1969, looks frozen in time, like a traffic jam on the way to a Grateful Dead concert. Rusting bugs, vans, and hatchbacks from the '60s and '70s, parked for parts, all facing the same way, are disappearing into the land, growing moss. Between them, forest-tree volunteers have grown to shade giving size.

I approach the barn holding out the box of Thin Mints.

"Can I interest you in some cookies?"

"Oh, yeah, appreciate that."

Dave chews thoughtfully in his blue, zipper-front jumpsuit, unfazed by my desire to get the Vee-Dub roadworthy for a trip across the country.

"Oil change, tune-up, valve adjustment, repack the wheel bearings, check the brakes. Go through everything," I instruct. "And anything you think I might need on the trip across the country, like a spare fuel pump. If I could buy those items from you and take them with me?"

"Yeah, that's a good idea," says Dave. "Make you a little go kit. Clutch cable and throttle cable. Fan belt."

"And take a good look at that starter engine, too. I wrote up some stuff, but you can do whatever you want, because this is going to be a three-and-a-half-month trip, and since it's through the Midwest, I'm not sure I can get these older VW parts there easily."

"You need this in a week or so?"

"No, I'm not leaving until the first of April. You've got a couple weeks."

I start back home on foot until my prearranged ride, daughter Abby in her Mini Cooper, beeps for me to hop in.

Three days later Abby shows up at my house ten minutes early, at 5:50 a.m. on Sunday morning, to accompany me on another prejourney errand, retrieving the One Square Inch stone, which I want to take with me as a visible symbol of the project. Abby agreed to join me for the first leg of the trip, which I've timed for her spring vacation from high school. The plan is to take her with me to various stops in Seattle and eastern Washington, then get her to an Amtrak station in time for her to make it back for the resumption of classes. I want her to come with me today because retrieving the stone marks the official beginning of the journey.

Last week we joked about a covert op by moonlight, using Abby's makeup to blacken our faces. After a series of major winter storms washed out the road in three places, the Hoh Valley was closed to all public entry, and I had been denied a special access permit twice. Fortunately, a last-minute plea to Bill Laitner, Olympic Park's superintendent, who has hiked with me to One Square Inch, helped set up a rendezvous with a park ranger who will let us through the locked gate.

I'm by the door when Abby pulls up. I don't want to be late. This is an occasion to savor. Not only will we be allowed into the Hoh Rain Forest, but we'll have it all to ourselves.

"We aren't taking the VW, are we?" asks Abby.

"No."

"Whew!" She plops herself into the passenger seat of my 2000 Jeep Grand Cherokee, flips on the seat warmer, then promptly falls asleep.

I drive up the Piedmont Road that winds past the mirror surface of Crescent Lake, one of the finest natural amphitheaters on the Olympic Peninsula, or at least it was, before Highway 101 was built. This deep, cold, and clear body of water stretches about a mile across and 15 miles long and is entirely surrounded by steep mountainsides. Its icy waters cool the air immediately above, allowing sound to travel much farther and clearer than normal. Sound waves travel at different speeds depending on the temperature, so the cold, thin air layer that hugs the surface of the lake puts an invisible ceiling on the lakeside amphitheater, enclosing the sound, with remarkable results. Sound echoes back and forth across the lake, connecting everyone and everything.

The call of a loon, a wilderness hallmark, travels far on Crescent Lake; unfortunately, so does the sound of traffic. Highway 101 circles the entire Olympic Peninsula and is the main corridor of transportation between Port Angeles and the logging town of Forks. At all hours you can hear log-filled trucks, passenger cars, and RVs. More than three million visitors come to Olympic Park each year, and many of them arrive via the shores of Crescent Lake, admiring the spectacular scenery, oblivious to the vehicular noise they inject into the equally stunning natural soundscape. When I taught a Joy of Listening seminar at the Olympic Park Institute in the late 1990s, I would take the class here and we would listen to the lake, and then move elsewhere in the park in order to listen more deeply and closely.

Not a peep from Abby as we pass snow-capped Mt. Storm King, then the turnoff to Sol Duc hot springs, and a man on his front lawn pointing an assault rifle at a mound of dirt. He's gunning for moles, using a flame suppressor, no doubt so as to avoid burning his lawn. Ah, springtime in rural Clallam County!

Abby lives with her mother in Port Angeles, some 20 miles from Joyce. Summers, she's known weekly backpacking trips, dating back to the days when she'd skip down the forest trail wearing a glittering fairy princess costume. Her teenage years, however, seem to have brought a change of heart. Recently, after a hike to the top of 5,700-foot Hurricane Ridge, Abby announced, "I'm not a nature girl anymore." The words stung. I can't help but hope she didn't mean it and wonder if I can keep her in the fold.

Her mother, Julie, and I met as bike messengers riding in downtown Seattle. On our first date, calling upon some survival skills honed on hobo jaunts, I roasted her a rabbit over a spit by the railroad tracks. Six weeks later we got married at the Chapel of Three Bells just north of the Aurora Bridge, paying two witnesses $10 each to seal the deal. Abby's older brother, Gordon Leland, nicknamed Oogie after his first utterance, arrived about a year later, in 1985. Now, weeks from graduating from the University of Washington with a degree in computer science, Oogie already has a job lined up at Microsoft. He built the One Square Inch of Silence website.

Light rain splatters the windshield when we arrive at the locked gate leading to the Hoh Rain Forest. Awaiting us is Park Ranger Mark McCool, standing by a fully rigged 4x4. McCool is on duty to inspect the day's catch of fishermen. His belt holds a Glock, two extra clips, pepper spray, and a radio. An assault rifle is braced inside his vehicle.

Sunlight breaks through dark, dense clouds, suddenly illuminating each raindrop that still hangs like magic from the tips of millions of forest twigs. A display as remarkable as Christmas. Steam rises from the black asphalt. The river is milky white; the name Hoh is an Indian word for the whitish water.

"Okay finding a parking spot?" McCool jokes. He is lean, fit, and friendly. After a quick handshake, he unlocks the gate to let us in. "Maybe you guys will get lucky and see a bobcat. I was up there a couple weeks ago. It was great. We were seeing all kinds of stuff. Cougar tracks right on the trail."

Abby is still asleep, so I nudge her to wake her up.

"How long is the hike?" she mumbles.

"About two hours each way," I answer.

"Goddamnit. Can I get back in time for work?"

"Yes. I bought a smoothie in Forks while you were sleeping. Would you like it now?"

"No," she says, before cueing me to her mood. "Don't take it personally if I'm bitchy, Dad."

Everything is drenched. The trail is lined with dripping trees. The moss drapes hang long and straight from the maples. The first mile of the trail has been cleared of storm-downed trees, but soon we come upon path blockers that require teamwork. One person helps the other find a foothold or handhold to get up. Then the one on high helps the other climb up. Jumping down is easier for Abby than for me. My knees bear the wear of nine years as a bike messenger in hilly Seattle. Here in the Hoh, the downed trees can raise a barrier as wide as a person is tall, but thankfully, most of today's challenges are much smaller.

Most of the path-crossing streams are fast-flowing and swollen. Our feet are getting wet. Abby falls behind. I halt at a long stretch of water; a stream has gnawed its way into the trail and claimed the next 100 feet. Abby's silence weighs on me. I wanted her to enjoy this hike. I offer my back, as in the old days, and shuttle her across the flood zone. She dismounts without comment. I remember carrying her down this same path when she was maybe six and wore a backpack; after the first couple of miles I boosted her up on my shoulder, where she sat, resting atop my backpack. We looked like a mini totem pole.

Soon we hit another obstacle, a pile of brush blocking the trail. A vine maple has fallen, creating a maze of snares that make wriggling our way through a slow, tedious task. I hear a distant winter wren let out a long high-pitched, fast warble from the secrecy of a low-hanging hemlock bough. A Douglas squirrel trills in the distance. Abby remains silent. Was it a mistake to bring her? I whip out my Canon Powershot, hoping to coax a smile. No way. She doesn't even pause for the camera. She's a pinkish blur in the shot.

We arrive at the One Square Inch turnoff at 10:55 a.m. The ambient sound, chiefly from the distant Hoh, which is running full, measures 41 dBA. About 100 yards farther, through elk-trampled mud, we arrive at the log where the One Square Inch stone rests in silence. I pocket the red, angular stone and replace it with a round, white, stream-polished stone that will stand in for the original until I return with it months from now.

I motion to Abby to hand me the leather, necklace-like pouch I've made to hold the One Square Inch stone on its cross-country journey. She misunderstands and blurts out, "Go!" To my knowledge, "Go" is the only word that has ever been spoken at One Square Inch. We retreat back through the muck to the main trail and begin the return hike.

"One Square Inch is the place . . ." I start to say. But she cuts me off.

"It wasn't even silent. It's supposed to be the quietest place on earth, right?"

Abby now pulls out the leather pouch from her pocket and I insert the stone, then place this amulet around her neck.

"I'm sure I was not the first person to say a word there," she snaps, removing the amulet from her neck, then winding it into a ball and stuffing it into her pocket.

"Why don't you want to wear it around your neck?"

No answer.

"I'd like you to wear it from now until you give it back to me when you get on Amtrak in Wenatchee."

No answer. Not even eye contact. Abby turns and starts back alone, leaving me at the sitting log, where I usually rest and take notes. Today I impulsively say a prayer:

> Now begins the journey of the stone—I will make a prayer for all things peaceful and my wish that even the smallest dream from the smallest person from the smallest place can still make a difference and be granted, and that no wish be considered too small to be important. And that I pay attention to the little things in my life and understand that someday my anger and my frustration will disappear and I'll be whole again. I want this—I want this so badly. It's not just about quiet anymore. It's about the pain

in my jaw. It's about the angst of teenage years. It's about faith and moral values, and it's about peace, the search for peace.

I eventually catch up to Abby and we arrive back at the Jeep together. She announces that the hike made her sick. She sleeps all the way home and a couple hours more before I wake her up to go to work. She is a bagger at the local supermarket from 6 to 10 p.m.

I return to Happy Motors on foot with the OSI stone around my neck. Reaching the grassy bend in the driveway, I spot Moose hunched over, oily rag in hand, putting the finishing touches on my Vee-Dub, two deer munching away at his bed of tulips.

"You like my van?"

"Yeah, I like it. We got to get it out of here, though. Everybody wants to buy it. That little stove in there is the greatest," says Moose.

"And it works, too. Just throw in a handful of wood."

Moose fills me in on all they've done and not done to the Vee-Dub. Bending over the fuel pump, he says, "See these clamps on here? We've seen lots of fires, so we added them. And on this lid here, it's not 100 percent tight, but it's completely frozen, and if I start screwing around with that we might be getting into a major project. It's not bad, it's pretty good. You can take a little piece of rubber if you wanted to tighten it up."

He points to the battery. "It didn't have a protective cap—that's kind of scary. It was coming real close to this edge. If you get it hot to ground, it will turn this strap red-hot. What I did was take some neoprene, it was only that far away," he says, holding up his forefinger and thumb a fraction of an inch apart. "It makes me so I can't sleep at night. I cut a strip of adhesive—that was one safety factor. You'd be surprised how many fires we see. The voltage regulator, we took your old one off. Changing this generator, it was a nightmare. The tranny was down a pint. Packed front wheel bearings and put in new seals. The lock plates—you can see how these things get all chewed up."

And that's how it goes. A pinballing mechanic's recap of all the work they'd done—some of which I understand. What I understand for sure was how much loving attention Moose and Dave had bestowed on my van.

"So, everybody wanted to buy this, huh? You know, I always tell anyone who asks, that it *is* for sale. Then I wait. When they ask how much, I say, 'Seven hundred dollars,' then pause, and before they reach for their wallet I quickly add, 'Seven hundred dollars for the bus and then another twenty thousand dollars for the Karma.'"

Moose chuckles.

I'd coveted this Vee-Dub for years and left notes on it with my name and telephone number. But the owner never called me. Then one day I was taking my son to the hospital because of a possible kidney infection, and I saw the VW bus drive by with what looked like a For Sale sign in the window. "Oogie," I said, "you're going to be a couple minutes late." I chased it down and got the phone number off the sign. Buying it was like undergoing a prospective parent interview at an adoption agency. "I'm not sure I'll sell it to you," the owner said. I had to be interviewed! When he did agree to sell it to me, we hadn't even talked price. My checkbook was out when he said "Twenty-five hundred." The next day, when he dropped it off, it came loaded with every spare part imaginable inside. It had 95,000 miles on it and a newly rebuilt engine.

Some 41,000 miles later, I'm writing another check. The Happy Motors bill comes to $1,279.12. I thank them and turn to go. Dave's voice spins me back around: "First couple of times you hit your brakes, hit them easy. Sometimes a little dust gets stirred up."

Abby is waiting for me in front of Julie's house in Port Angeles on April 1. I've packed the Vee-Dub with wood for the stove, fresh lamp oil in the lantern, extra blankets and pads for frosty nights on the rooftop sleeping rack. There are cans of food, too: Dinty Moore and Lipton soup and packages of Top Ramen, Balance bars, and granola. We've got five gallons of water in a corrugated Igloo with a large red-and-white label that reads "Industrial Water." I've stashed a case of STP lead substitute behind the driver's seat—an essential to keep the valves from burning—and added a cell phone booster antenna to the top of the sleeping rack for checking e-mail in Montana. And there's a bag of homemade candy Sue gave me. One piece for every day of the trip. She also handed me two cards, one to open after the first day, the other an anytime card.

I've felt a cold Alaska wind out of the northwest, foretelling harsh weather, but I see Abby has packed light. All of her stuff fits into a day-pack.

"Are you sure you've got enough warm clothes? What have you packed?"

"I have a couple shirts and pants, a sweatshirt, two pairs of socks . . ."

"What do you have for shoes?" I ask.

"Flip-flops."

"You can't wear flip-flops to the Seattle Symphony!"

She shrugs. I bite my lip.

"It's going to be freezing cold where we're going, Abby. You might want to bring a hat."

"I'm going to get my swimsuit."

She disappears back upstairs to her room while Julie and I exchange bemused looks.

Abby reappears dangling a bikini. Julie takes our picture.

With a flick of the switch and a push of the button, I fire up the Vee-Dub and we're off.

Before we reach the city limits, Abby smooshes her jacket against the car door, pillow-like, and announces she's going to crash. But the finicky door pops open. Instinctively, Abby grabs for the handle on the dash. She smiles proudly. As do I.

Our first stop is Sequim, the next town east on Highway 101, for cheese-burgers and milkshakes at DQ and earbuds for Abby's iPod at Radio Shack. Back on the road I ask her what she's listening to.

"I don't know."

"What do you mean, 'You don't know'?"

"Acon and Rick Ross," she answers. The names mean nothing to me.

The door again flies open at the sharp turn near Jamestown. Abby slams it shut and settles back in her seat. Her cell phone rings. Her one sentence answer, "I'm with my dad," redirects the teenage torrent of communication that ensues. It's all text messaging after that. And song after song after song on her iPod, music that I can hear from across the VW.

"You know, Abby, your iPod is set way too high. It hurts my ears when you use it, and if it doesn't hurt your ears, it's because your ears have a hearing loss."

"Okay, well, it's already happened."

"Yeah, well, it's probably temporary at this point. Will you turn it down, please?"

She makes a motion as if she is turning it down.

"Is it down?"

"I turned it down. Okay?"

A headache is beginning to blossom, sending needles down the right side of my face and causing my jaw to ache. I ask Abby if she's ever had a hearing test at school.

"What? I can't hear you. The car's making too much noise."

I've got to concede that point. The VW is no Lexus when it comes to a quiet ride. It rattles and groans as its antiquated, air-cooled engine strains with even the slightest suggestion of a hill. Plus I've added a set of wind chimes, which tinkle with every bump. In the rearview mirror I see a line of vehicles bunched behind me. I pull to the shoulder to let them pass. Abby's iPod has now moved on to rap music.

As I pull back on the highway, I think about an unwelcome silence: the silence between a parent and a teenage child.

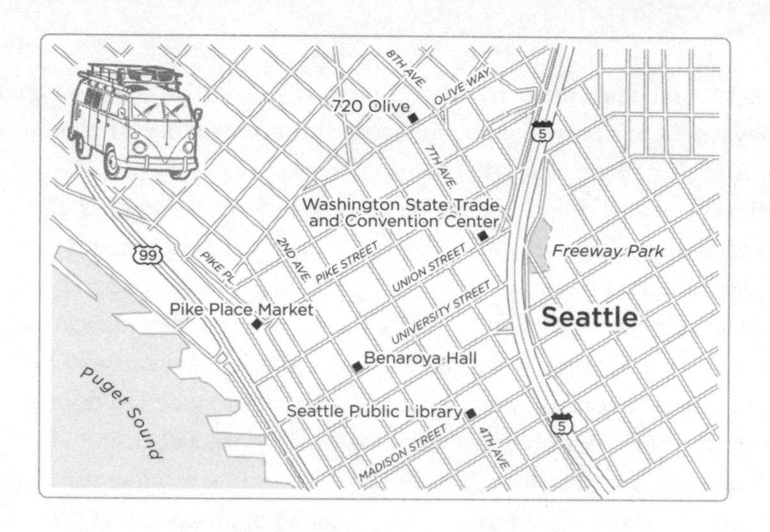

4 Urban Wilderness

And what is there to life if a man cannot hear the lovely cry of the whippoorwill or the arguments of the frogs around a pond at night?

—Chief Seattle, letter to President Franklin Pierce, 1855, regarding the proposed purchase of his tribe's land

From the deck of the Winslow Ferry, the city of Seattle rises like a towering mirage between sea and sky, both gray. Abby and I boarded, driving on 15 minutes ago, then headed upstairs for a snack and the Puget Sound view, which now offers the Emerald City skyline: Smith Tower, Columbia Center, Maritime, Coleman Building, Key Bank, Washington Mutual, Rainier Bank Tower, Financial Center. At one time I could name virtually every prominent building and, seemingly, every business on every floor, not to mention tell you which threw the best Christmas party.

I was a Bucky Bike Messenger, one of some two dozen anonymous key holders to the city. We delivered just about anything in hilly downtown Seattle in every kind of weather, usually rain. The name Bucky was synonymous with "Open sesame." Say it once at any door, and that door would open. Two-wheeled insiders, we got to know who did business with whom, who was on the rise, and who was on the outs. Our wheels helped the city spin. And sometimes our wheels stopped spinning.

It has been said that there are two kinds of bike messengers: those who have been hit and those who *will* be hit. I didn't remain a virgin very long. My second day at Bucky's I had my first accident. A car popped out of an underground garage, its driver giving little thought that a bicycle might be whooshing by. I got hit 11 more times during my nine years as a bike messenger, mostly because I was slow to realize that I needed to do more than stay on the lookout for trouble. My eyes alone couldn't possibly do the job. Only hearing can monitor every direction at once, even foretell what may lie around a corner. Eventually I became so good at listening that I found it impossible not to listen. Eskimos have a word for this, *seuketat*, which, literally translated, means "the ear of the animal." During my Bucky days, riding my delivery bicycle in rivers of moving steel, I was an animal surviving in an urban wilderness, a more dangerous environment than anywhere I have been since, including the Amazon and the Kalahari.

I learned the sound of car brakes that work, and those that don't. I learned the sound of an engine in tune and the sound of a diesel that's late for its oil change. I came to know the sound of a policeman's Kawasaki more than a block away. I know the thundering crescendo of rush-hour traffic, and how it suddenly diminishes when the cars slow to bumper-to-bumper. I know, too, the relative silence of a secluded private residence near the top of the Columbia Center, a residence that takes up an entire floor. I have heard the sound of street musicians in front of Nordstrom's and Bon Marché before outdoor music was used to drive them away. I recall the sound of steamed milk at the Starbucks at Pike Place Market when it was the only Starbucks in the land. I have heard the screams of sidewalk preachers just before they crack. I have listened to the sound of bus tires on flooded streets and the sounds of my bike tires on freshly fallen snow. I have heard the gutters beat rhythms below the Alaska Way Viaduct near the Public Market and the irregular *whack-whack—thump-whack* of cars and trucks passing over the overhead roadway seams.

More than a century and a half ago, before most of these urban sounds even existed, Sealth, the chief of the Suquamish Indians, also known as Chief Seattle, offered a Native American perspective on noise and quiet:

> There is no place in the Whiteman's cities. No place to hear the leaves of spring or the rustle of insect wings. But perhaps I am a savage and do not understand—the clatter only seems to insult the ears. And what is there to life if a man cannot hear the lovely cry of the whippoorwill or the arguments of the frogs around a pond at night? The Indian prefers the soft sound of the wind darting over the face of the pond, and the smell of the wind itself cleansed by a midday rain, or scented with a pine. The air is precious to the Redman. For all things share the same breath—the beasts, the trees, the man.

Today we live in a time of planetary fast-forward. We've unleashed a million years' worth of changes in a single human generation. We take from the Earth and return less than we greedily grab, and all too often we don't even know what we've lost. Like the sound of our own footsteps; I did not realize they were lost until one autumn evening in Venice, Italy, a city famous for the presence of canals and the absence of cars and other motorized vehicles. There to record the city and the route of the Orient Express to Paris, I listened spellbound from my open hotel room window. An older man and woman walked hand-in-hand in my direction from a block away. The sounds of their footsteps tapped on stone pavement, a slow, relaxed gait, leather heel and sole laying down two separate sounds for each step. The couple walked in perfect unison, and in my mind they, too, were aware of this, their very footsteps embodying their perfect match. Then a gritty pivot signaled a pause, a turn to intimate silence followed by soft conversation, during which the bells from a distant tower echoed through the narrow street, clanging in a many-layered wave.

I heard something very different, but no less wonderful, in Tokyo, where I was sent to record the sounds of Japanese pedestrians for a computer game. In the heart of Tokyo is Shinjuku Station, through which more than two million people pass each day. The urban din of cars and trains dominated the outdoor urban soundscape, rendering footsteps unintelligible above ground level. But deep below ground, away from traffic noise in the station's subterranean corridors, came a stunning surprise. The advent of soft soles, the short-shuffling strides, and the almost complete absence of conversation rendered the movement of commuters into a constant swishing of clothes, a sound comparable to the flapping of feathered wings.

The simple step is one of nature's basic sounds. Predator and prey alike are keen to its subtle clues: hoofed, clawed, or padded feet; quick or slow strides. And numbers: many steps or those of a lone interloper. The footstep, even when silent, can leave a sound print. I learned this important listening lesson one early morning in the Sinharaja Rain Forest of Sri Lanka the day I made this entry in my field journal:

> It was completely dark as I walked into Sinharaja to record the sounds of dawn. Occasionally a burst of stars would appear overhead through the forest canopy, and the air was warm and humid, ideal for listening. The surrounding tree frogs and the insects wove rich textures, unlike any that I've heard from temperate latitudes, and this extra treat made my prospects for a valuable recording high.
>
> I found my position, next to a little clearing at the crest of the hill where the first rays of sunlight would descend, and I set up to record. Suddenly I was overcome with panic! I wanted to run but for no apparent reason. As I tried to get a grip, I told myself, "You are halfway around the entire planet, you fool, and if you don't stay and record you may never be back again! I don't care what you're feeling—just stay put!" Then I realized, "You don't need to be here, not now! Just leave the equipment running, and in two hours you can come back and pick it up."

Four months later I was back in my Seattle studio listening to that beautiful morning. Yes, the sounds were wonderful, clear yet complicated, and I was eager to listen to the bird calls that I'd had missed when I left the recorder running unattended. Then I heard something odd, almost an imperceptible eclipse of direct sound, as if a shadow of a massive object had suddenly appeared. I rewound the tape and listened again. Yes, something was clearly blocking the sound. I heard the sounds of my departing footsteps—and then the guttural growl of a leopard, before it, too, departed through the bushes.

This brush with the leopard helped teach me what it means to become a nature listener. We are already perfectly prepared, by instinct, to listen to the world around us. Rather than ask How do we go about listening? I've found it more essential to ask How can we reduce the distractions of the modern world?

Our footsteps are self-defining, individually and culturally. Murray Schafer, the author of *Tuning of the World* (Random House, 1977), has proposed that the ability to hear our own footsteps should be a noise standard for cities, meaning, simply, that the places that we inhabit should be quiet enough that we can hear ourselves (and others) walk. That honorable goal, however, is far from obtainable in most cities, where machines now rule the acoustic day.

Abby and I have not heard our footsteps since disembarking into downtown Seattle's inescapable din, a kind of grime that cannot be seen, only heard, and is both difficult and expensive to erase through noise abatement, acoustic design, and behavioral change. At a Thai restaurant near the Seattle Center, with a view of the Space Needle and a statue of Chief Sealth, I lay out for Abby the aims of the next several days, the reasons for going to tonight's pro basketball game, for touring the much-talked-about new public library, for taking a walk through Freeway Park, and finally for attending a symphony orchestra concert at Benaroya Hall. After we visit acoustical environments designed and created by humans, we'll go to a very different place, Pipestone Canyon, one of my favorite listening spots. In rare places like this, I tell her, you can still step into a pristine natural amphitheater like that experienced by our earliest ancestors, the kind of place where human hearing evolved, a place where we can get more in tune with our humanity.

Abby doesn't get it. Or hasn't listened. And now she sits sullenly across the table from me, iPod on, earbuds in, head turned away. When I catch her eye and frown, she yanks out the earbuds.

"The whole One Square Inch thing—I think it's bullshit. I've always felt that way. It's stupid."

I guess I've got only myself to blame for this. Victim of that old expression "Be careful what you wish for," I had wanted Abby along for a stretch of the trip to stand in for today's youth, as she moved from the noise of an urban environment to the deep listening insights of a naturally quiet place. I hoped to gather her observations and insights, too. Well, she's not very shy with her opinions. So let's have them.

"Why is it stupid?" I ask, dutifully allowing Abby to give her voice to the project.

"I think it's a waste of time."

"Do you understand that if you preserve quiet at One Square Inch that it affects noise pollution for 1,000 square miles? That's not bullshit!"

"I don't want to argue about it," she says. "I don't care."

After 16 years, she knows how to push my buttons.

"You know, your not caring is really coming through. What *do* you care about?"

"Maybe what a regular teenager cares about. I care about friends. I care about having fun. I don't feel the need to go deeply into things I don't care about. I'm willing to take a train home or a bus. Right now."

Our chicken satay arrives in the nick of time.

I may not see her again for months, I think. We need to spend time together, one on one. What am I doing heading off on this trip? I feel caught up in the momentum of plans.

After dinner we cross Denny Avenue, which is clogged with cars headed to the basketball game at Key Arena, the Seattle SuperSonics versus the Denver Nuggets. We have seats in Section 216, Row 1, up high. I purchased them online, at a premium, from some guy in New Jersey.

Although I am not particularly a sports fan, I have attended my share of games. Microsoft has sent me to football, baseball, hockey, and basketball games and also to golf tournaments to capture the sounds of these venues for the company's video games. I was hired to apply the same wildlife and nature-recording skills that I employ on behalf of museums and films and my own CDs to sports stadiums filled with cheering fans. Different as they are, I've discovered that these two settings have a lot in common. Where else can we scream like a wild animal and still be socially accepted? But more revealing, I found that just as it is possible, by sound alone, to accurately tell the time of day in the Amazon rain forest, so, too, can I fairly accurately predict, merely by crowd noise, which team has the ball and how many minutes remain in the game or in overtime.

Soon after Abby and I pass through the turnstiles at Key Arena, my sound-level meter climbs to 74 dBA as it measures the babble of voices that reflect, unbaffled, off the concrete surfaces of the interior architecture. It is difficult to speak, or would be if Abby and I could think of anything to say to each other. When we arrive at our seats the piped-in pregame rock 'n' roll raises the meter to 83 dBA, louder than six-foot surf crashing on shore. But this perfect storm of manufactured home team enthusiasm has only begun.

"Let me hear you GET LOUD!" says the announcer at 90 dBA, a sound level typical of thunderstorms. The crowd dutifully cheers.

"Are you ready?" the announcer yells again, encouraging more cheers, and we see the arena's own "noise meter" buoyantly bounce up. The half-filled arena registers 98 dBA on my handheld meter. The storm is now overhead.

Boom-boom-clap, boom-boom-clap. Play has started. The first bucket: Sonics, 98 dBA. Second bucket: Nuggets, 90 dBA. Third bucket: Sonics, 102 dBA.

The person who controls the sound at the aptly named SuperSonics games sits high above the arena floor behind one of the baskets, up near the American flag, in a small room known as "the bucket." His name is Andrew Moren. Moren took the job in college for $25 a game plus a free meal, and 12 years later he is still at it. Even with a family now, and working as a contractor installing fire-preventing sprinkler systems, he still does sound for the SuperSonics because he enjoys the work, which is all about entertaining and stoking the crowd. He compares his job to creating the soundtrack for a Hollywood movie.

"I look at it the same way," he says. "A crowd plays a big part in the game. In my mind, they make a difference of a few points, and a lot of games are decided by a point. There's been games where I felt like I contributed to a win and there's games where I probably feel I should take responsibility for not getting the crowd more excited—and we might have lost by a point."

Moren uses software called the Game Ops Commander that puts all manner of sound effects at his disposal. His "made shots" folder gives him options like *baboom.wav* and *biojump.wav* and *torpedo.wav*. His "missed shots" folder enables him to razz the opposition with such sonic raspberries as *haha.wav* and *sucksuck.wav*. When Moren deems it time for a musical jolt he's got 5,000 choices in his song bank. Often, he says, his role is filling in the gaps in the game.

"For me, the loudest sound of all is silence. That catches my attention in a heartbeat," he says. "When there's no sound, it bothers me. I jump on that as fast as I can, and I fill it with something, some light background music, if nothing else. Non-recognizable instrumental, that's the safest.

You don't necessarily have to jump in with a very popular song. You don't want to pull attention from what the event is, but you do want to back it up so that it doesn't feel like it's been kind of stripped down or naked."

As for the arena's video decibel meter, he readily confesses it's not the real deal. "It's got ten boxes, three green, three yellow, and four red. The higher it goes, it seems to prompt the crowd. And we put a note next to it that says, 'Makes some noise.'" He identifies this as "a video roll. It looks the same every time." It would run identically if he were the only one in the arena.

Moren says his only experience with a real decibel meter "is when a network like NBC comes into the building. They have tight rules on the maximum volume, because they want to hear their announcers. We were fined, $10,000 I believe, because we pushed the volume too loud. That was during the big playoffs against the Chicago Bulls and Michael Jordan, so I believe the crowd was 90 percent of that volume and I was a very small part of it."

The National Football League, with its much larger stadiums, has clamped down on noise. Its rule book devotes a long section to crowd noise, noting, "Artificial or manufactured crowd noise in NFL stadiums has increased to the extent that teams have notified the league office that they have experienced difficulty communicating within their bench area as well as on the field." Expressly prohibited: "The use of noise meters or such messages as 'Noise!' 'Let's hear it!' 'Raise the Roof,' 'Let's go Crazy,' 'Pump it up,' and '12th Man.'" These and other noise offenses can be punished by a five-yard penalty or loss of a time-out. Still, the noise is so loud that players sometimes must count on visual signals, similar to soldiers surrounded by the noise of combat.

Hand signals have long been invoked on the frenetic floor of the Chicago Board of Trade, but lately a company called Sensaphonics Hearing Conservation, Inc., has come to the aid of traders. The company, which also makes special earpieces for Indy 500 drivers, astronauts, and musicians such as the Rolling Stones, has helped ease traders into the even faster paced world of electronic trading with a special earpiece it calls the ProPhonic TC—1000. The TC stands for telecommunications. "They used to have a guy sitting at a computer and a runner would run to the floor and give the trader the order. He'd make the trade, the runner would get the paperwork and run it back to the computer guy," says Sensaphonics president and founder Michael Santucci, explaining that the Board of Trade now facilitates internal-only cell phone calls between desk and trader.

"However," continues Santucci, "when it's busy, the noise level is extremely high."

"Do you know how loud it gets?"

"I do, but I'm under a nondisclosure agreement not to say. I measured it years ago. Let's just say it's at rock concert levels. It gets that high. Figure 100 people standing in a circle and you're in the center of it."

The traders, he explains, first tried to make do with sound-blocking earmuff units, but they weighed as much as two pounds and added to the already high stress load. So Sensaphonics designed a one-piece custom-fitted earpiece with a sound-canceling microphone that weighs less than half an ounce and plugs into the trader's cell phone. The device costs $650, a small price for helping to avoid multimillion-dollar miscommunications.

Noise also plagues the nation's pastime. In *The View from the Stands: A Season with America's Baseball Fans,* Johanna Wagner writes of her summer-long 2002 tour to all of the major league parks, observing, "The noise level between innings is intolerable to most." She quotes fans angered by the loud music between innings. And she names one ballplayer who made his own stand for quiet: New York Yankees centerfielder Bernie Williams, who is also an accomplished musician, requested that no music be played during his walk to the plate to allow him to concentrate.

Abby and I, too, feel assaulted by the noise at this sporting event. We don't even last through the first quarter. Basketball's not our game. We'd heard enough. We walk in silence back to our room at the Travelodge near the Space Needle. I sleep for nine hours. Abby logs nearly twelve. In the morning she helps me take a picture of the OSI stone with the Space Needle in the background. I expect to take more such photos as my Ambassador for Silence makes it across the country.

But I've got a couple more indoor spaces to visit in Seattle before pushing on. Abby has taken the path of least resistance, shopping, while I set off for an appointment on the 14th floor of a familiar address. Back in my Bucky Bike Messenger days, we called 720 Olive Way the Dajon building. Nowadays it's named after its largest new tenant, Marsh & McLennan. I know it as the site of my eighth mishap with a car. My bike helmet hit the pavement here on Olive Way. I suffered a concussion, learning the hard way what our human ancestors knew even before they spoke to one another: hearing is our most important survival sense. We evolved, after all, without ear lids.

I board the elevator to reach the offices of Yantis Acoustical Design. I've arranged to meet with Basel Jurdy, one of the acoustical engineers responsible for Seattle's new downtown library, a striking glass and steel edifice hailed as "exhilarating" and "thrilling," even "an ennobling space." How, I wonder, does a new-millennium library define quiet? Do its librarians still put their fingers to their lips and say "Shhhhhhhhhh"?

None of the glowing reviews of the $169.2 million library that I read referred to the acoustics. In advance of today's visit, that observation was put to Sam Miller, a senior principal at Seattle's LMN Architects, joint venture partner with the Dutch architect Rem Koolhaus and the Office for Metropolitan Architecture on the Seattle Public Library.

"To me that means it was successful," said Miller. "If it didn't work acoustically, you'd be reading about it. In fact, I think it works really well acoustically." He explained that Deborah Jacobs, the city librarian, said the goal should not be hushed quiet. "She said, fairly early on, that she wanted a library where people could be reading a book in and among other people and come to a passage that's funny and find themselves laughing out loud and not feel self-conscious, not feel like they're the center of attention."

Historically, Miller explained, the grand reading rooms were made of stone, terrazzo, and wood, and these hard, reflective surfaces defined the acoustics. In the past 30 to 40 years, carpeting and more absorptive surfaces helped deaden the sound in many libraries. Seattle asked for something in between, something, he says, "with a little life to it."

Tall and self-confident, Jurdy meets me in the reception area and escorts me back to a corner conference room. His soft, carefully chosen words bespeak an engineer who combines both art and science. Our initial get-acquainted chatter, about the local high-rise building boom and the number of cranes (Jurdy counted 13 in the nearby city of Bellevue), prompts me to ask about noise regulations for buildings. How loud can their heating and air-conditioning systems be?

"The code says not to exceed sixty dBA at the property line day or night," says Jurdy, explaining that that is the standard for a commercial district. "When the receiver is residential it's fifty-five dBA during the day and forty-five at night, so there's a penalty for nighttime."

"From my perspective, forty-five dBA is sort of a distant stream through the forest, still very present."

"It's interesting when you bring me this perspective, because when a client asks, 'What's thirty dBA?' it's hard to tell him, because we can't just be quiet and hear it. It doesn't exist in this urban environment unless we go to a specialty space."

Jurdy suggests we walk to the library because he wants to show me something on the way. That something turns out to be a women's clothing store, Anthropologie, which aided the acoustic design of the library. "We searched for a space to demonstrate the acoustics, not only to the architects but also to the librarians," he says after we step inside. "And we finally found this space, which had things in common with our future space: hard concrete, hardwood floor, a few low and high shelves." Music that Abby would enjoy pulses loudly. As I do a 360 he continues: "We asked the manager to turn off the music, came in here, and carried on conversations. We did the same exercise with the librarians. They all felt comfortable with it."

I can hear and understand him above the music, but not easily. When we are back outside, continuing on to the library, the urban din takes over. A truck roars by. Jurdy's mouth moves, but his words never reach my ears. Traffic noise now has the loudest say in our conversation. The Seattle Noise Ordinance limits noise to 95 dBA for any single vehicle, 35 dBA above property limits because the impact is temporary. How many missed words or misunderstood syllables can we tolerate before meaning is lost? And there's more than just the intelligibility of words. From a nature listener's point of view, I want to hear not only every word clearly, but also the heartfelt strains and tones in the speaker's voice. A touch of sadness. A note of irony. A smack of sarcasm. Noise rides roughshod over nuance with a constant flow of "temporary" impacts.

When we reach the 5th Avenue entrance to the library I take out my sound-level meter and measure passing vehicles from about 20 feet away. My meter pops up into the 80 to 90 decibel range. Changing the setting to average all values over 60 seconds, I see the level drop to 75 dBA. Even the lower value, I now know, is more than 10 times what a building owner would be allowed to emit into adjacent property in Seattle. Why the break for rolling thunder?

Jurdy leads me inside, to the main atrium, which the designers refer to as the Living Room. My meter hovers at around 55 dBA. Interesting. Although not as loud as normal human speech, which is commonly rated at 60 dBA, it is certainly competitive with it, meaning that you would have to stand close to another person to be heard. Yet my subjective experience is not noise, but quiet. I find this astonishing.

"In a classical library," Jurdy tells me, "you might want it to be dead acoustically, very absorptive, no response from the room. But here, as we started to talk through the function and the usage and how this room communicates with other spaces in the library, we felt as a team it can be lively, to really relate to the rest of the space as a gathering space, a living room. You hear people talking, but they belong where they should be. That woman's voice came from over there—it belongs to her. From a background noise standpoint, we introduced a little bit more background. You see on the floor there, nozzles are blowing air up, to keep this façade cooled. The majority of the cooling system is an under-the-floor system, which is more energy efficient and also quieter than a traditional overhead system."

He points upward some 100 feet to a structural component. "The architecture itself allowed us to do some absorption in this space. You see the black underside? That's a fireproofing material that is also sound-absorptive."

Riding up the escalator I comment on the appealing nature of the bright, fire-engine yellow of the moving stairway. Jurdy describes a very different use of color elsewhere in the building. "We should stop in the rest rooms. The colors are really nauseating." Intentionally so, he explains, to dissuade drug users from occupying them.

"Is it working?"

"I believe it is."

On our ride up an escalator I measure 62 dBA, as loud as normal conversational speech and even the brightest dawn chorus of songbirds in spring. Yet still, the overall sense of space feels relatively quiet.

The Mixing Chamber is a room full of computers, a reinterpretation of the traditional library reference room. I get a reading of 49 dBA, typical for a mountain stream that begins its descent in splashes and gurgles across moss-covered rocks. But rather than eventful, the acoustic energy, which is certainly present, is uneventful and the overall experience is quiet.

We travel on. In the stacks of books I note the quietest reading yet: 40 dBA. In the Reading Room, where only two of 17 people are holding books (the others peruse their laptop screens), the meter registers 44 dBA. Then a phone rings. From 30 feet away, my meter jumps to 58 dBA, equivalent to the snap of a cedar campfire in the dead of night.

Finally we arrive at the 4th Avenue entrance, where book processing occurs using a metal conveyor belt. There is an industrial sense of space here, the ceiling low, confining. Compared to the airy Atrium upstairs, this feels like being in a mine. With the ambient noise here of 63 dBA, you'd need to raise your voice to talk. Nearby is the Microsoft Auditorium, with its six-inch-thick, sound-damping doors and acoustically treated walls. It is empty but the doors are open, and outside "quiet" intrudes at 50 dBA, similar to a steady but soft rain.

My tour ends in the children's section, in the oddly shaped enclosed space called the Story Hour Room. I can happily recall reading books to my children when they were young, and I still enjoy watching children being read to or told stories. Particularly young children (the age group least likely to have impaired hearing) have not yet learned to listen in a modern sense, in which they focus their full attention on the teacher or parent. Instead, they remain instinctual animals, listening to everything, as if their survival still depended on it. Four- and five-year-olds make very good acoustic naturalists. Just put one on your shoulders and go for a walk at night.

My surprise is that there are *no* children here! About 100 feet away I see two adults carrying on a conversation, and what's more, I can hear everything they're saying at 52 dBA. Both the ceiling and the floors are highly reflective. I try to imagine the room with five or six, let alone two dozen children, and ask Jurdy about the inevitable "cocktail party effect." He's been told the children are well controlled, but my parenting experience makes me a bit skeptical.

I thank Jurdy before he departs, then return to the Living Room and settle into a cushy chair to reflect on what I've heard and seen. The ceiling 100 feet overhead is grand, no question about that, and the building's transparent skin, divided into diamond-shaped, triple-paned glass, is equally stunning. This visual grandeur inspires a quiet consciousness. It reminds me of a canyon area where public space is outdoors and natural. The acoustics here are very forgiving. I detect a soft babble of sparse voices with a granular texture, from how far or what direction is not easy to tell. I can see lips moving, lots of conversations, some as close as 50 feet, but I can't hear a single one. This is a gathering place with acoustic energy. I feel it.

The library is a remarkable achievement—a strong architectural statement offering a stunning landmark on the Seattle skyline and equally stunning views outward from the Living Room and Atrium, and a relative sanctuary from the noise of the city. There is quiet here, if not the stereotypical "Shhhhh" kind, a very functional psychological version. But can "quiet" really reach 50 dBA without surrendering something? When I measured the sound level in a "quiet" Boeing 737 as high as 81 dBA, I thought about the modern world's sliding scale on quiet and our ability to adapt, to tune out the world around us when we have no other choice. These modern, high-decibel versions of quiet, instead of offering places for heightened awareness, do the opposite. Like a bright, white room without detail, this loud quiet offers little. I hope my work will move people to listen more carefully and judge the true value of modern acoustic spaces.

The Odem Library at Valdosta State University has taken a different, much simpler approach to quiet. Visitors to this Georgia library can refer to a map that divides the 180,000-square-foot facility into equal parts, Blue Zones and Green Zones. In the quiet Blue Zones, cell phones are prohibited and only whispers allowed; Green Zones allow normal conversations and cell phones set to vibrate. The policy is similar to Amtrak's quiet car aboard the Acela, and it's had a positive effect. A circulation desk manager called the measure a success, citing 80 percent compliance.

Before literacy became the measure of education, aural traditions were. Our first libraries, if you will, were places of listening: the supper table, bedside, churches, town squares. While in Hawaii to record chants for the Smithsonian, I learned that until the nineteenth century Hawaiians had no written language. Their entire library was chanted. A single chant, such as the *Kumilipo,* or creation chant, tells of life arising out of the ocean and onto land. It runs more than 2,000 lines.

Sitting here in the cushy comfort of my overstuffed reading chair in the Living Room, I think maybe this is a stepping-stone to the library of the future, when the Information Age has shed Gutenberg entirely. I imagine the library of tomorrow will be a place of social intimacy, more like fine dining: an intimate place with soft lighting and a quiet atmosphere, where you can bring a friend or meet someone on neutral ground to exchange information with resources right at your fingertips. Book groups will gather there to discuss that month's selection. Authors will read from their books. Documentaries will be shown and discussed afterward. After all, I can't think of one thing currently at the library that isn't bound for the Internet—except the opportunity to engage with other people. The new Seattle library currently has two primary engagement areas: the Anne Marie Gault Story Hour Room and the Microsoft Auditorium. The rest of the library is practically engineered with auto-shhhhh, making any intimate conversation (or eavesdropping) impossible.

Our public gathering places, for sports, literature, learning, and music (my next Seattle stop), are intentional spaces, highly structured, and thus result in somewhat contrived experiences. Whenever I visit them I'm reminded of the vital importance of preserving places outside of human intention, unspoiled wilderness areas, places where we might regain sensory balance and learn from the unscripted, unedited, unenhanced, raw opportunity of nature. The wilderness is a place for recreation and certainly for learning and for the planet's original, purest music. The Earth speaks to us through the language of the senses. I don't begrudge $169.2 million spent on a library, but why not allocate even a fraction of that amount to preserve the living library that is at Seattle's back door: Olympic National Park.

Abby and I meet at Westlake Mall and retrieve the VW from the lot where I left it. We head for her maternal grandmother's house, arriving at about four o'clock. Yvette Keefe lives east of Seattle, in residential Bellevue on the opposite shore of Lake Washington. She's transformed her yard into a garden and a spiritual oasis. Columns of old Douglas fir trees rise bold and strong, and rhododendrons, azaleas, and primroses add greenery, color, and grace closer to the ground, near her many well-nurtured beds. Yvette spends most of her time either "down in the backyard grotto," by the fern-lined fishpond she dug by hand, or in her rocking chair in her kitchen listening to classical music or NPR's *All Things Considered* or reading.

We say hello. Abby quickly darts off down the hallway to her bedroom, where she always stays when at Grandma's, presumably to catch up on text messages. I stay with Yvette in the kitchen, where she sits in her rocking chair, hands clasping the sides of her teacup. Red Rose, always, with just a splash of milk.

She asks about Abby, already sensing that we're at odds. I explain that she really doesn't want to go to Pipestone Canyon.

"She's not anxious for that?"

I take a long pause, then sigh. "It has been like pulling teeth. I got this all cleared with Abby months ago."

"She said she'd go?"

"She said she'd go. Everything was cool. So I asked her, 'Now you're doing nothing but complaining. Why didn't you speak up back then?' And she felt like she didn't have a choice, that it was easy to say yes, back then, when it was months away. And now it's hard to actually do it. And I can understand that. This trip is hard."

"Well, do you actually have to have her?"

The phone rings and Yvette answers, "Hi, Jocelyn." She switches to fluent French.

My ex-mother-in-law, though born in Gravelbourg, Saskatchewan, grew up in Seattle, where her father worked as a steelworker, then as a maintenance man during the Depression. Her mother worked occasionally making cookies at a bakery. Yvette remembers the family taking in houseguests from time to time to make ends meet. She has lived more than 50 years in the Seattle area and alone for the past couple of decades. At 88 she is the matriarch of her clan and has provided me with wisdom since the day I was reintroduced to her with these words: "Mom, you know the guy you met last weekend? We got married." I made sure to return to the kitchen later and talk with her alone. "I know that you must be frightened and think that Julie has made a dreadful mistake marrying a man that she has only known for six weeks," I told her. "But I love your daughter and everything will be all right." Then I gave Yvette a big hug. Julie and I lasted 16 years. Yvette and I are still going strong.

With me sitting in her kitchen Yvette thoughtfully ends her conversation after a few minutes. As I have on other occasions, I ask her opinion. "What's your sage advice? Abby and I are here in Seattle to attend sporting and cultural events and then—then she gets to experience the exact opposite, a place that's natural, a place that's quiet, a place where you can see the stars at night and hear for miles."

"Wow. Sounds good to me."

"And I get to learn whether or not this kind of experience interests Abby, interests young people today. But I think she's got her mind elsewhere. So what do I do—drag her along?"

"Take her when she's thirty and has a couple of kids and she'll be glad to go with you," says Yvette with a sip of her tea. "I really don't know. I think it's up to you. You would enjoy it more, I think, without Abby. You're going to be under a lot of stress as it is."

I call Abby's name from the kitchen. Then repeat my call. Then walk down the hallway. She's on her cell phone. I invite her to join us for a family talk.

"My intent was never to torture you, you understand that, right? This is not a plot against you. Just give me this opportunity to explain your role in my journey, all right? Then you can decide."

"All right," Abby says quietly.

"If you come along you'll have the opportunity to listen to nature, something that you have not listened to in a long time, and I think that you'll be surprised."

"That's what I like," seconds Yvette. "You'll like it when you're eighty-eight."

"Yeah, when I'm older, maybe."

"The idea here," I continue, "is that when you're listening with an iPod you're listening only a quarter-inch away. If you come to Pipestone, you'll have the opportunity to listen to very faint sounds—sounds that have come from miles away. Wouldn't that interest you? Take a moment. I don't think you can even answer that question because you haven't heard what I am describing. This will be an entirely new place and experience. And I want to get your response to Pipestone. How does it sound to you?"

"Well, it doesn't interest me. I don't care. I'm not interested in stuff like that."

"You haven't listened to square miles before. You've only really listened to what's in your immediate room and your surroundings. You'll be able to hear a natural environment that is typical of what people used to hear—as long as a thousand, two thousand years ago. The question I would like to answer is whether or not, while there in person, you can connect with that part of yourself that is not just sixteen years old but thousands of years old. This was designed to be an experiment—I got your permission. I invited you and you accepted. And ever since, it's been tooth and claw, a bad experience. And I believe that if I drag you into Pipestone Canyon, you're not even going to be in Pipestone Canyon . . ."

Abby cuts me off. She's had enough of listening to Dad. "I want to go home *now.* Mom offered to pick me up tonight or tomorrow morning at the ferry. This experience is totally stressing me out, and I hate being stressed out. I'm going to start breaking out. This experience is like totally negative. It's not helping me in any way. I know it's not supposed to be so negative, but I'm really, incredibly, so not into this, that it's making it a hard time for me. I'd rather be at home grounded."

No, Abby says, she's not interested in the symphony. "I'm really, really not into that. Not into this whole experience. I'm willing to just catch a bus. I'll walk to the bus. I'll do whatever it takes. I just really, really want to get home. I feel like I can't relax until I'm at home."

And so I relent. Abby starts home immediately after dinner, getting a ride with her Aunt Jeanine to the ferry. Julie will pick her up on the other side.

After Abby leaves, Yvette counsels, "It's not an easy time for teenagers and parents. I'm glad I don't have to go through that anymore. Teenagers are so drawn to their own doings at this age. There's just nothing except their friends. They have to be on the phone. It's not just Abby."

Abby's parting words to me: "Thanks, Dad, for letting me go."

I remind myself of the difference between a trip and a journey. On a trip, you've got worry as a companion, for you're always concerned about what happens next and sticking to an itinerary. But on a journey you never have to worry. Something always happens next. It usually takes me two weeks to mentally transform a trip into a journey. Abby has helped me achieve this in just two days.

Long before sunrise I wake up at Yvette's house to remarkable quiet. Incredible: 20 dBA, the lowest reading on my meter. The ticking of my travel clock bumps it to 22 dBA. My breathing is loud by comparison, 30 dBA, due to a raspy throat from tree pollen. At five minutes before five, I detect the barest hint of distant traffic. The sound meter goes no higher than the ticking of the clock. Maybe I can hear Yvette breathing in the adjacent room. This is some of the acoustic energy that Basel Jurdy was talking about yesterday: something barely perceptible but still an essential ingredient of real places. Studio engineers mixing a Hollywood soundtrack would call it "room tone" plus low-level "background sounds." Wilderness walkers might call it the "spirit of place."

Later, the teakettle whistles. The old model has a pleasant sound. Yvette and I enjoy our Red Rose together—two cups apiece, each one right after the other. She taught me that routine. We resume last night's conversation.

"When I was a teenager," Yvette says, "there was no word 'teenager.' You were home with your parents. There was no subculture. Now it's a subculture. We never had that. You might have a few friends, but you didn't go around like a bunch of sheep like in the new culture. But how can you stop it? I don't know."

"Abby plays her iPod so loud that it hurts *my* ears!" I tell her.

"Somehow, all this noise protects them from what they don't want to hear. It's kind of like drinking. It's a dulling of the senses."

Teacup drained, my spirit raised by her wise words, I'm soon back in the VW, headed back to Seattle for today's listening sessions. I listen first to the sound of my wheels.

63 dBA at idling
70 dBA at 20 mph, second gear
84 dBA shifting from second into third
70 dBA 30 mph, third gear
75 dBA shifting from third into fourth
79 dBA doing 45 mph westbound on I-90 into Seattle
82 dBA doing 53 mph, top cruising speed with a slight downhill

No, this is not a quiet conveyance. But it is the reality of this air-cooled four-cylinder vehicle. Full of explosions! Without a schedule and with so much ground to cover I would prefer a different reality, my favorite mode of travel: walking. At the end of my trip I'll have my reward: a 100-mile walk along a portion of the C&O Canal. "It is a refuge, a place of retreat, a long stretch of quiet and peace at the Capital's back door," Chief Justice William O. Douglas described it when he hiked its length in 1954.

But for now, it's me and the Vee-Dub, which hits I-5 like a rock tossed into a fast-moving, shallow river. The traffic immediately piles up and streams around my comparatively slow-rolling wheels. Eventually I make it to the underground garage beneath Benaroya Hall. With lots of time to kill before my prearranged personal tour of the hall, I set off on a sound walk of Seattle, reacquainting myself with an urban center I once knew so well.

My first stop is Freeway Park, located behind the Park Place building on 6th Avenue and University. The park sits atop Interstate 5. I remember reading about Freeway Park "healing the scar" left by Interstate 5 when I was still in college. It was the first park in the world built over a freeway. Just part of the legacy of Lawrence Halprin, indisputably one of the great American landscape architects of our time. I plan to visit another of his achievements later on this journey, the FDR Memorial in Washington, D.C.

Three decades after it opened, the park's bold design begs a reassessment, particularly after the murder of a blind and deaf homeless woman in 2002 in broad daylight that has some questioning why Freeway Park has attracted a drug-using and drug-selling transient population. Apparently, as the trees grew, providing both shade and privacy, people began to feel another emotion: vulnerability. The number of visitors to the park has dropped substantially. To my animal ears, this had to be inevitable, considering one of the park's main design elements: fountains, most notably the Canyon Fountain, which Halprin included in his design perhaps as much for acoustical as for visual reasons. Halprin employed the sound of rushing water to mask the highway noise. But in doing so, he made it harder for someone standing near the fountains to detect approaching strangers. And when time added height and foliage to the trees, limiting sightlines, the situation only worsened. Wild animals, ever on alert for predators, would not choose such a place to linger. Why would we?

I understand that Seattle has hired a nonprofit organization called Project for Public Spaces to "develop a community vision" for the park with a "better balance between tranquility and activities and attractions that would begin to turn the park into a positive force for the neighborhood and the city as a whole." Pending their recommendations, here's a listener's perspective.

Numerous flowerbeds brighten the grounds and graceful hemlock trees lift our gaze to the surrounding buildings, whose mirrored glass surfaces reflect a hallucinogenic vision of the cityscape. The famous Canyon Fountain is dry, presumably still shut down for winter. Without its rush of water, here in the middle of what has been called a "five-acre urban oasis," the six-lane interstate that this park was built to tame is very much in evidence, though not a single car can be seen. The traffic noise peaks at 85 dBA. Later, with the noise hovering at 80 dBA, I have to shout into my digital recorder to have any hope of making sense of my oral note taking. Elsewhere in the park, near a sign leading to the Washington State Convention and Trade Center, a signs reads "Emergency Guard Call—Alarm Will Sound." Above it there's a gray panel with a red button and what looks like an intercom system. I am unsure if the square red box next to it is the alarm that will sound into this already noisy space, or if an alarm will go off someplace where someone may hear it. Or maybe both will happen. Other signs announce "No Loitering" and "This park is closed to the public 11:30 PM—6 AM (except for special events). Trespassers will be prosecuted."

With the sea of traffic below echoing steady urban surf, few people are congregating and nobody is really talking without yelling. Freeway Park is only a partial capping of the Interstate, a bold, though at best only moderately successful, attempt to make the downtown financial district more humane. I can't help but wonder if anyone has considered building a lid over the entire interstate where it passes through Seattle, thereby reclaiming, not masking, the acoustic environment for pedestrians. A project of such magnitude did occur nearby, where I-90 passes through the wealthy residential area of Mercer Island. There, the so-called Park on the Lid boasts two softball fields, four tennis courts, a picnic shelter, two playground areas, two basketball courts, and plenty of open space. The City of Mercer Island calls it "a great destination for the entire family. . . . The well-groomed open areas invite kids of all ages to romp around."

To my ears, there is not much at Freeway Park to enjoy, so I continue on, passing through the Washington State Convention and Trade Center.

I measure 48 dBA in a quiet corner across from Starbucks. Two Seattle policemen pass me riding bicycles on the wide pedestrian corridors of the Trade Center. I had no warning, the familiar click-click-click of their mountain bike freewheels silenced by an injection of grease, something I used to do on my personal bike before my knees blew out.

Soon I hear a familiar sound. The *Whirrrrrrr whirrrrrrrr whrrrrrrrr* of cars passing slowly over the brick-covered street at Pike Place Market, one of my favorite spots in Seattle. I liked it in my bike messenger days because it was just about the only place where you could get away with riding the elevator with a loaded delivery bike (in this case, to save the steep hill climb up Western Avenue). But I especially liked the unexpected events and charm and vitality of this century-old collection of vendors. Tables for merchants are still rented by the day. Most significant, this is one of the last domains of pedestrians. People walk freely from flower stand to vegetable shop, past hand-crafted souvenirs, and, yes, to the famous stand where huge whole fish sail through the air, enticing buyers. Strolling market-goers bring traffic on the adjacent road to a standstill. The loudest traffic sound I hear is from a motorcyclist revving his engine to keep it from stalling. I hear the *pop* of a merchant opening a paper bag with a quick shake, the sound of crushed ice shoveled onto seafood, the opening and closing of heavy, old-fashioned doors. It is rare, at least during busy hours, to hear a footstep, but you will hear people singing.

"Give me a song I may sing. Give me a dream that I may dream. Give me a mountain and the deep blue sea, peace of mind and serenity."

I've come upon the street musician Jim Hinde standing above his open guitar case, gray beard hanging way down onto his chest, backed by another singer adding harmony. They've collected about a dozen listeners, and I applaud with the others when the music stops. Then I dig in my pocket and contribute to the keep-Jim-coming-back fund. Fifteen bucks gets me a copy of his CD, *Shout Down the Wind: Songs of Peace, Protest, and Patriotism.*

It is an easy walk straight down 2nd Avenue toward "the Skids," thought to be the origin of the term Skid Row and so named because this is where, historically, logs were "skidded" downhill and into place for milling. I come to the Artist's Entrance at Benaroya Hall. Unlike Pike Place Market, where street musicians perform amid the cacophony of the city, Benaroya Hall has been painstakingly designed and built to exclude urban noise. It was completed in 1998 at a cost of $120 million.

I've got tickets to tonight's performance of the visiting Pittsburgh Symphony Orchestra, and I thought it a good idea to first learn more about the design of the hall and its acoustical properties. Mark Reddington, a partner at LMN Architects in Seattle, was lead architect on this project, working on and off for 12 years. He leads me straight to the sound of silence itself: the unoccupied 2,500-seat hall, to the heart of its ground-floor seating, close to what I'm guessing is the hot seat, that sweet spot just back from the front in the center of the seating area where the audience gets the best sound.

Boop! I let out a loud sound and listen to where it goes and how long it takes to disappear. *Boop!* Again I wait patiently for the last of the echo to fade from perception, then tell Reddington, "This is something I do when I'm out in nature to record. It's a fast and convenient way to judge the acoustics of a new space and decide if I want to record there."

My wristwatch times the decay at around two seconds in this unoccupied space. That's good. Reminds me of reverberation time at the Hall of Mosses above the Hoh visitors center.

"It's going to change a lot with people in here, too, isn't it?" I ask.

"The people will make it more absorptive in the room," Reddington explains. "In a symphonic hall, what you're normally trying to do is balance between that and a long enough reverberation time, which generally approximates two seconds, that gives you this kind of rich full blend of the sounds of the orchestra. The risk is that it can get too muddy if it goes too long or if you have other characteristics of the room that aren't providing clarity. So reverberation is important for the way the symphonic hall works, and then there are the characteristics of the room that diffuse the sound, which is also essential for symphonic hall acoustic quality. There are a number of characteristics of the room: the overall room geometry, the geometries of all the surfaces, the configuration of the balcony."

"So the entire performance will be the result of natural acoustics and not the result of amplification?"

"That's correct. There is an amplification system for lectures, but symphony performances are all natural acoustics," says Reddington.

I'm interested in how the acoustics of a symphonic hall might match up to a natural environment: the prairie, a clearing in the forest, a mountain amphitheater, someplace organic where our ears must have evolved. I ask if he knows of any book that gives this kind of information.

"Not that I'm aware of."

I'm disappointed but not surprised. Few audiologists stop to ponder the evolutionary significance of the peak sensitivity of human hearing, around 2.5 kHz, and ask What events in our evolutionary past required that we listen most carefully to this bandwidth? Few acoustical engineers stop to ponder any possible evolutionary origins in the ideal reverberation time in a symphony hall. But both these questions interest me. I can't help but consider the survival needs of our ancestors, to whom we owe our ears.

Sound travels a distance of roughly 2,200 feet in slightly less than two seconds at room temperature. Might this distance allow us to calculate the size of our ancestors' preferred living space? Might these two seconds provide ample time to flee or prepare to fight, in other words, sufficient surveillance time to relax? Might a reverberating natural space aid humans as echolocation enlightens dolphins and bats, reassuring us: The space is fully heard; all sound information has arrived. I have noticed a two-second reverberation time in certain large clearings in forests and hilly grasslands, typically in the early morning, before sunrise, because the air is often so still—not even a zephyr—and sound travels farther and more clearly than at any other time of day or night.

I reach for my sound-level meter and assess the silence of an unoccupied Benaroya Hall. I get a reading of 26.5 dBA. About the same as One Square Inch.

"And that," Reddington points out, "is with the door to the control room open and its two cooling fans blowing. See, these are eight-hundred-pound doors."

"This really is a vault."

The whole building is, in fact. The hall's urban setting and, more to the point, its specific location on top of the Burlington Northern train tunnel and next to a Metro bus tunnel made isolation from ground-borne noise the first priority, long before any orchestral considerations. "There was a consultant who went on site with instruments that went into the ground and measured the frequency of vibration that was generated when a freight train passed below the site," says Reddington. "Then he was able to design the isolation system to take out that frequency. You can tune it depending upon the density of the isolation pads."

The freight tunnel lies slightly below the bottom of two levels of parking garage. Atop the parking garage sits the auditorium, but not directly, thanks to site-specific acoustic isolation joints. "The auditorium itself was built as a self-supporting concrete box," continues Reddington. "It's constructed independently from the structure of the surrounding building. There's a clear gap between the two that is bridged only by flexible joints, so there cannot be any noise or vibration that's transmitted through the joints."

But the acoustical engineering didn't stop with vanquishing external ambient noise. The noise of the building itself—its circulatory system, if you will, fans, pumps, and blowers, all of which generate sound and vibration—also needed to be addressed and silenced for live performances as well as for recordings, to prevent background noise from being picked up by sensitive digital recording equipment. Building engineers and architects employ a metric for measuring mechanical system noise in occupied spaces, called an NC rating. A typical office building might have an NC rating of 30 or 35. The goal at Benaroya Hall was 15, equivalent to 22 dBA.

We walk on and Reddington points out and explains the absence of parallel surfaces. "You want to have very low background noise, and then you want to manipulate the sound in the room. You want to have a relatively high reverberation—the two seconds—and in order to achieve that the construction of the walls and all the surfaces are generally very dense, heavy material. Everything you see that's painted is full-depth plaster and then back-plastered to give it extra thickness."

"And this is to resist vibration? It's the opposite of a musical instrument, in other words?"

"So that it limits the absorption, yes. So that it keeps the sound live in the room," he says. "At the same time you want to scatter the sound. You will frequently hear symphony halls described as a shoebox configuration, and in the absence of any other articulation of the surfaces, that shoebox configuration will create the densest pattern of diffusion within the two seconds, scattering the sounds throughout the room. So that basic geometry is a starting point. Then all the surfaces you see are shaped with additional articulation.

"If you look carefully, there are almost always some sort of triangular sets of geometries, frequently tipped out of plane vertically, as well. The triangular geometry is a way of creating differently sized surfaces, wider at one end and narrower at the other end, so you're interacting with different frequencies of sound with those different surfaces. And then, likewise, you'll see all the soffits are folded in different directions. The dimension of the overhangs of the boxes is very critical for pushing sound back into the room after it comes off the stage and hits the sidewall and comes back off the underside of the balconies. Within the two seconds, you are absorbing sound. There is sound decay. So you have to control the absorption. So the other thing we did was develop ways to construct the walls so they absorb all of the different sounds evenly. You'd like to have all the different frequencies of the instruments decay over those two seconds at a fairly even rate."

Reddington steers me to one of the side walls. "Everything that you see here is first built out of concrete. Then, on top, is a series of wood battens spaced at different intervals." He raps his knuckles in several spots. It responds like a drum kit, with a different sound for each location. Indeed, when Benaroya Hall first opened, Reddington used to bring rubber mallets with him and play the walls for the groups he led on tour.

Nature, too, has many hidden surfaces and subsurface spaces that similarly affect the way sound behaves. I tell Reddington, "While recording at Yosemite, where there is a broad spectrum of frequencies being produced by waterfalls, I like to record the sound of a waterfall from inside the spaces between granite boulders, moving the microphone in and out, until I find just the right sound. Depending on where you position yourself, you can create quite a bit of mid- and low-frequency boost. And sometimes you can find hotspots where the beating rhythms are more accentuated. This is all achieved by using the hidden spaces out of sight but not out of earshot. The sound of the waterfall leaving these boulder fields is different from the sound leaving a granite cliff, for example.

"The same thing can happen in the pine forest," I add. "The raw sounds of a waterfall will enter a forest as a kind of white noise, but after it travels through the forest for a long enough distance, far enough so that the repetitive structures can selectively filter out some of the frequencies, you can hear a humming sound."

"Really?"

"At first, particularly if you just relax and let your mind sensitize itself, or, rather, desensitize itself to the idea that it's noise, just relax and let it find the patterns. Then what was originally a *SHHHHHHHHHHH* becomes a *MMMmmmMMMmmmMMMmmm.*"

"Is it a breeze passing through?"

"Well, a breeze can create that, if it's distant. Any white noise can create it, but I notice it particularly where either rivers or waterfalls send this out into a valley, where the valley and all the repetitions in the vegetation create that humming."

"Do you go into caves?" Reddington asks.

"I do. I don't like to go too far back into caves, because they get kind of creepy for me. In Hawaii I went into a lava cave near a beach where the surf rolls onto the lava boulders and causes them to roll and thunder, but the cave itself was dry and several hundred yards from the beach. The sound of the surf changed more and more the farther back I went, until it was just a very low mantra-like rhythm. I published *Back of the Cave* as a result. It's one of my favorite albums."

"I used to go caving in Kentucky when I lived in the Midwest," says Reddington. "The sounds when you get way down in there and you get completely detached from ambient outside sounds—everything is either rock, or it's water, or it's mud. So everything is very reflective and sound carries for a long, long ways. But it is so complex, all the shapes, it's very hard to tell where it's coming from. Often, it sounds like somebody is speaking in the distance, even though you know no one is there. It's just the sound of water, but your impression when you're back in those caves is really interesting."

"I've noticed that, too," I say. "In my perception of distant sounds or weak signals—a distant stream in a forest, for example, or a reflection of a stream off a forest tree—my mind sometimes turns it into human voice. I've actually stopped recording and tried to find the people to ask them to be quiet. But of course they weren't there. I'm all alone, miles away from anybody."

We agree there's some sort of human yearning at work here.

Reddington asks where I'll be sitting for this evening's concert. I reach for the tickets and hand them over. "Orchestra R, row DD 11," he reads, and then helps me locate the seat I'll take in just a couple of hours. I laugh, considering the curriculum vitae of my "usher."

With the tour coming to an end, I ask, "Can it get any better than this? Is this state of the art? Or is this where the budget stopped?"

"This is not where the budget stopped," he answers. "I would describe this as state of the art. You can spend, and people do spend considerably more money on symphony halls. For example, Disney Hall, which was finished in L.A. a couple years ago, is very sculptural and shapely and a lot of money was spent in making it that kind of a building, but acoustically I don't think it has any higher standard than this. In Philadelphia, the Kimmel Center was more expensive. What it has that this does not is a bunch of changeable acoustic chambers. In the upper part of that room there are some reverberation chambers that you can open and close with big concrete doors. Over the stage there's a big overhead canopy that is very heavy that you can move up and down to change the room acoustics. This symphony here did not want changeable acoustics. They wanted a symphony hall that really worked as a symphony hall. You go back to Boston and Amsterdam and Vienna and some of what are recognized as the world's great symphony halls, they don't have all the changeable stuff. They're more like this."

"Those older symphony halls, how did they design them? Did they set the standard? Are we just repeating what was originally arbitrary?"

"I'm sure it wasn't arbitrary. Many of those halls in Europe were built in the late part of the nineteenth century, and at that time, performance of this kind of music was moving out of its traditional place, which would have been in private chambers in the royal palace, to a performance hall where they were then designing rooms that simulated those older rooms, which had a shoebox configuration. I think they were likely reacting to what they discovered as they made them, and they had a series of them that were all fairly similar. I imagine they were re-creating things that they found to be successful. What's commonly described as the world's first scientifically designed hall is Boston Symphony Hall, built in 1900. There were actually calculations about the reverberation and absorption, although I don't know how much that actually played in the design of the room. It's still pretty much like the traditional European hall.

"This was finished in 1998, so the dollars today would be different, but the construction of this was eighty million dollars. Total project was about one hundred twenty million dollars, including land costs. The acoustic isolation here is state of the art. There are people who advocate for an NC ten or a five, but I've heard people question if you can even test it that low, because just to be in the room running the test causes more background noise than that. This is at fifteen, which for a room like this would be considered state of the art."

And so there we stood, with nothing more to say, enjoying a few unfilled seconds of the silence.

"Thanks for the tour."

"My pleasure, and have an enjoyable concert and a safe journey."

Five hours later I hand my ticket to a bona fide usher and, for the second time today arrive at Orchestra R, DD 11. In Abby's place is her aunt Jeanine, who, I soon discover, claps really loudly. The Pittsburgh Symphony Orchestra, conducted by Andrew Davis, is warming up. People are still arriving and talking in lowered voices. The sound level is now 45 dBA, about 100 times more acoustic energy than this afternoon, when the hall stood all but empty.

My seat is some 100 feet from the stage and five feet from the wall. Even so, I find the sound well balanced. In the early moments of the music, I see the sound level has fallen to 32 dBA. The decibel readings during the first selection hover in the 55 to 60 dBA range. Quiet moments give readings in the low 30s. I note how, in contrast to the audience, which is often motionless, as if in peaceful prayer, the musicians are swaying and nodding, some of them quite passionately, as they play. Most surprising are the amazing swings from quiet moments to very much the opposite. During one quiet pause before a new selection, uniformly respected by the audience, I see the decibel reading fall to 27, the same as earlier today with no musicians on stage and no one in the seats. But I soon take much higher readings, as these dBA notes show:

80, 81.5, 82, 89, 81, 91 and then 94: the crescendo
107.5: peak value for the applause
36: the next quiet interlude
121: another burst of applause
25 to 45: the next quiet interlude
116: standing ovation; hurts my ears (Thanks, Jeanine)

Noise can be powerful and socially expressive. This genteel setting reaches sound levels every bit as loud as a basketball arena or football stadium. The loud moments, of course, are few and fleeting, but isn't it ironic that we show our appreciation for perhaps the most beautiful human-made music with noise—loud as a chainsaw or a pile driver.

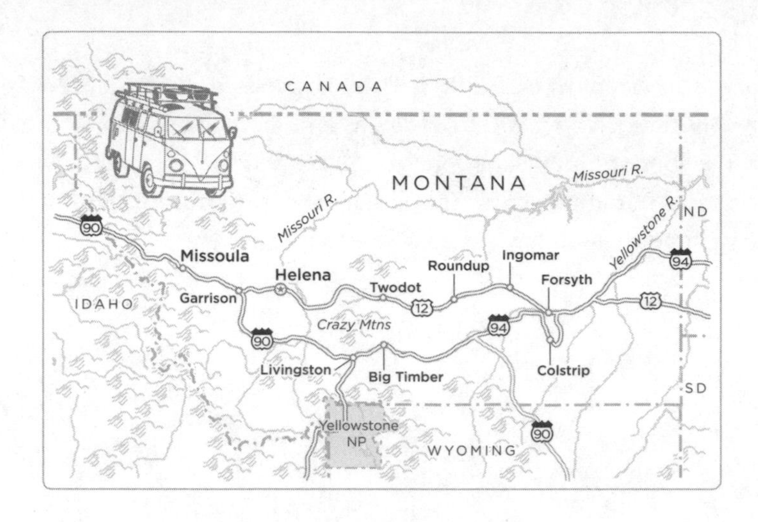

5 Endangered Quiet Beauty

> We the people of Montana, grateful to God for the quiet beauty of our state, the grandeur of our mountains, the vastness of our rolling plains, and desiring to improve the quality of life, equality of opportunity and to secure the blessings of liberty for this and future generations do ordain and establish this constitution.
>
> —Preamble to the Montana State Constitution

Montana references quiet in the very preamble to its state constitution. I've come to listen to the land and its people—as if the two were separable. My first stop is a retirement home called The Springs, in Missoula, where I'm eager to meet one of the state's pioneers, Bill Worf. We've talked on the phone and corresponded for more than 10 years, ever since he sent me a letter of encouragement after reading about my efforts to save quiet in Washington State and invited me to write an article for the newsletter of his organization, Wilderness Watch. A career U.S. Forest Service employee, Worf headed the development of regulations and policy for the agency's implementation of the Wilderness Act after its 1964 passage.

He founded Wilderness Watch in 1989 to look out for America's wilderness regions and keep them pure. Bemoaning the lack of attentive stewardship for lands designated as wilderness areas, he slammed $20 on the table at the Old Town Café (now the Two Sisters Café) and announced, "I'm the first paying member."

I spot him immediately in the living room of the retirement home by the description he gave me when I called to confirm the time of my visit: "Totally bald and wearing a red sweater." Worf is 81, with failing eyesight, recently widowed of the love of his life. A large man, he still bears a rugged demeanor forged by harsh Montana winters, but a gentle wisdom in his soft, deep voice is as captivating as a prevailing wind. While we wait for TinaMarie Ekker, Wilderness Watch's policy director, to join us, I tell him about my cross-country journey and my hope of finding and recording "some good old-fashioned Montana quiet."

"The noise of human beings is just reaching every place now," he says. "Not many places you can find."

I ask Worf about his boyhood in Montana.

"I was raised in eastern Montana on a homestead south of Rosebud. My dad and his brother homesteaded a piece of ground out of Reed Point, Montana. That's where I was born, out on the homestead, in 1926."

"Talk about a home delivery."

"When I was just two they moved to an irrigated farm near Big Timber," he continues. "Spent one year there. In 1929 they were convinced by an insurance company that had mortgages on a bunch of homesteads in eastern Montana to take five of those they'd foreclosed on and farm. That's what took us to Rosebud County: eight hundred acres. We farmed about six hundred of that, all with horses—mostly wheat and oats, some corn and some barley. We raised cattle, too. We starved out after about ten years. Just completely starved out."

From the nearby dining room I can hear the tinkling of glassware and utensils and the joyful voices of young children who have come for the upcoming Easter egg hunt. The in-house music system gently plays "Some Enchanted Evening."

"We just went flat-out broke. It was a lot of work and a lot of heartbreak—they went there with great dreams. They had a Farmall tractor. The insurance company that funded them bought a Farmall tractor and a

Model A Ford truck. The place where we lived, there was a pretty good barn there. That was one of the things that attracted my folks. And there was, I guess you could call it a house, a shack anyways. There were four children. This had one room. It seemed big to me but it probably was twelve feet wide and eighteen feet long. Upstairs was kind of an attic room, where the kids slept. Mom and Dad slept downstairs. The second year we were there, they moved a second shack in, drug it in with the Farmall tractor and horses. We connected it by a little hallway and built a woodshed in between the two, and until 1939 we lived in that."

"Back then, when you were working with horses and tractors, the sounds of the farm must have been entirely different. You didn't necessarily have to go to a wilderness to find quiet."

"Oh, no."

"Do you remember much about the sounds or the quiet?"

"I do remember we were seventeen miles south of the Yellowstone River. Highway 10 and the railroad ran along the Yellowstone, and I remember that on a good quiet morning you could hear the train and trucks and cars down on what we called the Trail. We referred to Highway 10 as the Trail. It was a gravel road at that time. Sometimes all you could hear on a cold winter morning was just the rumble of the train. Well, that tells you a little bit about the quiet."

"Do you think that it is still quiet, if I went there today?"

"Well, I would think so, because nobody lives out there. It might even be quieter than it was then."

My mind is racing ahead, rerouting my journey across Montana, when Worf suggests lunch in the café at The Springs. He has reservations and asks for a booth.

"We will have a third person joining us," he tells our waitress, whom he introduces as Howdy. Twice. I have difficulty hearing her unusual name because of the unaccommodating sound level in the busy room. The café is cheery and clean, but the surfaces of the walls, ceiling, floors, and furniture are all smooth and reflective. My meter captures the sound level of voices, added to the clatter of silverware and the delivery and busing of plates, at 67 dBA. That's simply too loud for a setting like this, where older people with compromised hearing gather for memorable conversations with family and friends.

"Sorry I'm late," says TinaMarie Ekker, joining us at the table just in time to order.

"The food is good," Worf says, "but one thing you can't do here is tip."

Over lunch he recalls one of the first issues that Wilderness Watch got involved in soon after its founding.

"The Air Force was setting up training areas and they were looking for a place where planes would not have an impact on population. So here they saw this big blank spot on the map that was called the Boundary Waters Wilderness, and they put this air operations area over the top of the wilderness. Of course, that was a pretty heavy-duty noise. Mainly they chose it because they didn't think it was going to impact very many people. Friends of the Boundary Waters took the Air Force to court. The attorney for the Air Force was trying to play this noise issue down. They came on with a noise expert who said that the noise of an F-whatever-it-was—fighter planes at a thousand feet—was no worse on the ground than a vacuum cleaner at five feet."

I nearly fall out of my chair laughing at the outrageousness of that conceit.

Worf says he was asked about that comparison when put on the stand as an expert witness by the lead attorney for the Friends of the Boundary Waters: "Mr. Worf, what do you think of that? Do you think that that is true?"

"Well, you don't go to the wilderness expecting to hear a vacuum cleaner."

Ekker chuckles.

"Did everyone laugh?"

"No, they didn't laugh much, but the judge, he started really paying attention after that. He understood what we were talking about. Before we were done, the Air Force folded its tent, said, 'We'll just move the whole air operation area off the wilderness area.' The reason I'm telling the story is because one of the first things Wilderness Watch got involved in, in a legal sense, was over the noise over the Boundary Waters."

"Is the public receptive to quiet today?" I ask Ekker. "Do they understand the need for natural quiet?"

"I believe so," she says.

"But," says Worf, "one of the things we need, we need some real scientific information that explains a little bit of the why and the what. What has changed? What is changing? If we are going to preserve any spots of quietude, I think that wilderness is the place to do it."

Worf does not, however, believe the government agencies are onboard. "The four agency heads pretty much do not support the idea of wilderness anymore," he says. "The Forest Service, the Park Service, Fish and Wildlife Service, and the Bureau of Land Management are the four federal agencies responsible for managing existing wilderness. They'll give verbal support for it, but deep down, they really think the whole idea of wilderness is just a bunch of damn silliness. The idea that we would want to limit sound, limit overflights—what does that hurt, you know, a plane is there for a minute, two minutes, then it's gone. What's the problem?

"For example, right now I'm working with some dam operators that have to do some repairs to the dam, and they believe, the dam owners and the local forest service believe, we're just going to have to do it with motors. Well, I'm trying to prove to them it can be done with horses. And of course one of the arguments that I get back is, 'You know, Bill, if you do it with horses, it might take you three times as long. If we get in there with motors we can do it quickly and be gone. Then it will be quiet again.' It's a tough battle. I think maybe there are quite a few citizens who don't understand this at all."

Ekker follows with a story about visiting government officials. "The wilderness governing counsel was out here a few years ago," she says. "The chairperson was from the U.S. Geologic Survey and she asked, 'What's the difference between backcountry park areas and wilderness and roadless areas, because they all look the same?' Nobody could answer. And here they all were active in wilderness management. They also wanted to know what was wrong with landing a helicopter, because it doesn't leave a scar. Then I said, 'The difference is the way you interact. If you insult your best friend or neighbor, nobody else can see, but the damage has been done. You've harmed that relationship. We need places that you treat differently.'"

I am thinking this is a hard thing to talk about, our spiritual relationship to the land. And Ekker is absolutely right. When we feel a certain way, we act a certain way. How can we profess our love of the wilderness and then abuse it on the grounds that the insult is temporary? This certainly wouldn't work in a marriage. Those who would charter a helicopter to perform research should think again. If they really love the wilderness they should be advocating for packing in and getting closer to the very thing they say they love. True wilderness, that is, pristine places far from the ever-present intrusion of humans and their noise, offers an opportunity to fall back in love with the Earth, teaching us ecological moral values.

At Olympic Park helicopters are used at treetop level to count individual Roosevelt elk over designated wilderness, including directly over OSI. When I asked the park's public information office about this in March 2006, I was told by e-mail, "Aerial counting is the only way that will produce a population estimate for elk." Yet, I wonder, might not older, quieter methods, such as track and scat data, be more appropriate in wilderness areas? Biologists studying bighorn sheep in the Grand Canyon found noticeable changes in feeding patterns and movement patterns in the presence of helicopters. The animals spent 14 percent less time feeding during the spring and 42 percent less time feeding in the winter, all the while traveling 50 percent farther while doing so. In other words, the helicopters made a big impact on their energy input and expenditure balance.

Worf stresses the importance of positive thinking. "The man who thinks he can is the one who can do it," he says. "That's one of the lessons I learned on the homestead when I was four years old, when my mother would send me out to get some wood for the stove."

With lunch over, Worf invites me into his residence. We ride the elevator to the second floor, and he leads me to a two-bedroom apartment for independent living. Once inside, he hands me a printout of something he downloaded from the Internet. (The poem, originally titled "Thinking," was written by Walter Wintle more than 100 years ago.) Worf keeps copies on hand to give to folks who may "need encouragement."

The Man Who Thinks He Can

If you think you're beaten, you are;
If you think you dare not, you don't.
If you'd like to win, but think you can't,
It's almost a cinch you won't.
If you think you'll lose, you've lost;

For out in the world we find
Success begins with a fellow's will;
It's all in the state of mind.

If you think you're outclassed, you are;
You've got to think high to rise.
You've got to be sure of yourself
Before you can ever win a prize.
Life's battles don't always go
To the stronger or faster man;
But soon or late, the one who wins
Is the man who thinks he can.

Bill Worf can. Even though he is legally blind he watches television using a pair of binoculars. He surfs the Internet, invoking inch-high type he views through thick glasses.

On the wall are pictures of his wife, born Eva Jean Batey, who was also his high school sweetheart. Worf carried a picture of her in his shirt pocket through three years of service in World War II. When he came home from Iwo Jima he was only 19 and, according to Montana law, had to ask his father's permission to marry. They were husband and wife for more than 60 years. Eva Jean, he says, lived just long enough—five weeks after they moved in—to decorate the place. She said goodbye to him, smiling, declining life support.

Very little has changed in the apartment since then, except for maybe one or two more pictures of children and grandchildren on the walls. There's also a photograph of the old Worf homestead. "This is the homestead shack I was born in," he says. "We rode horses the two miles to the one-room schoolhouse. That horse was old Bridgette."

The next is a school photo depicting nine children. "Mind you, this is not a class photo. This is the entire school."

He opens the bedroom door. "More space than I need, but I'm not about to move. Sleep with my window open most of the time—I like the cold."

"How would I find the old homestead?" I ask him.

"Little town of Rosebud, not much of a town left there anymore. Look south—about seven or eight miles you'll see some clay buttes with a bit of ponderosa pine. We lived south of those buttes, 'out in the buttes.' An area of about a hundred square miles."

"What's it look like?"

"Well, it's rolling sagebrush hills, and most of the stuff that we plowed up then has gone back to sagebrush. It has very few trees—wide-open country. Nobody lives in the homes that our neighbors used to live in along Sweeny Creek," he says, telling me that from here it's about 400 miles to Rosebud.

"Along the way," he adds, "Ingomar is a place you ought to see. Total town population of about eight. The only business is a restaurant called the Jersey Lily Saloon. Order the bean soup."

After our goodbyes, I sit in my VW bus for a long while before starting it up, enjoying the warm glow of a noble man who is as rare as the wilderness he wishes to save, a man I'll always equate with the quiet beauty of Montana.

Four hundred miles. That's more than the distance from Seattle to Missoula! I can't see the road very well at night because of my puny six-volt headlights, so I better get started. My journey has a new mission here in Montana: find the old Worf homestead and listen to the Yellowstone from 17 miles away. Can't wait.

It is rare that I actually take the advice of someone when seeking a quiet place in nature. Ninety percent of the time such advice leads me to the noise that they didn't notice. Even people who should know better lead me on unnecessarily. Like the time I followed a wildlife manager nearly 20 miles on a deeply rutted two-track road and, only seconds out of our vehicles, tried to give him an out: "Are those oil wells new?"

Much more typically, my hunt for quiet starts with remote-sensing data available from NASA; the best indicator of quiet is NASA's image of the Earth at night. The U.S. portion looks like the Milky Way in the East, and then it thins out to nebulae and constellations in the Midwest and West. Black space, of course, is where I want to be. Next, I'll check air traffic corridors between major cities, instantly eliminating most of the black spaces. Then I'll study state highway maps and rule out other possible places. My cartographic acoustic audition continues with USGA topographic maps.

The list grows smaller. I may see power lines (that crackle), mines (that roar), oil wells (that boom), gas pipelines (that hum), navigable rivers with tug and tow (that groan), and possibly ranch houses or other dwellings. I try to visit what's left, always encouraged when I lose cell phone reception. Yet, invariably, even these well-screened areas fail to provide at least 15 minutes of continuous, uninterrupted natural silence, my gold standard for quiet. "There's always something" could have been coined by a silence seeker like myself. Most often it is transportation noise that intrudes, usually jets and planes. High up they create a wide wake of disturbance. And increasingly there is more than one noise source. In the absence of laws, protected only by their anonymity, I can only wonder how long it will be—a few years? a decade?—before the last dozen disappear in the encroaching din.

Once I strike true quiet, I'm like a prospector with a gleaming nugget in his pan. I'm glued to the spot. I'll start recording and won't stop for days, sometimes weeks.

I'm hoping that Bill Worf has saved me all those steps. I head east on Highway 12 struggling to do 50 miles per hour, even on what must be the world's smoothest, most beautiful asphalt, a mix of what appears to be tire- and wind-polished semiprecious stones embedded in a fine-quality tar. All these materials are native in this land rich with mineral deposits.

It's calving season. I see mothers licking their newborns and cottonwood trees pushing out swollen buds. The smell of the cottonwood buds heralds the beginning of warmer days, but warmer is relative in this part of America. It's probably going to snow tonight. I plan on stopping at a roadside rest area to enjoy the last rays of sunshine and take the chill off. I'll fire up the woodstove, peel a fresh orange, and savor a cup of Red Rose steeped dark brown. That's my home remedy for my nagging cough.

I make it through McDonald Pass (elevation 6,325 feet) and enter Lewis and Clark County at the Continental Divide, dropping steeply and ever downward into magnificent valleys with rolling mountains and snowcapped peaks. An enormous view. Montana is living up to its state name. And barely a house in sight. Beyond the last snowcapped mountain ridge lies Helena, the state capital (population 27,885). Too cold for the VeeDub's roof rack tonight.

I spot a Days Inn, pull in near the front door, boot up my laptop, and I'm in. An unsecured Wi-Fi network has allowed me to use the Internet from the parking lot. I log into my account at Priceline.com and put in a bid for a room at $45. Might be the Days Inn, but maybe not. I'm anxious to find a warm place because my cough is getting worse and my toes have been numb all day. Let's see what happens.

"Congratulations, your price has been accepted!" The Red Lion Colonial Hotel on the corner of 15th and 12th has accepted my bid of $45. Another mouse click, and the map shows me I'm just a couple of blocks from warmth.

"Checking in. Hempton."

"You don't have a reservation?"

"Yes, I just made it through Priceline.com. I can go out and park my rig if you need a little time."

"Actually, it just showed up."

"Is there a part of the hotel that's quieter than another?"

"Normally, the second floor and the back."

I can hear the cold wind blowing over the roof and country western tunes floating in from the jukebox in the lounge.

My $45 room is Montana-size. Wow. Double Wow. Two huge beds. And, yes, a bathtub. In the lounge I order a double Chivas on the rocks and call my dad. I move on to a hot meal in the dining room. Then, back in my room, I let the hot water roar into the bathtub. "Ahhhhhh," I say, slipping into the steamy water, "thank God for unsecured Wi-Fi!"

When I awaken during the night I notice immediately how quiet it is. Peeking through the curtains I see a light dusting of snow. My meter registers 23 dBA, then falls to 20 dBA, its lowest limit. I fall back asleep.

Breakfast is behind the wheel: organic carrots, some bread ripped from a crusty loaf, and some chunks of cheddar cheese. Even with the heater on high I'm wearing gloves and a sweatshirt underneath a down jacket, admiring last night's fresh snow in the mountains behind me in the rearview mirror. Out my windshield I see rolling hills and grasslands and a sliver of sunshine poking through dark gray clouds. A cold horizon. I pass the exit for Good Earth Campground and a sign announcing "No Services." Signs like that are inviting to nature lovers like me.

The only towns for miles belong to the prairie dogs. They dot the land on both sides of the road. I pass what at first looks like a large tan squirrel munching grass by the side of the road. Then another prairie dog, this one brazenly bent over right in the middle of my lane, his snout pressed into a large crack in the asphalt as I roar right over top. In my rearview mirror I see him unflustered, still on the roadway.

Here the willows haven't even changed color yet. No swollen buds either. One day till Easter, and spring is still a long way off. Past the town of White Sulphur Springs I pull over outside of Elk Peak Ranch and climb on top of the Vee-Dub to shoot a panorama photograph in the bitter cold. Turning 360 degrees, capturing the view, I think I've figured out what's up with that prairie dog with a death wish. Just as gold travels downstream and fills the cracks in the stream bottom, seeds blow down the road and fill the cracks in the asphalt. Now, on the far end of winter, food must be getting scarce, pushing the prairie dog to find new harvests. But what a deadly dining room.

My toes are numb and the snow has started to fly. It's 38 miles to the next town. I'm actually enjoying this—sure beats staring at a computer screen! A sign warns "Rough Road," which, after all the cracks, patches, and potholes of late, is like saying You ain't seen nothin' yet! I stop for a break and to shoot another panorama. The sky is white; the ground is white; a black ribbon of asphalt passes from one vanishing point to another. The only sound I can hear is the faint whistling of the wind through barbed wire, then a distant vehicle—only one—coming closer and closer until, finally, it thunders by and vanishes to the horizon. This primo Doppler effect lasts an incredible seven minutes before the last rumble falls below my auditory threshold.

In this delicate soundscape I can detect the cooing, croaking sounds of migrating cranes somewhere up there in the snow clouds. I glance down at my cell phone. "No service" is displayed on my Treo 650. These days, it's rare to see those words staring back at me. Right now it's pleasant. Farther on, I pass an eagle feeding on an antelope carcass, then a cow licking its steaming newborn, born directly onto the snow and ice.

Maybe it was the song "Twodot, Montana" by Hank Williams Jr. pushing up from the recesses of my mind, or just my curiosity to find out who lives here that caused me to take the turn off into town. I can't be sure. But the important thing is, I did stop.

Down the center of Twodot's main street comes a bearded fellow in a wheelchair pulled by a dog on one of those retractable leashes. The man is smiling proudly, bundled in a camouflage jacket, hood up, wearing winter boots and gloves.

"That's a pretty nice setup you got going," I tell him. "Do you mind if I take a picture?"

"Sure, go ahead."

Did I hear a feminine voice? I lean to pet the dog.

"What's your dog's name?"

"Tootsie." I raise my hand to pet Tootsie, and the dog's unflinching response speaks well of its owner. "Yeah, I found her lost. Don't know if she fell out of somebody's truck or what. So I brought her home to give her away. But nobody wanted her. I'm thinking, 'What am I going to do with a dog? How am I going to walk her—take care of her?' But she winds up pulling me in the wheelchair and saving me miles. So it's a good deal."

We continue the conversation where it began, right in the middle of the road. I explain that I'm from Seattle, that I'm hoping to find quiet south of Rosebud, and ask, "Do you have any favorite quiet places here?"

"Well, I used to hang out down underneath the bridge and watch the little fish. I'd walk down to the river with my cat. I had a cat that would walk on a leash. It depends. There's a lot more traffic coming through, depends on the day. What do you do?"

I explain about my recording job and the need to escape all man-made noise.

"Well, it's mighty important, I think. I don't know what you think about God and all that stuff, but I know that I got sent here by the Lord because of stress-related disorders. He stuck me here where it's very, very removed from stress—and I'm healing."

"Do you care to talk about that?"

"Well, I just got a disorder that was caused by too much stress and it made the chemical system in my body turn on itself. It started eating itself."

"You look like you're about my age. Was it Vietnam, by chance?"

"It's way more bizarre than that. It freaks a lot of people out. I'm a woman. I started out a woman, and I've given birth to two kids. Had a real stressful childhood and then married into a real stressful situation, and it made my endocrine system flip. I had a dad who tried to kill me. It was like living with a bear. You didn't know when the next attack would come."

I point out what I think is a mule deer nearby and he, or rather, she corrects me. "It's probably a white-tail."

Tootsie's owner introduces herself as Judy and wishes me well on my journey.

"You never know where life is going to lead you," she says. "I hope that you find the quiet you're looking for. If you believe in God, I've learned to go and sit on His lap and sit in His presence and it doesn't matter what the environment is. If I quiet myself and just spend time with God, that's where I get my healing."

A flock of Canada geese passes overhead.

"I find my quiet in nature," I say. "Where there's no escape and no distraction. That's where I find myself and my true needs."

"You've got to get to a quiet place," says Judy, "because God whispers."

During our goodbyes, Judy asks if she can pray over me. When I agree, she places her hand on top of my head. "Heavenly Father, I just lift up Gordon to you. I don't know what his needs are. I know he's got a good heart that's searching for peace and joy."

Days later, I could still feel Judy's thermal handprint.

Past the Crazy Mountains, Deadman's Basin, and the Chapel of Hope, I reach Roundup, fittingly, at the end of my day of driving. Here, $37.50 buys me a room at the Roundup Motel. I cough a couple times pushing open the door of my motel room, quickly crank up the heat, draw a hot bath, and balance a six-pack of Moose Drool on the edge of the bathtub, ready to celebrate my newfound warmth. It takes 30 minutes in the hot bath for me to stop shivering. There's no stopping my cough. Each time I lift my arm up out of the tub to take another swig of Moose Drool, up rises a pillar of steam. When I'm still, the only sound is the ticking of the timer on the bathroom heat lamp. I hope it never clicks off.

Easter morning is cold and calm, low 20s for both temperature and decibels. The loudest sound is the Coke machine outside: 55 dBA measured one foot away as I head out to the car. My organic carrots, left in the dashboard holder once used for the ashtray, resemble a bouquet of icicles. I'm anxious to get an early start, mindful of quiet recording opportunities before folks start their pickups and cars, bound for church, relatives, and Easter dinners. Expecting natural quiet below 30 dBA, I break out my ultra-low-noise recording system, a $10,000 proprietary system that is so sensitive it expands the listener's horizon for more than a dozen miles in every direction. Like a mountaintop telescope that allows astronomers to examine distant galaxies, this system enables me to listen far and wide, to the spirit of a place. When that place is quiet and the conditions are right, I almost always hear nature's voice, its many layers of rhythms, as music—music conducted by the rising and setting of the sun.

The morning dawns clear and blue with a gorgeous sunrise that reminds me of a flower with petals of light. Just outside of Roundup I hear the call of the Montana state bird, the western meadowlark, coming from the cemetery. There are several, actually, their bright, ringing notes perhaps settling some Easter morning territorial dispute. From 150 feet their songs measure 47 dBA. I go no closer, having discovered a nice ring similar to the two-second reverberation of concert halls. The quiet between the notes measures 27 dBA. That is, when I am not coughing.

I've learned to dress in natural fibers when I'm recording to avoid the ruinous swoosh of synthetics. Normally, I can be as noiseless as the One Square Inch stone around my neck. But not today My cough is getting so much worse that I am having difficulty breathing, and I decide that my only chance for a good recording is to leave the equipment in record mode and walk away. But by the time I do, the Easter morning traffic has begun to stir, one or two cars every five to ten minutes, crashing my recording party at 46, 54, and 55 dBA, and a truck at 66 dBA. Another lost opportunity.

Back on the road, I see something's been hit. I stop, throw the Vee-Dub into reverse, and find a mangled ball of feathers with outstretched talons and frosted wings. I've seen owls succumb to the temptation of headlights before, swooping down in their glow to snatch a scurrying rodent, only to be hit by a vehicle coming in the other direction. That's my guess here.

Most owls hunt at night using their keen vision, enhanced hearing, and silent wings to successfully nab their prey. Owl hearing is about 10 times more sensitive than human ears, made possible by asymmetrically set ear openings, where one ear is higher than the other, and by a feathery facial disc that helps to guide sound waves toward these openings. Furthermore, the owl can change the shape of this facial disc at will, gathering even more information by focusing the sound image. Owls can also discern the difference in arrival times between left and right ears down to 30-millionths of a second—useful information for putting a bull's-eye on its target. The owl's brain processes even the faintest of sounds and converts all the auditory data into a mental image, enabling it to spot its prey through fallen leaves, vegetation, even thick snow. I have no doubt that traffic noise interferes with the owl's perception of faint sounds (for it confuses my ears) and affects the efficiency of its hunt. After an extended winter, when food is especially scarce, extreme measures might have to be taken. This owl, however, was not as lucky as that prairie dog miles back.

With no radio in the Vee-Dub (that six-volt antique was yanked for good after it broke), I'm in tune with the surrounding countryside. Sandhill cranes leap in the air as part of their courtship. Sharp-tail grouse cluster on a bare patch of ground called a lek; I hope to get in close to their unbelievable courtship later, in Nebraska. A herd of deer browse on tender shoots down near the river, and a hawk works a marshy area with cattails, executing low, fast, jagged movements, probably hoping to snatch an unwary red-winged blackbird. During a stretch break, a ring-neck pheasant blurts from an unseen distance its raspy, crowing call.

A small group of antelope gracefully graze on endless rolling grasslands. Stopping right in the road, I decide that I am going to give these rolling grasslands a listen. I see no visual evidence of people except for this road and two lines of barbed wire fence, until I spot something on the horizon to the north that might send me scores more miles farther on.

Getting out my binoculars to take a closer look, I confirm my suspicion. Looks like some kind of petroleum-related structure, either a gas pipeline compression station or oil derrick. Though I cannot hear it yet because my ears haven't shed the noise of the Vee-Dub, I know well the hum that's surely hovering over this remote land, having heard it many times. It's born of oil-well math. One oil well = one oil field = many more oil wells. A silence seeker doesn't just have to escape that single oil well. He has to escape the entire oil field!

I take a spur road to get closer and catch a glimpse and a sound-level reading of this particular impediment to quiet. My tires roll over a cattle guard. *Burrrrrrrrr.* Several antelope bolt to lightning speed, jumping across the road in front of me. The spur leads to an oil well powered by electricity. Between hacking coughs, I get a reading of 69 dBA, much quieter than the unmuffled, older bangers burning crude oil that I've come across in national wildlife refuges like Quivira National Wildlife Refuge in Kansas. But even this newer, quieter extraction technology emits a narrow, low-frequency bandwidth that will spill out for many miles in every direction. Oil field, oil patch: by either name, the noise blight is always difficult to escape.

Unsuccessful quests for quiet like this one make me thankful for my camera. I can point the lens away from the oil field and let the frame silence the outside world. I take lots of photographs and can well imagine prairie schooners plowing these grass-covered swells that surely indicate a sandy soil. But I'll know the noisy truth outside the frame. As Bill Worf reminded me, "The sounds of human beings have reached just about everywhere."

I hear Worf's gentle voice again a few hours later when I see a sign announcing his must-stop on the way to his old homestead: "Jersey Lily Saloon and Eatery, Ingomar, MT—gas, phone, RV camping, live buffalo, bunk and biscuit—only place to sleep in 100 miles." Any place advertising "bunk and biscuit" gets my vote. From the highway Ingomar looks like a sizable town: more than a dozen utility buildings and homes of all kinds, a stranded railroad car, and, of course, the Jersey Lily Saloon, a single-story brick corner building on the main street with a covered wood walkway circling the front, which is adorned with bleached animal skulls and a pair of wagon wheels.

Looks closed, but it's open. A fellow comes to the window and sees me, and then steps back. I let everybody adjust to my arrival, then head over near one of several abandoned trucks to set up my recording equipment and give this place a closer listen.

My sound-level meter hovers around 24 dBA, daytime ambience. Have I found the quietest town in America? Maybe, if I disregard the commotion of a dog being kicked out of the saloon with a shout and a slam of the door. The few people I see approaching the saloon appear to be old, and they move slowly. With that dog dealt with, the loudest sound in Ingomar is now an American flag clapping in the wind. My sound portrait of the town carries the following to my headphone-covered ears: the wind lightly whistling down main street while a few distant meadowlarks come in and out of clarity and that whipping American flag. For 20 minutes Ingomar intones silently, singularly, in my ears. Then, after just a few more minutes, I'm satisfied and ready to step inside, meet some people, and order up a large bowl of bean soup.

The squeaky door to the Jersey Lily Saloon opens onto a high-ceilinged room with a magnificent oak bar, an open kitchen, and five oak tables, each currently unoccupied. The walls are covered with photographs of bygone days and heads of moose, elk, white-tail, grouse, pheasant, antelope, even a stuffed turkey—all seeming to stare at an old man seated at the bar with the bushiest eyebrows I have ever seen.

"All right if I take some pictures?"

"Go ahead, just don't take any pictures of me. I'm having a bad hair day," says a woman from the open kitchen. She is furiously preparing an Easter dinner feast of prime rib, lamb chops, and rack of lamb that will attract diners from the far reaches of the county, filling the saloon's tables and, for a couple of hours, more than triple the local population. The bathrooms, I'm told, are outside. My wood-toned xylophone footsteps down the boardwalk lead me to a choice of freestanding outhouses: "Bull pen" or "Heifer pen." More wooden notes mark my return, and I settle in at the center of the bar near the old man, just as a much younger fellow wearing a black felt cowboy hat appears.

I order a Chivas on the rocks from the guy in the hat. It's still short of midafternoon. The scotch, I tell myself, is for my cough. But it's also antishyness medicine. In a town like this, a new face in the door is good for a case or two of whiplash among the locals. I shouldn't have worried. A younger woman comes over, introduces herself as Marnie, and asks me where I'm from. She tells me it's her husband, Todd, who is pouring my drink. Tells me she grew up in this town and that she and Todd drove here 70 miles on gravel roads to join in on the Easter feast.

"When I was in eighth grade there were three of us in graduation. Maybe ten families lived here when I grew up," she says. "There's only a few of us that know the history. There's only three people who live here now."

Eric Ericson, the old man next to me, is Marnie's father. "I've lived here all my life," he tells me. "This used to be big sheep country, then, after World War Two, it went to cattle. Used to be wide open. Never used to see a fence."

I tell him that Ingomar might have a claim on the quietest town in America. "I measured it at twenty-four decibels," I say. "That's three decibels below the silence of Benaroya Hall, the symphony hall, in Seattle."

"I'll be darned. I never thought of that. A lot of tourists, the hunters who come in the fall, they comment on that. They can't get over the quiet. They're so used to hearing traffic and whistles. They listen, and they don't hear anything."

"Everything sounds louder through the recording. That flag sounds like a fire burning because everything is so quiet."

"Brand new flag Monday. By Wednesday it was torn. That damn wind came up. Pretty strong winds this year."

"And in a VW bus, it gets slow."

"Which way you headed now?"

"A zigzag route all the way to Washington, D.C."

"Now only four of us live in this town overnight," says Ericson. "Me and my wife. Morris, the guy who works here. And Kathy, who just started. But some people from California are building a two-story house, for retirement, and a trucker bought the lot next to me. Says he's going to bring a trailer in."

Back in the 1920s, I learn, Ingomar had 250 residents.

"I never thought the railroad would pull out," continues Ericson, "but it did. Then we ran out of kids for school. Pulled the rails and the ties all the way to Miles City. County bought the right-of-way gravel and graveled these roads, which is the smartest thing they ever did because there just isn't any—it's all gumbo, heavy soil. If it rained you just didn't go anyplace. You just stayed there until it dried. Hunters still get in trouble, those who overstay their time in their vehicles, anyway."

While I'm eating my greens from the salad bar, a car alarm goes off. "Uh, oh," says a recent arrival at one of the tables.

"Taking the back roads and seeing things?" asks Marnie, delivering my order of leg of lamb.

"Back roads and listening," I correct, and then dig into a meal like I haven't seen since I left Yvette's house in Bellevue.

"Welcome to Forsythe, Home of the Doggies. Stay a day."

I have arrived at the Yellowstone River and the railroad tracks and make my way to Rosebud, then south, as Bill Worf instructed. Some things have changed since his boyhood. Interstate 94 now cuts across the creek valley, and I suspect the highway's massive embankment blocks some of the railroad noise and also cold air drainage out of the creek, possibly altering the thermal air layer stratification that allows sound to travel great distances. But mostly the land matches Worf's description: ponderosa pine and the buttes right on the mile mark, seven or eight miles down. I stop at a historic marker informing me that General Custer camped here on June 22, 1876. In the background I hear a *Yit-yit-yit* that reminds me of a Hoh Valley Douglas squirrel. But following my ears, I spot a prairie dog sentry announcing my arrival.

Worf's eyes would detect another change in the land. In just the 10 miles south from Rosebud I've passed maybe 20 ranches running cattle, not growing wheat. Approaching the 17-mile mark, the gravel road turns to gumbo. Good time, I decide, to knock on the door of a new ranch house and ask the occupants if they know of the old Worf homestead.

Three dogs intercept me getting out of the Vee-Dub. "Are you guys friendly?" I inquire, spotting their owner, an elderly rancher, eyeing me from his porch. "Hi. Sorry to bother you. Happy Easter."

"Same to you."

I explain my mission: find the old Worf homestead and listen for the train that Bill Worf remembers hearing along the banks of the Yellowstone River, that is, discover if it's still quiet enough to hear 17 miles away. The rancher isn't familiar with the name Worf. Or the sound of that train. But he knows of another train. "Well, you hear a train, but it's that one over there that you hear. The one you hear, that comes off the mine over there. She's really a chuggin' when she comes out with a load of coal."

"How far is that coal mine?"

"Oh, it's about thirty miles south of here, but I don't think it's more than fifteen miles to the tracks."

Several gunshots echo through the valley. They bring no comment from the rancher.

"I haven't heard it in quite a while, but we used to hear it all the time, especially in the morning. I can't remember hearing it lately. I'm getting pretty hard of hearing. Ten to twelve years ago, when we first come here, we could hear it."

Maybe the old Worf homestead has disappeared into the land. In any event, I'm obviously close to where it stood. So I get the rancher's permission to return early tomorrow morning with my recording equipment, bid him goodbye, and then point the Vee-Dub back to Forsythe. I've ignored my nagging cough longer than I should. In the past few hours I've been hacking up some nasty hues in the green spectrum. I need to see a doctor.

My only option is the hospital ER, where I'm the only patient this Easter evening. The doctor asks me a few questions, tells me to take a few deep breaths, and listens to the sound of my chest with his stethoscope. He makes his diagnosis by listening. In what seems like only an instant, I have not just a scrip but erythromycin. The antibiotic is pressed into my hand; no need to go to a pharmacy, if one were even open today.

I spend the night there in Forsythe and wake naturally with the first hint of ambient light. I'm off in search of a sound that may no longer be audible. My headlights are so dim on this rig they're like running lights on a ship. Thankfully, there's a beautiful sliver of a moon. The chiaroscuro landscape is breathtaking. A few scattered lights over the vast plains, shining as bright as any star or planet, remind me of a wilderness lake reflecting the sky just before dawn, when only the brightest stars remain.

I seem to be responding well to the antibiotics. Not only has my coughing subsided, but I can enter notes into my voice recorder unstrained for the first time in days. It almost feels like the Vee-Dub's heat is working; I'm actually sweating. Out of the starry landscape the first silhouettes emerge, and I can now detect individual grass stems, perfectly still. I smile. Calm conditions are crucial to long-distance listening; even a ripple to the atmosphere will reduce the listener's horizon from many miles down to only two or three.

The rising sun is painting the undersides of the local cloud cover crimson. This is not only beautiful to look at, but the clouds should aid my sonic quest, reflecting the train sounds back down toward the land, while also helping to exclude air traffic noise above it. Just past the 16-mile mark, where the road turns to gravel, I hear the loud clear ring of the meadowlark's song pierce the racket of the Vee-Dub. Earth's morning song has arrived: the dawn chorus, the acoustic shadow of the rising sun, an orderly sequence of songsters that reassert their claim to the new day.

I stop and record for about 10 minutes, enjoying a lively chorus of meadowlarks tossing a song back and forth from the tops of sage. Then one of them flutters swiftly upward to a new playing position while making a chortling sound. There are also some red-wing blackbirds, ducks, a woodpecker drumming, and American robins. But a bellowing bull steps all over this music with loud bursts of simple thought. When I'm bent on silence, I hate bovine bellows. Why? Because they're unintelligent and out of place. We have done such a good job protecting our livestock that their communication is no longer essential for survival. A bull sounds clueless compared to a buffalo or an elk. Until this yahoo bull barged in, I heard music, orderly, sequential, developing toward a crescendo still an hour away. Yes, I consider all forms of domesticated animals noise when I am attempting to record Earth's living music!

In the background I pick up a deep rumbling that might be a passing train, it is hard to be sure. At 29 dBA, averaged over a minute, a deep rumble is hard to discern, but three minutes into the recording I hear a chugging: a locomotive. But there seems more to this deep rumble, something even more apparent months later when I listen again, back in my studio, something eerie and unchanging.

Bill Worf might call this quiet. Most people would call this quiet. But I would not. I say that because what I cherish occurs in the absence of human-caused noise intrusions. In this case, I heard the train, the distant passing of a truck, and the domestic cattle.

I drive on to the 20-mile mark and there, on the western horizon, see plumes of what looks like steam and smoke. Maybe it's the mine the rancher was talking about. I drive farther and record again. The meadowlarks sing wonderfully, but now the background rumbling is even clearer. And now there's no train. I press on to the west. The road turns to gumbo, thankfully dry, but its washboard surface turns the Vee-Dub into a bolt-rattling percussion instrument, sending the wind chimes into hurricane mode. Happily, it's back to gravel and then asphalt. And then I see it, like Oz in the distance, and brake.

In the town of Coalstrip, Montana, "Tomorrow's Town Today," four tall stacks and a huge coal mining and power complex fill the air with man-made clouds and noise that radiates for miles and miles and miles.

I mentally erect a headstone: Quiet RIP.

But I'm not ready to give up on Montana and its quiet beauty. My next scheduled stop isn't until I get down to southeastern Utah. So this has me backtracking, over the Big Horn River, past the town of Custer and the turnoff for Little Big Horn battlefield, heading west on I-90 in Sweetgrass County. I'm instantly captivated by the Crazy Mountains, which stare down at me from a most marvelous and ever-changing cloudscape of scudding grays and whites. "Because of the spiritual power of these mountains the Indians were able to seek refuge," says a historical marker. That's enough for me. I'm going.

But not today. A squall line blows in, completely obscuring the Crazies. And then the pelleting rain reaches me, too. My already noisy ride becomes a racket, and my antiquated, slow windshield wipers prove no match for the deluge.

This is one more reason I love driving this bus and put up with its rattles. The Vee-Dub forces me to slow down, as now, and smell the sweet earth as the rain turns to snow. The land here remains largely unfenced, affording a rare vista, almost a look back in time. The snow shower switches to flurries, making it possible to turn up an unnamed gravel road, but it quickly turns to gumbo, the thick stuff I've been warned about. I'm not that crazy for the Crazies. I'll let them go—for now. I turn around, contemplating a hot bath in Big Timber.

But on the way, I venture down another side road that leads me past an abandoned church with a toppling spire, to a sheltered valley with no visible signs of settlement except the road—no fences, no cattle, no dwellings. No noise that my ears can detect. I see horned larks fluttering about after the storm, promising that dawn, if calm, will feature their amazing twinkling sounds. And I see western meadowlarks. Their classic bright call will echo off the slightest hill with a ring. This is where I will record tomorrow morning.

I check into the River Valley Inn in Big Timber, take my next dose of antibiotics, eat dinner, and fall over, sound asleep by 7 p.m.

I'm up before the sun, brew a pot of coffee in the room, and check my e-mail, opening a message from a buyer of my recordings, Dr. Samara Kester, an emergency room physician in Valparaiso, Indiana. After her stressful, exhausting, noisy stints in the ER, Dr. Kester takes refuge listening to some of the soundscapes I've captured. Having read about my journey to Washington on the One Square Inch website, she's been corresponding with me and is eager for an update. I write her:

> Listening for 17 miles with the naked ear apparently is not hard—very natural. What is hard is finding a place to do it. The Worf homestead was not found but I was easily close and I spoke with a neighbor there who was a newcomer—he moved there only 10 or 12 years ago. He was confused upon thinking that Bill could hear a train for 17 miles on the Yellowstone River and he said that he was probably hearing the train coming on down from the coalmine and pointed in the other direction. Okay, I asked, and how far was that? And he said 15 miles. So I poked around and set up to listen: meadowlarks, cattle mooing, skeet shooting booms, dogs barking—all very far away—and yes a rumbling train and yes, something else. I went around and finally located the something else. Miles distant (I haven't measured it yet) is a new American town built around what must be one of the largest coal mines in the world with four large stacks and lots of cooling towers. Looked more like a power plant than a mine. The town is Coalstrip, Montana; besides exporting energy, it consumes a thousand square miles of natural quiet. I don't have the heart to tell Bill Worf at the retirement center in Missoula that his childhood memory of natural quiet on the homestead has vanished.

I shut down the computer and gather my belongings and stow them in the VW. The wind is truly whistling. With the mercury at 27 degrees Fahrenheit, the wind chill is certainly below zero. I can't imagine there's a pocket of still air anywhere for 100 miles. My odds of a spectacular dawn chorus have grown very long. Turning in my keys at the front desk I ask, "Does it ever stop?"

"It did stop—yesterday evening—for about five minutes," says the woman.

"Will it stop again today?"

"Well, over there in the Crazies, I think it is. Could be it's still."

At least I think that's what she said. The cigarette hanging out of her mouth and her nervous head twitches distracted me a bit. But maybe she's right. Why not give it a try? So I climb inside my Vee-Dub and the gusty tailwinds rocket me through town and across the Yellowstone. This is DUI: driving under the influence of road bumps and wind gusts. Each one-two punch has me concentrating not on staying in my lane, but on the road. As crazy as this might sound, today really seems to be coming together. The intensity of the wind feels like the intensity of my search.

There are so many different kinds of wind. Years ago, I was asked to provide wind recordings to be used for the soundtrack to the movie *Alive*, which tells the true story of a rugby team whose plane crashes in the Andes. I inquired, "What are the emotions—what are the actors feeling?" before I started to plumb my catalogue of wind recordings. Each species of tree makes a different sound in the wind or rain or snow, and to my ears, each evokes a different emotion. John Muir recognized this and used the different sounds of pine wind to navigate up Yosemite Valley at night. His favorite was the yellow pine, also called ponderosa pine. With its especially long needles, it produces deeper tones than its shorter-needle cousins. When I visited his grave in Martinez, California, years back, I noted that someone had planted a yellow pine next to it, a conspicuous addition to the native oaks, at least to an astute Muir fan.

If the wind doesn't settle down, I will look for a place that has both pine and sweet grass, seeking deep roaring tones from the pine needles and the faint, almost hallucinogenic, swirling sounds from the tall thin stems of sweet grass. Imagine listening to a roaring river at the point where a stream trickles in.

As I drive up to the Crazy Mountains, a herd of cattle blocks the road. I rev my engine to generate enough juice to honk my horn. The Vee-Dub makes a sound like a party favor from the 1950s. The herd splits. About 10 more miles down the road I'm back to yesterday's valley and the abandoned church in the rolling grasslands. Framed by snowcapped peaks, its steeple rises above the entrance but, humbled by time, slumps forward. By the looks of the missing windows, the peeled paint, and the age of the large wood stove where you'd expect the pulpit to be, it's easily 100 years old and probably hasn't seen a service in 50 years.

The wind has begun to die down, and as it does, insects emerge, which adds a nice humming to the soundscape. I like insects in my compositions—as long as they're not the biting kind, since I can't swat. With my recording light on I'm reduced to a defenseless piece of juicy meat. But when the wind dies down and the insects come out, so do the ranchers, with their pickups and private planes. Many large ranches have their own landing strip and take a Sky King accounting of their four-legged holdings. As it happens, I can hear the fading trail of a jet, the first of the day. Then the *Craa-ooo* of a faraway crane. The wind, now just a breeze, blows through the well-weathered church, creating beautiful wood tones. I feel as though I've been ushered in.

I set up my ultrasensitive gear in this quiet place of worship and begin to record from the position of the congregation. My sound-level meter rests at the bottom of the scale and doesn't move during the session, except when I hear hooves beat on the ground, not a desperate bolt but at a sprightly pace. Apparently a deer had been hiding nearby, no doubt listening to my every move. Earlier it chose silence (a good strategy if I had been a poacher), but now that I was here to stay, it had to move on.

The wind and the quiet take me to a profoundly peaceful place: aural solitude. Words fall short of capturing this deep listening experience. Even a recording does not do it justice. Emerson got close when he advised, "Listen to what the White Pine sayeth." He did not say what the white pine said. You have to listen to the white pine for yourself.

For 13 minutes I listen to this peaceful place, until the wind subsides entirely and a fixed-wing plane approaches. Even at a distance, the plane sets the room into a lovely, violin-like vibration, so I'm thankful for that rancher's plane. Here in the church, I'm recording a cultural sound portrait, a piece that echoes human intention. The perspective is quiet and tuned and heard, but barely. This delicate expression of place tests the limits of my auditory sensitivity, my spontaneous thoughts fading like ripples across a mountain lake, revealing the depth of my existence through clear water. I listen with more than my ears. And I eagerly take it in. This plane did not take my peace away. I am completely still, inside and out, and I know this journey has finally begun.

Feeling cleansed by this solitary, early-morning service, I head farther along the road toward the mountains, passing a road grader. Bad choice? Beyond it, the road gets so bumpy that 18 miles per hour is soon top speed, and I notice a disconcerting assortment of car parts in the road—metal parts, rubber parts, some plastic—and can only wonder what I'm adding to the collection. I'll keep my eyes open on the drive out for any VW car parts that look familiar—should there even be a drive back out.

The closer I get to the Crazy Mountains, the crazier it gets. A snowstorm is raging across the mountain peaks several miles distant, releasing long trails of what look like sifted flour that end in windblown curls. I hop outside to have a listen.

Yeeoooowww! It's blowing so fast and hard that the little pieces of dirt and grit feel like BBs on my exposed skin. A magnificent willow stands alone in a field beside a small creek and I walk closer to hear the local rendition of willow-wind. On my last cross-country listening trip, in 1990, I found a willow in Kansas, swaying and singing, that held me spellbound for more than two hours with a wide range of fine-toned vibrations. The longer I was there, listening, taking it all in, the more I heard. At first I noticed only the larger patterns, simple gusts and lulls, but then my mind dug deeper and discerned the individual wind torrents weaving through the branches. After 15 minutes the details were countless; the tree was a congregation chanting a hymnal to the sky.

This Crazy willow has not yet leafed out, which I like, because the deep tones will be pure, not cluttered with leaf slaps. Instead of something akin to a waterfall, this will be clearly a wind driven event. The roar reaches 75 dBA during the strongest gusts. I set up to record low to the ground, near its massive trunk, to escape the wind that would distort the microphones. There I spot one of nature's marvelously subtle achievements: a windblown blade of grass dips its tip into the creek and rises up again, a frozen drop of water at its end. This must have happened again and again last night to create this tiny marvel. Dip. Freeze. Dip. Freeze. Like candle making. I watch this bejeweled blade of grass do its little dance while the grand trunk of the willow—which has to be three feet across—actually bends with this wind. The power of the wind surely continues on, down through the roots of the willow, for I swear I feel the very ground move, too.

Once back in the Vee-Dub, headed still farther up into the Crazies, I realize I've been up for six hours. I realize, too, that I forgot to eat. So I stop and build a fire in the stove inside the Vee-Dub and put on some tea water. I brunch on my usual: piping hot tea, some chunks of bread, slices of cheese, and oranges—a simple, quick meal that is the equivalent of a hot bath (from the inside) and provides enough energy and nutrition to fight back fatigue without becoming a culinary distraction. I developed this on-the-job habit as a bike messenger, sometimes eating my working lunch one-handed while still in the saddle.

Back behind the wheel, I hit what I call black snow: a blanket of white over a thick layer of black mud. Uh oh. Got to make it through. Wheels spinning, I cut deep ruts. Whew. Safely on the other side, I decide I've gone far enough. Of course, now I've got to make it through a second time, and this time my ruts lie ready to trap me. I continue up a small hill and turn around. With a running start and a deep breath I plow back through the rutted muck in second gear (an old Wisconsin snow-driving trick from my graduate school days), wheels spinning, inching forward, finally jerking free, like an eager fan through a turnstile at a sold-out ballgame. I might still be there if the mud had been just a tad slipperier or deeper.

Before I leave the Crazies I take time to appreciate those moments of quiet that I've found here. Silence seems to make music from everything, simply by isolating individual sounds, allowing the sounds time to form temporal relationships. Music is made out of rests and notes. Quiet times and exciting times, silence and sound. We need them both. More than any other sense, hearing unites everything.

Back in Big Timber I fuel up for the push to Utah. The guy at the other side of my fuel pump says he likes my rig. Tells me he, too, used to own a "splitty"—an apparent synecdoche for the split or two-paneled windshield that graced VW buses through the '67 model year—and now owns a '68. I tell him of my difficulties driving in the wind. "You aren't from around here, are you?" he says, telling me the wind can last for days and that I should expect worse in the direction I'm headed. "When you get to Livingston, there'll be a blinking light. Exit there. Otherwise you'll get blown off. A semitruck has been blown off."

Livingston is another 50 miles, nearly two hours away, considering the best I can do is 31 miles per hour, and that's taking advantage of the draft created by every passing semi. I resign myself to the shoulder of the highway. Good thing. I'm down to doing 29 miles per hour, more than 45 under the posted limit. If there were an exit ramp, I'd take it. But there aren't any, just warning signs: "Gusty Crosswinds" and "Severe Cross Winds Next 12 Miles. Use Caution." Finally I see a sign for Livingston and a billboard for the Del Mar Motel that advertises "Quiet" in big yellow letters, with quotation marks. I make that my destination.

I find the Del Mar, and it's closed. But that's not the reason for my chagrin. The motel is right next to the rail yard. Quotation marks or not, how could they possibly call their motel quiet? I'm just sitting in my Vee-Dub, shaking my head at my recent stretch of bad luck, when my cell phone rings. It's an old friend, Jay Salter, confirming that we are still on for hiking into Canyonlands National Park. "Gordon, where are you?"

"I've ground to a halt in Montana. Headwinds made driving impossible."

"While you're still in Montana, there's somebody you should look up that you're going to want to meet. Doug Peacock. I heard him speak about how he came back all screwed up from Vietnam and worked things out by living with the grizzlies in Yellowstone. I think he might be someone who knows what quiet is all about and also someone who knows how to listen."

"Well, I'm not going anywhere, I'm stuck in Livingston."

"That's where he lives!"

The next morning, after a café breakfast of two over easy, home fries, coffee, and secondhand smoke, I pull in behind a quickie oil change business off one of Livingston's main streets and get to work with a screwdriver, one of those double-enders with a Phillips on one side and straight edge on the other. The process of tightening up the bus is like patching a leaky roof with flour paste; it's never, ever over. It's just a matter of how much time to give it. But I've learned a few tricks, like carrying a coffee can full of screws and a box of wooden toothpicks. When you come across an empty screw hole, just insert the toothpick, then replace the screw and break off the end of the toothpick. It fits real snug; might actually manage another 500 miles before it drops out again. About 20 minutes in, I see three oil-soaked workers staring at my rig and laughing at my un-routine maintenance. So I go inside and ask them if it'd be okay for me to change my oil out back. They even let me use their nifty spill-proof oil catcher that slides underneath the Vee-Dub's drain hole.

The Vee-Dub is running like a kid in a new pair of sneakers. I'm equally happy. It's not every day that you might get to meet the inspiration for a fictional hero in an American classic. When creating the character of George Washington Heyduke, the protagonist in his environmental romp, *The Monkey Wrench Gang,* Edward Abbey looked no further than his good friend Doug Peacock.

I find Peacock's number in the phone book and give him a call. I introduce myself and my reasons for being in Montana. "Sure, come on by," he says, providing me directions. I've now crossed the Yellowstone River and I'm climbing up toward the mountains on a dirt road. There's the jog around a red house, and here's the two-track road he mentioned. This must be his "old ranch house," a white two-story house with a steep pitched roof in the middle of large trees, still leafless, that provide a sunning place for all the birds attracted to his bird feeders and water bath. Peacock's house is set in grassland backdropped by hills patched with grass, pine, and snow, then backdropped again by mountains that tower into the clouds. The entirety of the view speaks of bear country.

I'm greeted at the door by a big man wearing a fur cap with a leather brim, several layers of Patagonia, head cocked slightly to the side. His eyes look directly at me. His walk is ambling, slow and deliberate. His wife, Andrea, offers to make us some coffee. I explain about being blown to a standstill on I-90 in my Vee-Dub. Peacock tells me he once "drove across the winter landscape in a VW bus with no heater."

Out the living-room window we see a big pheasant strutting through the yard. "He's the only one left," says Peacock. "If we had a thousand pheasants here, I would cook one or two, but he's the only one. We feel obliged to protect him."

I explain in more detail about the purpose of my journey: searching for true natural quiet and speaking to people along the way about quiet and the significance of quiet in their lives. Ultimately, I tell him, I hope to speak to government officials, including those at the FAA and the National Park Service, to try to have Olympic National Park recognized and removed from the FAA preferred flight plan.

"Yeah," says Peacock, standing near the wood stove, his back to a wall of books and photos of bears and a portrait of Andrea. "I forced the government to take a helicopter pilot to court to stop him from buzzing grizzly country in Glacier National Park. Sometime in the eighties. But they wouldn't do it. It was their pilot. They lost the case on a technicality, but the guy quit flying that route over the wildest part of Glacier, where there's no trails."

Peacock rolls his glasses between his fingertips. I see he's got books about travel, natural history, and Native American life. One bright yellow book says "Edward Abbey" in large red letters.

"I never use trails in the backcountry," he says.

His Vietnam ghosts, Peacock tells me over the next two hours, drove him deeper and deeper into the western wilderness. Mortar concussions assaulted his eardrums, leaving him for a time with tinnitus, a maddening ringing in the ears in the absence of external sound. His prescription for self-healing? He bushwhacked the Continental Divide nearly to Canada, seeking refuge in remote grizzly bear country.

"Since 1968, until I had children, I lived in the wilderness. I've written four books. They're all about wilderness. Two have 'wilderness' in the title. I spent twenty years with grizzly bears, ninety-some percent of it by myself."

"You mean, camped out on their turf?"

"Yeah, in places like Yellowstone and Glacier, you sit and wait. I started when they were at their lowest point in Yellowstone. You wouldn't see a grizzly for seven or eight days. But I'd just sit there waiting. I'd go at this time of year, when the roads were all closed. They were then. They weren't plowed. Nobody was in the park. And there wasn't a sound. Maybe every other day you could hear a commercial jet. Absolutely no human sounds of any kind."

"That extent of natural quiet, could you find that today?"

"You've really got to work at it. I'm all gimped up this year, but I spend most of my time in the most remote places in the continental United States. In season, it's grizzly habitat, till they hibernate in November. Then it's southern Arizona. I take solo hikes across the southwest corner of Arizona. From Organ Pipe to the Colorado River, with Mexico as southern boundary, is this giant wasteland. There's nothing in it. It's a bombing range. It's a wildlife refuge called the Cabeza Prieta. I've hiked across it alone, more than any living human. Seven times end-to-end and once north-to-south. These are one-hundred- and one-hundred-forty-mile walks. They take about ten days. If you don't know where the water is, you die."

I can't help but ask. "How do you hike through a bombing range?"

"Well, carefully," he says. "The last chapter of *Walking Off* is called 'The Bombing Range.' It's a hike I take by myself, from I-8 all the way down toward Mexico, to where I buried Ed Abbey in 1989. It's a very beautiful place. You don't see a human sign—except for unexploded two-hundred-fifty-pound bombs, shit like that. I walked through an area where there is a lot of live fire. I know what they do out there—and it got to me. However, when it's quiet, there's no place on earth like that. It's absolutely quiet. You can't hear anything."

Peacock's wife arrives with the coffee. We talk a bit about people's perception of quiet. He acknowledges he's very atypical in his wilderness needs. "A lot of people can get what they need in their own backyard. I have friends like that. I need to be alone in the middle of a huge wilderness, big enough where it's four or five days to walk out in all directions. Where it is so absolutely quiet that you hear the inner sounds of your ears. You can't believe how much noise they make. The silence takes your breath away. It's absolute silence."

Over coffee he tells me more about living with the grizzlies and filming them and doing an ABC-TV show on grizzlies for *The American Sportsman* with Arnold Schwarzenegger.

"You track these animals, it takes three or four days, and you've really got to know what they're up to or you'll never catch up to them. Or else I set up in a really good place and wait for them to come to me. Either way, when I go into the woods, I don't talk. You're essentially hunting the whole time. I use birds a lot to find bears, to know what bears are doing. They're the best informants in the land," Peacock says, taking a big, noisy sip of his coffee.

"I probably spent the equivalent of four or five years of my life sitting, totally quietly. I'm talking about hours."

"I know the experience," I tell him.

"Just listening. I've got a favorite place up in Glacier where they come in for huckleberries. I almost always hear a bear before I see it. You can hear little sounds in the day bed. A little squeal. It's a mother disciplining her cubs. They messed up a little bit. You can hear a bigger boar come down through the timber. There's other animals in there, too, like moose and elk, but they travel differently. Bears are noisy when they're not aware of people around them."

"I have a funny story," I interject. "Years ago, I was in Kodiak. I was working as a deckhand on a halibut boat and I got off there for just a few hours for a little in-port. While everybody else headed into town, I went out to pick some berries. So I'm a couple miles outside of town. I'm enjoying the walk, and I say to myself, 'Boy, I wouldn't think people would walk this far to pick berries, but the paths in this berry patch are just so crunched down.' I'm not even feeling the brambles."

Peacock and his wife are laughing along with me.

"Not even thinking bear. But, of course, I was in a bear berry patch."

"You sure were."

We talk some more about bears, how the Yellowstone grizzlies at one time were habituated to feeding in garbage dumps, about dominance hierarchy, which kicks in at about 300 yards in open terrain, and how he can quietly get much closer. "My goal, when I'm shooting movies, is to get in and get my shot and the bear never knew I was there. My last book, which Andrea and I wrote together, is called *The Essential Grizzly.*"

Andrea goes to track down a copy and Peacock returns in his mind to the 1960s. He is still standing, legs braced, arms arched outward to the side with hands hanging from the wrists like, well, bear paws. It's irresistible. I can't help but compare Doug Peacock to his longtime object of study. So much of the bear's spirit seems to have seeped into him.

"I crawled into the wilderness after two tours as a Green Beret medic in Vietnam," he says. "The war was ultimately too much for me. My homeland happened to be the Rocky Mountains. I'm most comfortable camping out. I like being alone in the wilderness, so that's what I did. I went looking for grizzlies, up in the Wind River Range. Except for the wind and the lightning storms, it's pretty damned quiet, too. It's such lousy weather, that's going to be the last grizzly refuge in Yellowstone because it's so cold. After shivering around the fire for about three weeks I had a malaria attack and thought I'd better go to an easier place. So I went to Yellowstone. I wasn't thinking about bears, but they were there—and they certainly get your attention. Once you're in bear country, your self-indulgence vanishes. All your senses are directed outward. It's such a healthy attitude. It's really enforced humility. And the ambience in which it happens, happens to be very, very quiet. They only live in the most people-less, remote habitats now."

"I know when I'm in a quiet place, the dimensions of the space I'm in are known to me," I say. "Nothing's going to sneak up on me, if it's truly quiet, because I can hear pretty well, and just about every footstep on a hoofed animal, even those with pads and claws, makes a little bit of a sound with twigs. It's a combination of feeling more aware of the animals around me, but also more vulnerable—because any activity I make also sends out sound. There's a heightened sense of place for me when it's quiet."

"When I'm sitting listening, I don't feel very vulnerable at all," Peacock says. "I feel I'm probably at my safest. I have enough confidence and experience in being there, and I know what sounds mean, and I can hear what's happening before it happens. Nothing's ever going to sneak up on me. People used to look for me at night with flashlights firing blindly into the bushes. Paranoia is a way of life. I'm a total wacko. Hardly ever without weapons on hand, waiting for the worst. That's where I feel the safest.

"Once you realize how good your senses of smell and hearing are—we modern people have no notion of how good they are and what they can do for us. There's a total acoustic and olfactory universe out there that we totally shut down to. I think it's because of all the racket."

Our heads turn simultaneously at the *Aaa-aak-ek* of the pheasant, which doesn't derail Peacock's train of thought.

"I also think," he continues, speaking to the value of preserving natural quiet, "it's the closest way to really get in touch with what I consider your innermost humanity, because that's how we evolved, listening and smelling in ways that aren't imaginable today. We're the same species. The human mind, our intelligence, our consciousness, it all evolved from a habitat, whose remnants here in this country we call wilderness. The issue I'm continually raising is this: We evolved from that, which is essentially a wilderness, a wild habitat, using our senses, and that which evolves doesn't persist without sustaining the conditions of its creation. That's a giant argument for silence right there."

We're rightly quiet for a moment. Then I mention the great disconnect at the National Park Service. "On the one hand, they have in their management policy the duty to preserve the natural soundscape, and they define natural quiet as the absence of human-made sounds. And yet an even bigger, much more elaborate section deals with air tour management plans, because air tours operate over many of our national parks. As a listener and collector of sounds for twenty-five years, I look at something like that in a management policy, and I think these guys can't possibly know what they are talking about."

"They don't know what they're talking about. No one needs to fly over wilderness. They should ban all flights over wilderness. Just like that. You know Doug Tomkins?" he asks. "He founded both North Face and Esprit. This guy never finished high school. He's a climber. I've been on polar bear trips with him, to Siberian tiger country with him. He's got all this money now, owns about a fifth of Chile. That's his current project. But he also published a wonderful book, *Clearcut,* that just shows you pictures of clearcuts. His next book is on ATVs. It's just amazing how powerful that lobby is—and they're really hard to take on. And given this administration, the pressure to have ATVs—anyplace a horse can go, they want ATVs—and they're largely getting their way. We're doing what we can to stop it. I recommended 'The Moronic Sport' for a title, but they didn't use it. They used something more prosaic. Yeah, between airplanes and ATVs, those intrusive sounds are audible miles and miles away."

I decide it's time. "I think you'll understand this," I say, retrieving the leather pouch from under my shirt. "This stone is from One Square Inch."

"Yeah, that's somewhere in the Hoh."

"You know of One Square Inch?"

"Yeah, I know of it. I know about it. Yeah."

Holding it, he laughs gleefully. "It is. That's a square inch. It's beautiful. It looks like ardulite from an early Cambrian date."

"It's taking the tour with me."

"Good," Peacock jokes. "It will probably end up knowing more than you'll know."

He calls Andrea in for a look at the stone, telling her, "This square inch takes miles and miles of insulation."

They agree to have their picture taken with the stone. Peacock giggles as I snap the shot. "Yeah, I like that," he says.

What had seemed like a winding down of our conversation proves anything but. He tells me he's scheduled for knee surgery in six days. Tells me he just got a Guggenheim grant to write his next book. "They gave me more money than I asked for. They gave me fifty-five grand. I can light my cigars with hundred-dollar bills," he laughs. He signs and gives me a copy of his most recent book, *Walking It Off: A Veteran's Chronicle of War and Wilderness* (Eastern Washington University Press, 2005), with the chapter on his passage through the Arizona bombing range. When I tell him I'm headed next to Canyonlands, he tells me about a natural arch where the wind sings passing through it and who to ask for directions at a bookstore in Moab, Utah. "You should go there," he says. "Ed Abbey was there before he died."

I thank him again for the book and get ready to go. But before I do, I take a sound-level reading here in his living room, six feet from the wood stove. I look twice to see that I've read the number correctly.

"The base level is 27 dBA," I tell him. "That's the same as the silence at Benaroya Hall in Seattle, after they spent eighty million dollars."

"Out here at night," he says, "it's really nice."

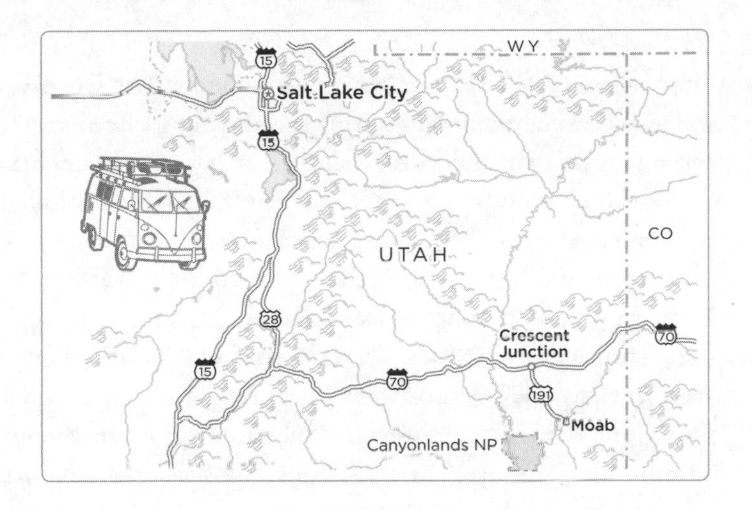

6 The Earth Exposed

> I am 20 miles or more from the nearest fellow human, but instead of loneliness I feel loveliness. Loveliness and a quiet exultation.
>
> —Edward Abbey, *Desert Solitaire*

My flight from Seattle has begun its descent into Salt Lake City, where I'll resume my journey after a brief trip back home to catch up on pressing projects for clients and spend time with Abby. For the moment, I am one of the thousands of people who make up the airborne community over the United States. The view from my window seat shows the raw, exposed Utah landscape, the immensity of Great Salt Lake, the rugged Wasatch Mountains to the east, and the converging interstate highways that have become our nation's circulatory system.

"Where civilization is most advanced, few birds exist," said Charles Lindbergh, who died of lymphoma on Maui in 1974. "I would rather have birds than airplanes."

Lindbergh became an instant aviation hero on May 21, 1927, when he landed the *Spirit of St. Louis* at Le Bourget Field near Paris, completing the first transatlantic flight and uniting two continents. The world became smaller that day and has continued to shrink, becoming ever more in need of Lindbergh's vision, which lives on today in annual Charles A. and Anne Morrow Lindbergh Foundation grants to "improve the quality of life through a balance between technology and nature." The foundation seeks to support present and future generations in working toward such a balance, that we may "discern nature's essential wisdom and combine it with our scientific knowledge" (Charles A. Lindbergh) and "balance power over life with reverence for life" (Anne Morrow Lindbergh).

In 1989 I was awarded a Lindbergh Foundation grant in the amount of $10,580 (the purchase price of the *Sprit of St. Louis*) to help preserve the nature sounds of Washington State. When I answered that exhilarating phone call, my son Oogie, then three, got to watch me jump up and down like a monkey. Afterward I shouted, "Daddy doesn't have to be a bicycle messenger anymore!" One Square Inch was conceived that year, but not established until 2005, after I decided I could no longer wait for the National Park Service. I would do it alone.

Our final approach circles the airport. I'm hoping to spot the Vee-Dub in the long-term parking lot not far from the toll booth, where I parked it 15 days ago with $36,000 worth of recording equipment hidden beneath a sleeping bag in the back. I can't help but worry, because my pristine antique auto attracts a lot of attention, and anybody curious enough to peer in through the windshield may notice that a key's not needed to start it. Just a push of a simple button on the steering column will do. Moreover, those who know this vintage also know you don't even need a key to get in—just the right touch to slide open the window.

We land. I board the parking shuttle. We roll by acres and acres of cars. Finally, I spot my Vee-Dub. But as I walk up to it, I can tell instantly that it's been opened by the half-cocked door handle. My heart starts racing. I spot a business card, of all things, face up on the driver's seat. "Salt City Air Coolers, Volkswagen Club, since 1986." On the back is a handwritten invitation to attend the next meeting, first Thursday of the month. The brotherhood of old VW owners! I check the rear. No surprise. All my equipment is undisturbed. Even the chocolate cookies on the shelf behind the driver's seat are untouched. Gotta love the VW karma.

I am tooling south on I-15 in heavy traffic, after the turnoff to go east on

I-70, when everything gets spacious. "Warning: No services for 109 miles. Next services Green River." This is my kind of country. Another 40 miles down the road and about an hour later, I pull off at a roadside rest stop. The sun has set, there is a bare hint of colors, the moon is half full, the sky yet to sprout its first star. It is going to be a gorgeous night. Surrounding the viewpoint, the canyons, with their horizontal seams of black, glow reddish tan. The juniper trees are ancient, twisted by time into dance postures. A semitruck idles nearby. Even so, I can hear the silence out there. Stepping over the edge and below the lip of the canyon onto a ledge, I am treated to something unfamiliar. The relative silence seems to destroy the noise, and the atmospheric attenuation, the rate at which a sound wave weakens, is much greater here in the dry desert than in the moist northwest air. I like working with the unfamiliar and eagerly await my time in the Canyonlands of Utah.

As I look out over the moonlit landscape it appears as if an aeolian storm has sandblasted away everything except the hard and unmovable, which here includes the trees, contorted monuments to life's enduring patience. In the absence of stars, the darkened landscape shows a line of pearls stretching for maybe 10 miles. Like luminescent ants, they roll determined and purposeful down the interstate.

The first planets have popped out of the night sky to join the half-moon. I can feel this great expanse, the incredible canyon and the celestial sky. It's humbling. Silence is humbling, and I crave it, I think, because it releases me from responsibility, this burden, this sense of, Gosh, there's so much to do in this world. In silence I feel God's presence and ultimate control. "Save the Earth" has become the battle cry of some environmentalists, but the Earth does not need to be saved. It has been doing fine, life evolving, for billions of years. It is we humans who need to be saved—from ourselves (by ourselves) and our myth of "Super Abundance," as Stewart Udall, former secretary of the interior (1961–1969) and a key figure in the Wilderness Act of 1964, once put it. His book *The Quiet Crisis* (Avon Books, 1963) warned us, "America today stands poised on a pinnacle of wealth and power, yet we live in a land of vanishing beauty, of increasing ugliness, or shrinking open space, and of an over-all environment that is diminished daily by pollution and noise and blight."

I awaken inside the Vee-Dub, right here at Ghost Rock viewpoint. How could I not spend the night? A second regimen of antibiotics is kicking in. I feel good. I can breathe again, and the day is dawning majestically. The horizon is turning amber. The view stretches maybe 30 miles—subdued drab greens in the canyons, the rock still pale—but with an implicit voice. The sheer antiquity of the rock speaks of deep and timeless silence, reminding me of my fragile existence. I'm used to marveling at trees in the Hoh that are 400 years old. Here, the exposed rock is millions of years old.

Craving my morning cup of coffee, I begin the 40-mile drive to Moab. The sun is rising so fast that the first line of light descending down the canyon slope outpaces me, the rock turning a mystical, deep red. I flip on my blinker after a sign for the Spotted Wolf Canyon viewpoint and am rewarded with incredible views of Navajo sandstone crowns on tilted flatirons, a sedimentary rock from the Paleozoic that makes up the San Rafael Reef, with cliffs over 200 feet high. An interpretive sign identifies shales, siltstones, sandstones, and other layers in the geologic formations, attributing the typical red rock of the Colorado Plateau to the presence of iron oxide. The depth of the color depends on the amount of iron oxide. "Here," the sign continues, "water has sliced and sculpted stunning narrow canyons and formations in the sandstone—a paradise for hikers and rock climbers." Past the sign, I-70, two lanes each way, divided by a concrete barrier, snakes its way right through the canyon where the ancient riverbed once flowed.

No information explains why this canyon is called Spotted Wolf or which Indians once called this place home. But a nearby bronze plaque reads:

> In 1957 the decision was made to increase the nation's interstate highway system and I-70 was engineered to bisect the San Rafael Swell. Here at Spotted Wolf Canyon, workers could stand and touch both walls of the canyon before construction began in October 1967. Engineers and surveyors used body harnesses and ropes to work as high as 400 feet above the canyon floor. Crews excavated 3.5 million cubic yards of rock from the area at a cost of $4.5 million for eight miles of road. On Nov 5, 1970, the Utah Dept of Transportation opened the 70-mile section from Freemont Junction to Green River to two-lane traffic. Two more lanes were added in the mid-1980s.

Somebody has scratched on a coda: "TOO BAD!"

Exit 182 puts me onto Highway 191 South toward Moab. A few miles farther a sign advises not stopping on the roadway because of possible dust storms. "Dead Horse Point, 40 miles." Well, that's convincing! As I approach Moab I'm getting the notion that it might not be the rustic mining town of bygone days that I imagined it to be. It's Friday, and though early, I'm joined on the road by weekenders in big-wheel 4x4 pickup trucks pulling supply trailers, fifth-wheel travel trailers so large that the hitch is a pillar, and full-blown motor homes pulling ATVs. I note, too, that the raw, arid wilderness of Utah has sprouted a species of sign new to my eyes: "Entrance Station." These, of course, attract the legions of off-roaders.

One final bend, then the drop to Moab and its verdant valley. Snow-capped peaks to the south, rock walls almost all the way around, and the magnificent Colorado River. But Moab has clearly outgrown its quiet, small-town past. It's become the Aspen of off-road adventure, and by off-road I mean up, down, and sideways: raft trips, kayaking, ATV adventures, hang-gliding, parafoil, skydiving, rock climbing. There's all manner of custom modifications of 4x4s, beyond anything I've seen in the Australian Outback, the setting of the Mel Gibson film *Mad Max*.

I'm supposed to meet the poet and fellow nature sound recordist Jay Salter, who wants to share with me his favorite secluded spot in Canyonlands National Park, a place he's visited annually for the past 10 years. It's hard to believe that anything within 100 miles of here could be one of America's last great quiet places. But since I trained him, I know he and I agree on the quality standard for noise-free: no audible human noise intrusions of any kind for a minimum of 15 minutes. Unfortunately, a phone message tells me he's been delayed by vehicle problems in the Sierra Nevadas. But he'll make it. No way would he miss his annual retreat. Even Superman needed his Fortress of Solitude in remote places like the Arctic, the Andes Mountains, and later the Amazon rain forest. Salter's natural fortress of solitude lies deep within Canyonlands. His only condition, in taking me to it, is that I must keep its location secret, a stipulation I'll steadfastly honor until that day that laws are in place to adequately protect such quiet places.

I return the call and leave my own message: "No worries, Jay. Take your time. I have another stop scheduled. Just let me know when you get to the turnoff for Moab. That'll give me plenty of time to get my gear ready."

Meantime, I'll be meeting with Skip Ambrose, for nearly 30 years a Fish and Wildlife Service raptor expert in Alaska, who, nearing retirement, switched government agencies to help found the National Park Service's Natural Soundscape Program. These days, enjoying more hospitable weather and living conditions, Ambrose resides in ranch country about 20 miles outside of town, where driving directions go au naturel, as in "Take a right, then go about a half-mile to a large cottonwood tree."

Ambrose lives in an idyllic setting: log cabin, 10 lush acres with fruit trees, surrounded by mountains. Birdsong fills the valley on this crisp, clear, blue-sky day. Ambrose greets me outside his home and introduces me to his wife, Chris, whom he met doing falcon work 12 years ago in Glen Canyon. He says he, too, owned VW vans, even in Alaska. I tell him I managed 19 miles an hour, tops, on the steep climb up here. Standing on the front porch he steers me over to a spotting scope pointed toward the opposite canyon wall.

"Take a look," Ambrose says. "Golden eagle sitting in a nest. She's got a baby, probably about a week old."

I spot the nest, right there on the cliff wall about a mile away. "Am I actually seeing a head there?"

"The golden head you see is an adult. The baby should be bright, bright white. The baby can't yet lift up its head above the bowl. That nest is probably six feet deep."

"How did you ever find that?"

"We knew there were eagles around here," says Chris. "We saw them flying, so we started watching them and searching the cliff."

Whitewash streaks, Ambrose explains, pointed to the nests. Though he hasn't seen these up close, he's spent considerable timc around other nests, especially in Alaska in the late 1980s and early 1990s, when he was installing the equivalent of peregrine falcon YouTube systems. Peregrine falcons were then on the endangered species list, and the Fish and Wildlife Service, among others, feared ill effects of a major increase in Air Force training exercises over their Alaskan habitat.

"We were trying to assess the effects of low-level jets on peregrines. I had done avian surveys, but I was new to acoustics," Ambrose says. "We designed this system where you put a tiny camera in the nest of a falcon. Peregrines are really tolerant. Eagles might abandon the nest if you go into it. We could transmit the signal down to a notebook computer that had a receiver and a program that ran, and when a jet went by it would take a before-and-after video of what the peregrine did. The same camera also took a still frame every five minutes.

"At first we used those things," he says, clapping his hands, "like you use to turn off the lights in your bedroom. It was really crude for noise. Then the Air Force wanted to get decibel levels for other studies, so I started using a sound-level meter, and that ran through the computer and gave us a more precise trigger. We could put it at forty-eight dBA or fifty-two dBA and save the video. Plus it was saving decibel data and making recordings of what went by so we could listen and know what the noise was."

"So, did the peregrines react?"

Essentially no, Ambrose explains, but he adds a couple of qualifiers. One, the birds, which are among the most adaptable creatures on earth (known to nest even on Manhattan high-rises), had been habituating to the noise for 15 years, even in the egg before hatching. Two, the study was hard to control because the Air Force pilots often ignored the study parameter of staying 2,000 feet above the cliffside nests. Ambrose's equipment often registered noise levels in the mid-80 dBA range and bursts as high as 114 dBA.

"The good news for the peregrines," he says, "they habituated. There was no difference in productivity, their activity, their feeding, or incubating. The Air Force probably liked it. The Air Force didn't want to do it at first. This was their symbol. So good for the peregrines, but they're mostly visual. I don't think other species are necessarily that way. Species like owls that hunt in total darkness and depend on hearing, they're going to be more affected. Birds need quiet to find mates, defend territories. At some point, the noises do overtake an animal's ability to communicate and catch prey, but it can be very difficult to prove."

Still, researchers have amassed more than suspicions about the harm of man-made noise on wildlife. The National Park Service's own Nature Sounds Program website has an annotated bibliography that lists more than six dozen scientific papers exploring the impact of noise and overflights on the likes of red-tailed hawks, spotted owls, elk, caribou, mountain goats, even humpback whales. Among the findings: caribou experience negative reproductive effects; bighorn sheep exposed to aircraft noise become less efficient foraging for food; communication between whales suffers. Studies in the United States have shown that songbirds in noisy habitats sing louder than birds in quieter spots, forcing them to expend greater energy to do so. Many investigators have demonstrated that animals spooked by jet and helicopter overflights flush from their hiding places, which increases stress hormones and exposes them to potential harm.

Chronic noise has also been fingered in the population decline of animals. Researchers studying the impact of noisy compressor stations in Canada's boreal forests have documented a decrease in successful mating among native ovenbirds, a beautiful, insect-eating warbler. More than 3,000 such industrial noisemakers pump oil and gas in the vast Alberta wilderness around the clock, radiating a low-frequency 75 to 90 dBA noise impact on the surrounding forests. The authors of the study report ovenbird pairing success of 92 percent in quiet habitats, compared to 77 percent at compressor sites. Writes Lucas Habib, one of the investigators, "Female birds are attracted to males by the strength and quality of their mating song. Because of this, loud background noise could affect the signaler-receiver relationship. If a male's song is distorted, or doesn't travel as far through the forest, females may not be attracted to him. This could have severe consequences for him—if he can't mate with a female he won't be able to produce any offspring that year."

After his peregrine studies, Ambrose and his wife wrangled a deal to winter here in Utah. The superintendent at the time at Canyonlands and Arches National Parks, Walt Dabney, brought Ambrose in to start measuring noise levels. "He said there's going to be an issue some day with air tours. This was 10 years ago. He was ahead of his time, and he was right, though Canyonlands and Arches are still pretty low key compared to Grand Canyon."

Currently, 675 air tours are authorized to fly annually over Arches; 1,039 flights are allowed at Canyonlands. The number soars to 28,441 at Hawaii Volcanoes National Park. But Grand Canyon stands alone, by far the national park most affected by noise from sightseeing helicopters and fixed-wing aircraft. Each year some 800,000 individuals crowd onto some 90,000 flights over Grand Canyon. That comes to 246 a day, or 20 flights per hour, if spread evenly throughout the year and over 12 hours. In fact, flights tend to be heavily concentrated in summer, so in prime visiting months this natural wonder becomes a grand collector of aircraft noise.

These grandfathered flight numbers have essentially been frozen since the federal Air Tour Management Act of 2000, Congress's third pass at addressing the noise from sightseeing flights over our national parks. The original bill, the 1987 National Parks Overflights Act, declared that "noise associated with aircraft overflights at the Grand Canyon National Park is causing a significant adverse effect on the natural quiet and experience of the park" and directed the National Park Service and the FAA to provide for substantial restoration of natural quiet by managing air tours. More than 20 years have elapsed. *Not a single national park has implemented an air tour management plan.* Little has changed in the air, while on the ground the fight between the air tour operators and environmental groups and the internecine administrative battle between government agencies rages on. One of the goals of my journey is to better understand this contentious ground noise, so I'm pleased when Ambrose provides some firsthand overviews and observations.

"The 2000 bill basically said the FAA and Park Service have to work together and develop these plans and said the FAA will be the lead agency, but both agencies have to sign the document," he says.

Representing the Park Service, Ambrose worked with the acoustical engineers tapped by the FAA, who hailed from the Department of Transportation's Volpe Transportation Center and brought a highway and runway perspective to the study of noise over America's most stunning and pristine natural landscapes. "They had never measured in parks, where it was down to zero dBA. They'd start at sixty and go up. We agreed on how to collect data; the big issue was how you interpret it. In the early days, the FAA would say, just like at an airport, if it's lower than sixty-five DNL [day-night level], which is what they use around airports, then it must be okay.

"We said, 'People sleep in tents. The ambient is ten dBA. We can't start off at a sixty-five day-night level.' All this was new to them. And they started coming around. They progressed a long way. But it really comes down to the interpretation. The key words in the act are 'no significant adverse impacts.' How do you define 'significant' and 'adverse'? In the beginning, if it wasn't above sixty-five dBA it wasn't adverse to the FAA. Of course, it was to us."

What's evolved is this: In the Grand Canyon, ground zero for the battle over air tours, the standard strives for half the park to be quiet at least 75 percent of the time. Ambrose points out that if you took that literally, you could have three minutes of quiet and one minute of intrusive airplane or helicopter noise *around the clock*. "Would you call that quiet?" he asks. "It's like the night sky. No one thought you could lose it, that it would become so noisy that you would lose your natural soundscape. Well, it's happening."

For many, I point out, all the measurements and debate and glacially evolving standards are beside the point. "How is it that air tours are even allowed over our national parks?"

"The air tour industry answer would be, 'We have to give people with special needs access to the parks,'" Ambrose says.

"That's a consideration," I admit, but hardly a reason to fill the sky over many of our crown jewel national parks. "If a handicapped person wants to acquire a special use permit and be flown over an area on a particular day, that might make sense," I suggest. "But it's another matter to offer air tours to anybody who pays the money, and, as in Hawaii Volcanoes National Park, have them listen to the rock music of AC/DC over headphones, as my aunt and cousins raved to me at a family reunion, telling me about their flight."

This doesn't even surprise Ambrose. "People are so oblivious," he says. "From the air it's a wonderful sight, but from the air they don't realize that people on the ground aren't wearing headphones." Even the magazine published by the Wilderness Society, he says, ran an article by a private pilot lauding the view from his small plane as the perfect, no-footprint way to see America's wilderness (see appendix A). "I'd like to kill that author and the Wilderness Society," he says. He's overstating his anger, of course, but there's an edge to his voice.

Ambrose tells of taking off from Denver some years ago on a commercial jet to measure noise in Hawaii Volcanoes National Park. "We flew over Colorado National Monument, Arches, Canyonlands, Bryce, Zion, and Mojave. We had noise meters in every one. I thought, God, I'm flying over all my noise meters. And it's such a line. If you moved that flight line twenty miles, you could avoid all those parks."

I offer a commercial jet story of my own. When I was studying John Muir's sound descriptions in his journals and recording in Yosemite, I decided to fly to San Diego to visit my brother. Because I'd had to pretty much limit my recordings to nighttime to avoid the noise intrusion of high-flying commercial jets, I asked the fight attendant to ask the pilot if he would fly around Yosemite. So I was surprised to soon find myself looking down at Half Dome and El Capitan. After we landed, the pilot was standing by the cockpit, so as I was getting off I said, "Well, thanks for at least trying not to fly over Yosemite." He said, "Not fly over Yosemite? I thought you wanted to fly over Yosemite." I said, "No, no, it's because of the noise of the aircraft." He told me, "At 36,000 feet you can't hear the aircraft."

"More people say that," Ambrose snorts. "And they believe that. But how could a pilot say that? One of the things I've done, though it ruins a lot of people's experience, is tell them the next time they go to a national park, to write down everything they hear in ten minutes. You don't even need a recorder, just a piece of paper and a pencil and a watch. If they're really paying attention, they'll realize that things aren't that quiet. Sadly, in our society, urban noise is so pervasive that when people get to a park, they say, 'This is nice compared to L.A.' They'd be correct, but it's probably not quiet at all. We're so accustomed to noise we don't think about it. We've got to make people think about it."

Near the end of our conversation, Ambrose tells me that he tries to buy a CD of every park he's been at. But few of these CDs, he says, are filled with the pristine sounds of the parks. "There's like thirty seconds of a bird and maybe thirty minutes of a flute and a piano. You might have a thunderstorm, but there's always music in the background."

And I know why. More than 30 seconds of uninterrupted birdsong isn't easy to find in America anymore, not even in our national parks.

Back in Moab, my overnight happens to coincide with a classic and collectible car show. All the hotel and motel rooms are booked, so I end up at the only place available, Slick Rock RV Campground, the Coney Island of camping. I'm lucky to get a spot, which puts me closest to the highway. But the campground has Wi-Fi, so I fire up my computer and catch up on business before heading into town.

Main Street is jammed with slow-cruising classics and hotrods. A souped-up Pontiac GTO with purple metal flake paint and a supercharger protruding through the hood ready for a challenge. A '40s-something Buick with a black cherry high-gloss paint job and chrome everything. A lowrider '50s Chevy pickup with two-tone paint and flame accents with moon caps. I fit right in, my Vee-Dub prompting some hoots and hollers from the sidewalk crowds lining both sides of Main Street. I grab dinner at one of the restaurants, pizza and a pint of Heffenwiesen that comes with a slice of lemon. My sound-level meter displays a peak reading of 105 dBA: dishes clattering over the babble of voices. The music plays at 75 dBA, loud enough, were I dining with a friend, to make it difficult to have any kind of intelligible conversation. After dinner, looking for a Laundromat, I come upon The Wet Spot. Not a name worthy of my patronage. My dirty duds can wait.

Back at Coney Island, no fit with the RV crowd, I listen from the top of my sleeping rack because I don't know what else to do. I'm lonely for the first time since my journey began. I listen to a stew of noise: people's voices, dogs barking, lots of engines, some birds chirping, people laughing, a motorcycle pass-by, and a baby wailing. An unpredictable sonic recipe for a restless night of sleep.

The next morning I'm treated to a botanical snowstorm of fluffy cottonwood seeds loosened from the trees. Shovelable drifts accumulate to several inches. I reach for my camera and snap away, freezing the graceful flight of these silent aircraft. Then, removing the OSI stone from around my neck, I place it upright in one of the "snow" banks and take one last picture.

I park the Vee-Dub at a Days Inn for free. Well, not exactly free; I ensured my parking spot right in front of the main office by booking a reservation for next Saturday night, a week from today, when I'll return after hiking out from the backcountry.

Jay Salter pulls up in his '95 silver Jeep Cherokee, climbs out, and stretches. Tall and lean, he looks like he has lost weight since our last recording adventure two years ago in southern California near the Mexican Border, gathering sounds for the San Diego Natural History Museum. Bandanna around his forehead, sunglasses hiding his eyes, and a Scottish goatee anchoring his face, Salter looks straight at me and says nothing. I sling my pack into the back of his Jeep and we hit the road.

He's concerned because his soon-to-turn-moonshot Jeep (he's at 236,000 miles) has been acting up, and we're running late. It's already after 3 p.m. and we have to get to the Canyonlands visitors center before it closes to get our backcountry permit. He's got the Jeep's heater roaring at full blast, hoping the engine will stay cooler, but only five miles outside of Moab we grind to a halt next to an abandoned vehicle that offers scant encouragement. All of its windows are smashed out. Jay thoughtfully positions the Jeep so that I get the shady side, then pops his hood while I sip on a cool Pepsi bought at the fuel stop on the edge of town.

"So, I'm not trying to be a nag, but we might not make it?" I'm teasing.

"We might not make it," he answers solemnly. Cars and trucks are swooshing by. "If we get there by six it shouldn't be a problem."

While we wait for the engine to cool I ask him how long he's been coming to Canyonlands.

"About fifteen years. I was hired to teach a class for Prescott College. I taught poetry and art. And I would take my class to this place for two weeks, long enough for them to get homesick. I didn't want to leave."

"Does this place have a name?"

"It's called You-Can't-Share-the-Name, Utah," he says, and then tells me more important things about it and why it's become his personal Mecca. "What it is, Gordon, is a place where I can return to myself. It's enough of a scramble to get to . . ." he waits for a loud truck to pass ". . . that the energy expended is significant, and it translates into a change in my body chemistry and my psychological chemistry and my heart chemistry. I'm often the only person out there. I might not see a person for days. I'm out at night a lot. You don't see anybody, just the critters. I have to be on my toes—you can get hurt out there and just lie there. You have to be self-sufficient. That's part of it. You have to plan well. You have to know the place well. But once you're out there, it's just yourself and the place. My mind has to let go."

We do make it to the visitors center before closing. Inside, it feels like a hotel lobby during the off season without the Muzak; there's the hum of ventilation and a whine of something electric with a soft babble of voices coming from a corner of the exhibit area.

"Two of you?" asks the disinterested voice of the park ranger as she enters our data into her computer, which uses Wilderness Tracker software to control backcountry travel and keep a record of each person's history in that area. "What you can do next time when you make reservations is ask them to mail it to you and you won't have to stop here." Her voice is a dead ringer for Nurse Ratchet from *One Flew Over the Cuckoo's Nest*. She then recites the list of wilderness rules: no bathing in streams or waterholes; no fires; no walking on the living soil, just bare rock and washes. Then she mentions that there have been problems with people entering archaeological sites and removing artifacts and leaving graffiti behind. "You are going into bear country—haven't heard of any reports, but we have heard of ravens tearing holes in plastic sacks. You guys have headlamps with you?"

Finally, we have our permits (one to take in with us and another to leave on the dashboard of the car). In the parking lot, Salter sees there's still cell phone reception and walks away from me for one last call. I peruse the park brochures we've grabbed. The literature lists the activities—hiking, four-wheel driving, mountain biking—then goes on to remind visitors that this is mountain lion country and that the dirt is alive. "Cryptobiotic soil, a knobby black crust is a living ground cover found throughout Canyonlands National Park and the surrounding area. Protecting cryptobiotic soil will insure that Canyonlands will remain ecologically healthy."

This is what the ranger was talking about: the living soil, a.k.a. desert glue. It is made up of bacteria and fungi that secure the desert soil from erosion by wind or rain. This soil is so fragile that footsteps are enough to destroy it; regrowth can take five to seven years under favorable conditions. Well, I think, you can forget about regrowth anywhere near those desert entrance stations outside of Moab.

Back in the Jeep we ford a 30-foot-wide river in four-wheel drive and pop out on the other side—a baptism and entry gate to a stark, mesmerizing landscape. The vegetation, a kind of pelt or fur or skin, has been pulled back, revealing the planet's raw flesh. It's as if the earth is unearthed, and we're headed to the center of being. By the time we reach our jumping-off place, a trail at the top of a large butte that descends into a series of canyons and the creek below, the Sunday evening rush hour has begun high above us.

Four large white jet trails scratch an otherwise peaceful sky. I pull out my sound meter. At the moment, I hear only a light wind through the juniper trees, a rustle so faint that my sound-level meter cannot measure it. I see one passing jet, so high it's practically out of sight, but don't yet hear it. The cone of noise it drags behind it hasn't reached us. When it does, it measures 40 dBA, unnoticeable in an urban environment. But we aren't in a city. Here, the jet sound is easily 100, maybe 1,000 times more evident than the ambient sound of the rustling wind. It feels deafening. More intrusions loom, for the darkening sky is becoming crisscrossed by jet trails: five . . . seven . . . eight . . . nine.

The nearly full moon rises bright enough that we don't need flashlights to get ready to bed down I hear *Poorwill, poorwill, poorwill,* then the *Whot, whot, whot* of a nearby owl. Then another jet intrudes at 50 dBA. God whispers. Man shouts. I have come to this national park wilderness as a pilgrim in search of aural solitude, but the noise reminds me instead of what I am trying to escape. But hey, I'm tired, tired enough to roll out my sleeping bag onto the powdery dry dirt that covers the flat rock at the top of the butte. I fall fast asleep.

I awaken early, well before sunrise, and listen. A light wind fans the juniper and pine trees ever so lightly in slow breaths of 20 to 30 seconds each. For more than an hour it remains genuinely silent. No sound of an insect or a bird or a footstep of any animal—only the breeze. I'm feeling peace. A shooting star blazes across half the sky in less than a second. I follow the slow path of a satellite. I conclude that the wind has waned. Nope. Here comes another gentle breath, another chaser of all but essential thoughts and hard questions—the buttes of my stay. This is why I'm here: to find out what I really care about and who I am. During times like this, it is possible to exist, even to think without words. "Listen to what the white pine sayeth." I'll just sit here for a while on this rock, as a rock, wearing a rock, and watch the stars fade and the day begin.

Like salmon milt released into a clear mountain pool, the dome of daylight builds in the east, slowly erasing the stars. It's going to be a beautiful morning.

5:25 a.m. *Berlew, berlew*. The call of a bird that I cannot name.

I undo the drawstring at the base of my tubular down bag and stick out my legs, slip on my pants, and lash on my boots. Because of this simple feature that I had sewn to my specifications, I am able to stay cozy inside my down bag while walking around, able to get up before I get up, on this occasion, to head to the edge of the canyon to witness sunrise. The dawn chorus precedes the first rays of light, a sonic roll call of bird species as distinct for every locale as a thumbprint. I confess that I have never studied the names or identified all the species whose calls and songs I listen to. To my ears, they're all different, each individual bird. So I prefer not to take roll, but rather, just allow the dawn chorus to wash over me, as I do a symphony, without segregating the orchestra into oboe, cello, flute, and drum.

Across the canyon I spot a light, evidently from a campsite lantern, and then make out two silhouetted figures. I don't know if, like me, they're awaiting the sun's early show of color. But I do know that if they are, they've done themselves a great disservice by lighting their lantern, for its light casts visual noise. Just as it would limit their enjoyment of the night sky, it will strip away some of the subtleties of the sunrise. Amazing: they've just lit a second lantern. Like headlights on a car, their campsite stares at me. Unnecessarily, for once your eyes adjust to these early-morning conditions there's plenty of light to move about, even prepare breakfast, if that's their aim. Camping for some people is a Hammacher Schlemmer experience; all too often the nifty gadgetry separates them from the immediacy of the place they've traveled so far to visit.

I am staring down into our gaping opportunity: a huge crack in the Earth framed in juniper and pine boughs that will soon swallow us up. The canyon appears like an inverted wedding cake with alternating layers of translucent cream and pink sponge that turn to amber and red in the direct sunlight.

At six o'clock I hear a flutish sound. *Loo loo looo loo. Loo loo loo.* Is it the sound of the canyon itself? To my ears it sounds like the faint resonance of gurgling water as heard through a long tube. Later, I hear the *Aw, aw* of a passing raven. Then a bird's song rings true through the morning air. I recognize the *Whurrrrrrrr-inggggggg* of a hummingbird moving into position for its unique aerial aria. Ascending high into the sky, the minute bird plunges Kamikaze-style toward the earth, pulling out in the last few, death-defying seconds to rocket straight up again. The high-speed turn causes its wings to sing with a loud ring that can be heard clearly at a distance of 100 feet or more.

These feathered flyers are soon replaced by man-made aircraft. The day's first jet intrusion comes at 6:20 a.m. A fixed-wing plane passes overhead at 6:55, and a few minutes later, a helicopter roars in low to the ground, descending into the canyon, temporarily out of sight but not out of earshot. I get a reading of 45 dBA for the claps of its blades before I can see it again, hovering halfway down and near a wall of one of the canyons. It's a small chopper, the kind often used by television and film crews. Salter and I assume it must be on some kind of search and rescue mission, for at that elevation it's against park regulations. A later call to the park's chief ranger, Denny Ziemann, prompts a check of that day's records. There were no hiker rescues, and the only helicopter permit issued was for after noon, five hours later, and not even over the park but over adjacent Bureau of Land Management terrain. This flight was apparently unauthorized because it operated below the canyon rim. FAA Advisory Circular 136–1 states that all flights conducted for hire must maintain a minimum 5,000-foot altitude above ground level while operating over national park lands. Unauthorized or not, this intrusion is an assault on resident wildlife and human visitors alike, for it came at an especially sonically delicate time of day, precisely when songbirds send their messages most efficiently and wilderness seekers can listen to a place at its most expansive.

Readying our gear for the descent, Salter and I make sure to fill our canteens and water bottles. He asks if I want any ibuprofen, holding up an enormous generic container that must hold 500. I stream some into the pocket of my backpack, barely denting his supply. I'm happy he hasn't offered me aspirin, which is one of the ototoxic drugs found to cause temporary tinnitus and some hearing loss in some individuals. Then he pulls out his iPod, pushes play, and hands me the headphones. I hear male elephant seals near his home in Santa Cruz, then pups being weaned. The sound quality is excellent. I'm transported to the Pacific. But I find the experience almost hallucinogenic, for here in the Canyonlands there's no visible water, only an ancient sea to imagine. I hand back the headphones, telling Salter I can't listen anymore. Not that his work isn't fantastic, it's just that I need to be here right now.

After Salter signs us in at the trailhead, we hoist our packs onto our shoulders, cinch our waist straps, and take the quiet path ever downward, shedding layers of clothes with the rising sun and heat and letting our animal bodies find their natural stride.

Contrary to what some might think, a hike down is always tougher than a hike up. On a steep downgrade like this, unrelenting knee strain can leave you with two bum wheels at the bottom. Slow is good, especially for my bike-messenger-taxed knees, and we each carry a hiking staff to absorb some of the shock and steady loose footing. I made mine from titanium tubing that I salvaged from Boeing's Seattle surplus outlet, where, with enough spare change and enough spare time, you can acquire all you'll need to build your own 747. I fashioned my lightweight hiking pole in two pieces that I joined with a quick-release seat clamp hacksawed from one of my broken messenger bikes. The pole has served me well for 20 years. It's doubled as a monopod for camera and microphone. Filled with water and set across a roaring campfire, it has helped me brew many a decent cup of tea. And now it helps support me on the six-hour descent.

Once we're down on the flats the trail disappears, washed out by a recent flash flood. Although we know which direction to head, we don't know exactly where to go. By the looks of it, the flash flood happened many months ago, but the trail has not been re-marked, so instead of one real trail across the fragile desert floor, there are many, forming a crazy maze of continuous choices. At every new junction we choose the most popular and well-worn option. At a shallow rock ledge, this trail dead-ends. We were wrong—as was everyone else who came this way. That realization is sobering, for I'm suddenly aware of how unaware I had been while strolling easy street. I can see why Doug Peacock doesn't hike on trails. They're like roads; somebody's already done the thinking for you. We backtrack. Paying more attention, we find our way across the washout and rejoin the trail.

The only source of water is a wonderful cascade that drops into a deep pool in which tiny tadpoles wriggle about. Though tempting, we cannot go for a dip. The salts and oils from our bodies would pollute the water and possibly endanger wildlife. We do refill our canteens. Salter uses iodine tablets. I call upon my ancient ceramic filter. We pour water over ourselves and wring out our T-shirts away from the pool, spreading them over the sagebrush, where they dry quickly in the hot afternoon breeze. Thanks to the ibuprofen I popped, my knees feel fine, albeit a bit warm to the touch from the stress on the joints during the long descent.

Late in the afternoon we pass an abandoned, hand-hewn log cabin around a bend, and then head into one of the side canyons. Soon we arrive at Salter's longtime campsite. It's a beautiful spot, offering a spectacular view of the canyon from a nice little rock amphitheater shaded by a spreading oak. My sound-level meter doesn't budge from 20 dBA, its lowest reading.

When we're still, I hear only a faint, residual whining in my ears from all of my recent travel. I know this should soon disappear. Aaaaaaah. The immersion in silence, like a good soak in a hot springs, relaxes my tired spiritual muscles. Right now it is all about letting go of my ingrained thoughts and being open to the moment. Everywhere there is beauty to see, music to hear, desert holly flowers and sage to smell. Wispy gray and white clouds accent a deep blue sky. The canyon rock glows fiery red. I feel invisible, almost selfless. A hot rock in the desert is what I am, dissipating conscious thought like the stored heat of the sun. To the west, backlit by the setting sun, darker clouds are dropping curtains of rain. This is a magical place that breathes inspiration.

The moon rises full and glorious. My dinner is a few handfuls of granola. Salter and I part ways soon after, each to attempt to record the silence. I position myself at the base of a nearby cliff that forms a semiparabola that reflects distant sounds down to my listening point. Not only do faint sounds sound closer in a spot like this, but I believe that the sound portrait contains an added sense of place by using this naturally occurring acoustic feature.

Good spot. Bad luck. Instead of natural silence, I hear distant voices from somewhere above, a quarter-mile or more away. At one point, I even hear a zipper. So I turn in for the night.

At 1:35 a.m. I wake with the passing of a jet and measure it at 35 dBA from the inside of my tent. After it passes, the desert night is immeasurably quiet. Then another jet passes. I try to ignore it. Counting jets instead of sheep will hardly soothe my mind toward slumber. Then another jet intrudes, so loud that I again reach for my sound meter. I measure it at 41 dBA at 1:50 a.m. After it passes I listen to a delicate breeze pass through the oak, juniper, and pine that surround me. The first breeze barely touches the sound-level meter at 25 dBA. The second breeze registers 27 dBA. I nod off to sleep.

Edward Abbey wrote about the stillness of this region in *Desert Solitaire*. "I wait. Now the night flows back, the mighty stillness embraces and includes me; I can see the stars again and the world of starlight. I am 20 miles or more from the nearest fellow human, but instead of loneliness I feel loveliness. Loveliness and a quiet exultation."

At 1:58 a.m. a jet wakes me again. Its impact on the canyon floor registers 54 dBA. That's 9 dBA *above* the maximum permissible nighttime noise level of a residential area in Seattle. Former Canyonlands superintendent Walt Dabney would surely be saddened by these nighttime ruptures of the natural quiet, but he would not be surprised. Ten years ago he told a reporter for the Dubuque *Telegraph Herald*, "What I hope the American people will recognize before it's too late, is that there are a few of these places where we want to recognize natural sound as a national resource." Notice that Dabney said "the American people," not the government agency he worked for, the National Park Service, whose very mission is to preserve our parks in their unspoiled condition. I've come to agree with him. We the people will have to raise our voices if we're going to preserve the few naturally quiet places that remain in America.

Clearly, Canyonlands is far from perfect as a quiet sanctuary. But between the noise intrusions it's wonderfully silent, and in extended, pristine acoustic moments, the sonic equivalent of a stand of old growth forest or a stretch of unplowed native prairie, it offers aural glimpses of our nation's disappearing natural soundscape. Come the dawn, today will be a day of sound tracking.

My eyes open to the sound of wind-blown birdsong, one of my favorite compositions to work with! There is something very romantic and optimistic in the voice of a bird carried by the changing rhythms of wind. A light rain dots the sides of my tent. Still bundled in my walkabout Worm, I move into a protected position with my recording equipment beneath a rock ledge that should be among the first places to receive sunlight. Not only will this shelter my microphones from wind distortions, but the smooth, concave rock surface will reflect the birdsong, strengthening its presence against the firm wind. My hope is also that this position will soon be visited by a bird who likes to sing in the morning light (many do).

I hang my microphone system in a bush so it remains fairly inconspicuous, then peel out 30 feet of cable to another bush, behind which I'll sit like a statue and listen.

A bird flutters in, then departs. More gusts. Gradually, with the onset of daylight, the canyon reveals itself as home to hundreds of songsters. The air comes alive with a chorus of birdsong that ebbs and flows with the eraser-like action of the wind for a good 20 minutes.

Then the spell is broken. A jet passes high overhead, louder than any natural sound reaching my ears or my microphones. My sound-level meter measures it at only 25 dBA, a very low reading, unnoticeable in a city and easily dismissed by some as a no-consequence event on park goers. But here, in this early-morning quietude, it's a severe noise intrusion, as spell-breaking as a cell phone ringtone butting in on a symphony's concerto.

The jet noise fades. The wind subsides. A sonic window opens for a timeless call: a desert songbird in search of a mate. A lone winged insect buzzes over a yellow flowering desert holly. A pine sighs. I lose myself in the varied symphony of the canyon. A raven calls out from a cliff top, the echo arriving in multiple layers.

I gather my gear and relocate to a distant rock parabola I hope to use as a big ear for listening to the entire valley. I set up my microphone on top of a tripod I've placed in the focus, the natural sound-collecting point of the rock structure. The ambient dBA is only 20.5, barely measurable, a superb natural amphitheater for upcoming solos. A hawk sails by more than 100 feet away, yet I can hear its very feathers cutting through the air. Then a hummingbird appears on stage, whirling its wings into a resounding buzz. The wind picks up steadily after this brief lull, but not enough to knock out the noise of a passing jet at 50 dBA.

The wind forces me to switch subjects, from recording birdsong to capturing the wind itself. I move on once more, settling in a tall stand of tamarisk. The windblown stems knock about, reminding me of bamboo. The wind subsides again, and I measure 26 dBA with the sounds of distant running water and insect wings.

Another jet. I move on, discovering a small oak grove in a secluded ravine sheltered from the wind, a wonderful, intimate, natural recording studio. Another hummingbird passes, but instead of a low-toned whirl this one gives off a clear, high-pitched ring that traces its every move through the soundscape. But my recording opportunities are shrinking. Jet overpasses now occur every four to five minutes. With each one lasting about three minutes, my noiseless listening and recording windows rarely exceed a minute. These are sips, not thirst-quenching drafts of quiet, but I'm thankful for them. Salter is right. He's discovered one of the last great, if not endangered, quiet places.

Eager to explore it further, I stash my recording gear, grab my camera, and set off—initially, led by my nose. I smell the sweet pungent fragrance of big sage and the brilliant yellow blossoms of desert holly. The rain-pocked desert dust seems almost lunar. The overlapping, translucent spring leaves appear neon green when backlit, creating a stained-glass effect. Beauty in all its sensuous forms nourishes the soul and sweeps away despair. I come across a small barrel cactus in bloom with the most incredible tropical flowers, scarlet petals with yellow hues toward the center, with a cluster of eight lime-green pistils and hundreds of anemone-pink anthers in a subtending fluff—a perfect work of art. Seeking the perfect picture from a position low to the ground (as a small mammal might see it), I crouch and shoot. Three frames into the study, I inadvertently back right into another cactus, making an instant pincushion of my posterior. Two hours later I'm still yanking needles out of my ass.

I have yet to see many mammals up close. From a distance, I have seen a herd of deer; a coyote pup without a parent—that worried me; a dead kangaroo rat, still in perfect condition. The only four-legged creatures I've encountered up close have been two rabbits that showed an unusual comfort at my presence, even at ten feet, munching on the spring flush of greenery. They remind me of how much I miss my wild-running pet rabbits back home in Joyce, where I have more than 20. When they hear me drive up, they approach, only to be petted, for I do not reward them with food. But in the wild, this kind of tame behavior in the face of larger creatures is suspicious. I wonder if perhaps the coyote population has cycled low for several years.

When we're back in camp together, Salter and I rarely speak, but we often exchange happy glances. Our mutual delight in being here extends to respecting each other's need for solitude. Even carrying our canteens to a nearby spring-fed stream, we're wordless. Here, too, we must abstain from the great temptation to jump in. So we do the next best thing: fill our canteens, step back from the water's edge, and then pour waterfalls over our heads.

A loud single-engine prop plane makes a tour of the canyon above the rim, a flightseer, no doubt: 58 dBA, louder than Salter and I will speak during our entire stay. Before its noise fades, a jet intrusion starts. It is now nearly 10 o'clock. I pull out the sound-level meter and raise it to the sky: 50 dBA, then 54 dBA. Then another, and another—almost all in the same east-west line of travel. To and from LAX is my guess.

The sun climbs higher in the sky, and the warming of the air creates enough turbulence to make the aircraft noise much less obvious. This is one of the reasons many tourists report on park exit surveys that they were not annoyed by aircraft; they visit during the time of day when plane noise doesn't reach them. Just as we can see farther to the bottom of a pond when the water surface is not rippled, sound travels best when the air is calm. That's why birds sing and other wildlife vocalize mostly at dawn and dusk; that's when calling is cheap, meaning that they expend less energy to send a message over the same area. This is the best time not only to send a message, but also to listen for one. The canyon's echoes—the voices of a listening land—are best heard then, too. Most likely the ancient people who inhabited this land centuries ago realized this and capitalized on it, possibly for hunting or defense or religious purposes.

The winds are picking up again and a thunderhead looms on the horizon. Salter and I exchange knowing glances as we rustle up our recording gear. He chooses to record from his "listening rock," his sweet spot of solitude within his chosen canyon. He comes to sit at this single spot every morning and every evening, just to listen, record, learn, commune, and become changed, and thus refreshed. I understand and leave him to his solitude.

There is hardly a more dramatic sound than thunder for revealing the physical dimensions of a place, either live or in a recording. Muir wrote this description from Yosemite: "Presently a thunderbolt crashes through the crisp air, ringing like steel, sharp and clear, its startling detonation breaking into a spray of echoes against the cliffs and canyon walls." In deeply forested areas the thunder echoes warm tones, as in Joyce Kilmer Memorial Forest in North Carolina. In a canyon like this, a distant rolling thunder will resound with a multitude of echoes. A thunderclap powerfully sets everything into vibration (putting my Benaroya Hall test *Boop* to shame), including the abandoned hermit's log cabin we passed on the hike in.

Whoever lived there may have also built it—it's obviously hand hewn, still showing adz marks—from native trees. Long before Frank Lloyd Wright drew up the plans for Fallingwater, John Muir built his cabin in Yosemite Valley over a small stream just to hear its music. Did the hermit build an intentional listening place? What will it sound like in a thunderstorm? There is only one way to find out.

The cabin is about 15 by 20 feet, fashioned of horizontal logs with a dilapidated wood shingle roof and mudstone fireplace and chimney. The low front doorway is within 10 paces of what has become a huge cottonwood tree more than nine feet around and sheared off at only about 25 feet from the ground, but with a lot of life left, offering a bit of shade, some shelter from the wind, but, more important in my mind, a constant, ever-changing weather report. This living wood chime registers the slightest breeze with wide, spade-shaped leaves extended on long petioles. Even when the pines are subaudible, the cottonwood will sound like a misty sprinkler in the slightest breeze. Quaking aspen is a first cousin to the cottonwood and well known for the shimmering leaves that give both visual and auditory accents to home gardens. Judging by the assortment of feathers at the base of the huge cottonwood, this old tree has a long history of attracting birds, offering both song and company to the hermit. Perhaps, like Edward Abbey, the hermit was not filled with loneliness but loveliness.

Though the thunder in this thunderstorm never amounts to more than a few distant rumbles, I'm not disappointed. I'm completely in the moment in a very special place.

Day two in Canyonlands dawns fragrantly, the overnight rain having stirred up the smell of sage. Wearing my down jacket and hood, sipping my first cup of Red Rose, keeping the cup close to my face for warmth, I notice how I'm settling in deeper and deeper here, becoming one with my surroundings. The nearby gamble oak tells me its story, which is written in its bark and outstretched craggy branches. I learn of good years and bad, fast growth and slow, old fire scars. The cottonwood also speaks to me. As its trunk swells with age, its bark actually bursts apart, revealing the younger layers beneath. Over the years the bark has become deeply furrowed, especially at the base, but interconnected like fishing net, allowing it to persist. Both trees are so different yet have so much in common; each has fresh sprouts near old wounds.

The ear never sleeps, but I did. Last night I slept without waking, but as I was going to sleep it started to rain and I grabbed my sound-level meter: 30 dBA. Then a jet came roaring through, and I thought it was the loudest one yet, but it measured only 35 dBA, much less than I expected. I'm becoming aware of the slightest changes in sound, shifting from the crude, unawakened sensitivity that mostly serves one well in the city. I am fine-tuning. My senses are coming back into survival mode.

I haven't talked much about food—with good reason. During my John Muir recording project in Yosemite, I tried to eat what Muir ate because I thought it might help me better emulate his experience, maybe even think like him. So I baked up a big batch of Muir biscuits, based on what scant evidence I could find about his vegetarian eating habits, and I existed off them while I hiked and camped and recorded. Muir didn't write much about what he ate while he was out on the trail because it's not really that important. Eating is a distraction; there's so much else to do. This trip, I brought along about 10 pounds of snack material: granola, bread, and cheese. Not one thing requires cooking. Whenever I get hungry I have a little bit. Last night I wasn't hungry, so I didn't eat. This morning I ate a single handful of granola. My body weight is also fine-tuning.

Salter has his food organized into six packages, one for each day. His stomach is beginning to go sour, and I suspect the iodine water treatment. I would never swallow a poison that is supposed to kill one-celled animals and not think that it wouldn't be killing me, one cell at a time. My water is perfectly clear; his looks like tea after the bag accidentally breaks. I offer to let him borrow my water pump, which works quite simply. Just toss the intake tube into a pan of water, then a bit of brute force on the pumping arm pushes the water through a very fine ceramic cylinder. Out dribbles perfectly clean water.

Nourished and ready for the evening, we move to Salter's "listening rock" and look out onto a beautiful sunset beyond the verdant and songful presence of the spring-fed creek. We've barely spoken since we hiked in, and now seems a good time to ask him about this special spot of his and about his annual pilgrimage to Canyonlands.

"It's actually my listening overhang," he says. "I met this place fifteen years ago. I was team-teaching a class through Prescott College called Landscape Perception. We were both very interested in landscapes and wild landscapes. My specialty was and is poetry and art and how artists and poets connect with a place and how they make art. My own poetry is very much involved with place. And with sound. So we spent two weeks here and a place nearby with about ten college students. It was the middle of May. Every day I worked out a writing exercise for all the students. We'd go out for a certain part of the day and write and then get back together and read our work. There was a real communal feeling, sitting on this big rock listening to each other's work. I did the exercises, too. And I actually tapped into something that was very amazing. It was a transformative experience for me. Going out day after day to the same place and experiencing it with all of my senses without any distractions.

"Toward the end of our trip—and I remember this very well, for it was a landmark in my life, really. We were all camping outside. Sometimes we needed tents, sometimes we didn't. I was lying in the sack at about midnight, just savoring everything and listening. It was the yellow-breasted chats' mating season, and they were calling and responding with their vivid mating and territorial songs, and the toads were singing—this shrill singing—up and down this creek. And other birds were contributing. It was a moonlit night. It was so vivid against the silence that night that I felt like I was listening to what the people who lived here one thousand years ago or more heard. Maybe they wouldn't have heard the tamarisk rustling, but they would have heard the willow and the cottonwood and would have heard these animals. That was an extraordinary experience for me, to realize I was hearing the same sounds as the people who had lived here that I had studied, whose art I had been spending a lot of time looking at. That moment of appreciation felt like a petroglyph or a pictograph on my consciousness. The sound itself just changed my consciousness."

Salter's voice is soothing. There's an unhurried pace to his story and almost the equivalent of musical rests as he pauses, in search of just the right word or for emphasis, or perhaps to draw upon the invigorating power of these surroundings for inspiration. Yesterday, of all things, we both heard (how could we not?) from different spots in the canyon a bagpiper playing, a stunning, unexpected musical performance—inappropriate, to be sure, but implicitly interesting to Salter, who has been a piper since he was a boy. Over the years, he explains, and especially after coming to Canyonlands, he's found his three passions in life—piping, poetry, and the outdoors—converging and enhancing and illuminating each other. But first he enlightens me about the origins of Gaelic piping.

"Scotland was a clan society, tribal really, people wedded to very particular territories, and of course there was a lot of contest over territory, a lot of warfare. The music became important to the clan to establish a sense of themselves. The more well-to-do clans had their own piper who was responsible for composing, like the bards did, tunes that celebrated the prowess of the clans, especially the prowess of the sometimes mythic progenitors.

"These tunes are passed on aurally, and they're long and characterized by the melody against the drone. I always liked to feel that the drone was the land, the constant keynote of the land, and the melody moved through the land like a trail, which would give meaning to and take meaning from the land. By that I mean there was a tonal expectation, which was aroused by the melody against the constant drone sound, a sense of dissonance and reconciliation. I heard this music when I was a little boy and I said," Salter snaps his fingers for emphasis, "that's the kind of piping I'm interested in. It's meditative, the pieces are long, you can get into it, get lost in it, you can be transported. It works toward a kind of ecstatic crescendo. Then it goes back to the basic theme, which has been elaborated through a series of variations. Each variation is expressed with an increasingly complex kind of grace noting. The old pipers would often call them warblings. They're very much like birdsongs. They're very complex. They happen very quickly."

"Can you give a sample?"

"Sure. Can you sing a drone sound?"

He's never heard me sing, I think, and laugh self-consciously.

"So what you're going to do, Gordon, is sing the drone sound, the constant sound, and I'm going to sing a melody against that. You won't change your tone."

He has me hum a mantra-like *Ommmmmmmmmmmmmmmm*. And then

joins in, chanting a wave-like string of melodious, ancient-sounding syllables, braiding them around my hum.

"Those tunes," he says reverently, "are like landscapes to me. It turns out the pipers in Scotland were very in tune with the land. A lot of their compositions celebrated aspects of the land. There are tunes about the quarries. The quarry is a hill and often there's a hollow place on top of the hill. Sonically, a very interesting place, an incredible natural resonance, and, of course, as a piper, just like a rock musician, you want to be able to hear yourself and also amplify it and give it even more volume. So these places were beloved to the pipers, and they felt a spirit out there that was conducive to the music.

"And, to touch on a parallel with my writing, I've always been interested in sound in poetry, not just rhyming, which can become singsong, but working with the vowel or consonant shape of the sounds and how you can string them together. How you can make a music, how the throat has its own yoga when you're reading out loud a well-made poem and you realize physically and through the ear the resonance and the potentials of the language, which we don't always access in our day-to-day, business, kind of get-it-done speech. When we were an earlier culture we were more in tune with the sonic aspects of literature, of poetry, of song."

"I hear you describing how a place has a sonic structure that affects the people in it."

"Right. When I taught this class, I was selected to teach it based on work I had done on landscape and poetry. I'm very obsessive about finding poets who really honor the experience of the land. Patrick Cavanaugh, an Irish poet, for one. I became obsessed and looked at their work very closely, studying their reflections on the this-ness of a place, the solidity of a place, the reality of place, which might be experienced differently by different people, but there is some fundamental truth that people find individually in a particular place.

"I think a lot of that is missing from modern poetry because we've lost a connection with place, a connection with land—the kind of connection we're feeling here tonight, with the wind rustling here through the tamarisk and the willow." Salter shifts his gaze from me to our surroundings. "The light's starting to die and we're becoming aware that we are going to be in this place without any light, without any streetlights. We'll have moonlight, but it becomes a totally different world that we have to be attentive to and listen to. It has its own rhythm."

"Is that part of why you come here again and again?"

"We all have places that somehow speak to us. Places we feel most whole in, where we can discover more and more pieces of ourselves. I go to the Sierras for solitude. I live in a redwood forest by a creek. I've known a lot of solitude there. But what I experience here has to do with that life-changing experience I had listening to the sounds that first night I was here. This is very hard to describe, but—this rock is so old and it's so beautiful. There are paintings on it done by people who lived here, connecting with the land. The rock is red, the color of flesh, and it was flesh for the people who lived here. They made their homes in this rock. It was a haven, shelter. They built their granaries here. They lived very simply, at the mercy of a harsh environment. They had to leave several times as the weather changed, and then they would come back. And they left this amazing art on the walls here.

"I find it inspiring to return to a place where I can find art, I can find evidences of a people who lived in a very direct way, and where I can find few evidences of the modern world. Few people come here. There's no mechanized apparatus allowed here. No loud machinery. I can drop down out of the busy everyday to my apprehension of this place and what it's telling me. It's also a place that's wild. I can break a leg while I'm out here. It's enlivening, it's awakening to be here, to be relying on these capacities which we don't often get to use in the civilized world. Especially our senses. We're using our senses all the time out here. I live in a sonic reality, that's how I make my living. And I live in a beautiful place, which is sonically beautiful. There's something else here that draws me.

"So when I started recording here six years ago, I went back to try to capture, as if I was doing a petroglyph, my experience of that moment, with all the things that were funneling into it: the cycles of the weather, the day and night, the season, the animals' movements. All of these things that are increasingly rare and more and more endangered, I find here in plenitude and I'm renewed, apprehending them, in all senses of the word apprehending."

"So is it fair to say you fell in love with this place?"

"It's obvious, isn't it?"

"When you did, fifteen years ago, it was a place considerably free of all mechanized noise. Have you noticed a change since then?"

"I have become more aware of airline traffic in the past few years. This visit, the number of planes coming through has been extraordinary, until they were chased away by the thunderstorm for a short time. I have seen a change, yes, in the frequency of air traffic. What does that do to your experience? As a recordist, it fucks everything up for you. You're getting a sonic picture of the landscape, the bird is singing, it's a priceless moment, and then a plane flies by."

"And if you weren't recording?"

"That's a good question. I've talked with people who say it doesn't affect them. I've talked with people who say they're outraged by it. Yeah, I'm affected by it. Before we came into this canyon we were sitting on the brink and looking down and a plane passed through. Somehow, from above, looking at the landscape, seeing it all laid out there, that plane passing seemed ephemeral, a temporary disturbance in an endless matrix of natural quiet, but when you're down in it and plane flights are more frequent, it's a different experience. It's more difficult to weave it in aesthetically. For me, being here is a meditation. I come here to renew myself, to remember what it is to be human, which is not to be listening to a lot of things telling me what to buy or who I am or who I could be, but to find out who I am, to remember who I am, and to take that back into the world. What I'm hearing when I hear a plane is the contents of that mind that I'm coming here to get away from, because that mind is reflecting a purely, I'll say a simply human world. And that's insane. How do we know what we are if not by what we're not?"

Thursday, 3:30 a.m. The night breeze sighs through the trees in long drawn-out breaths, like glassy sea swells hitting a distant sandy beach. The full moon is gorgeous, illuminating everything. All is still, silent. Then a soft breeze whispers. Then back to nothing. Everything is immediate. Time is washed away, immeasurable, and unremembered. I simply am—inseparable from place. I am here.

I wake again, naturally, to the first calls of desert songbirds echoing off the canyon walls. Quickly and quietly, I gather up my recording gear and set it up at the base of a large juniper tree (which I expect will attract birds as a song post) nestled at the base of a rock wall (which will help brighten the sound by reflecting it). Then I push "Record" and retreat back to camp to boil water for a pot of tea. Squatting like a peasant, I admire the passing wave of birdsong, coaxed off the surface of the Earth by the first morning light: the dawn chorus. This is the same wave of birdsong that I have recorded more than 1,000 times on six continents. It just keeps circling and circling, one long Earth song that has evolved with life itself, including our own. This morning feels like it once was, music untainted by noisy human irreverence for nature. I feel not merely 54 years old, but in tune with millions of years, still evolving, still listening, and becoming changed by what I hear.

Dipping my Red Rose tea bag, silently, in and out of my Texaco thermal mug, I appreciate that the dawn chorus is a perfect match to the landscape: sparse, expansive, dry, bright, and jubilant—full of optimism, hope, and a sound of prosperity. Out of a crack in the desert rock I can see a green sprout. It's springtime, even in the desert.

My unit of time has grown to a day, not hours or minutes or even seconds. I'm quite content to watch a dry blade of grass ripple in the wind. Somehow its action is a statement, and in some ways a poem more interesting than anything a person could write. The scene is not being described to me; this blade of grass is my reality. Everything is direct, immediate, and uninterrupted even by my thoughts.

Another blade of grass is a different poem. Both are different; both are true. The truth is self-evident and unexpected; it is not what I think but what I feel. My body is my brain. The one supreme truth is that I simply exist. And I have a right to lay claim to this day and this place equal to the claim of any other creature living here. I am subject to the same laws of survival. It's thrilling.

In late morning the canyon is filled with the drone of winged insects. Then the largest insect of all, a fixed-wing plane, appears high above my perch on a large sunlit rock. I do not reach for my sound-level meter but dip my hand into a bag of granola, scoop up a handful, and pour it into my mouth. What are they thinking? Certainly not about me down here on the ground.

Our experience here, Salter's preferred method of enjoying his Canyonlands stay, differs from the more common practice of wilderness-seeking backpackers, including my own when I first took to the woods. I used to plan to stay at a different place each night, always moving on, noticing the changes from place to place but not noticing the changes in one place. I've learned the wisdom of staying put. There is so much that goes on in one place, and though largely repetitive from day to day, if you look and listen closely, you discover subtle variations, even minute to minute. It takes a minimum of several days in one place to know it, in my opinion. Think of how different it is to visit someone for just a few hours versus moving in for days and really getting to know him or her. Few people know what it feels like to be accepted by a wild place, to be known by a wild creature and have it trust you enough to resume its natural patterns of behavior. Always on the go, most people walk through nature never stopping to escape the wake of their own disturbance.

The wind has gotten blustery and the jet traffic is almost nonstop. The jet contrails are immense. Returning to camp I see the swirling dust has driven Salter inside his tent. I do likewise.

During a lull Salter and I emerge for more birdsong recordings. The birds have been waiting, too. I see Salter about 100 feet away, patiently waiting for a songbird to land on the branch right in front of his microphone. Then, just as it does, a jet intrudes. Salter points his middle finger at the sky.

He says he'd like to head farther down the canyon to one of his favorite recording spots. I follow him down the trail that leads through fragrant, tall sagebrush and flowering desert holly, past my rabbits and a presumed coyote den (somewhere near where I saw the untended pup), then another mile or so. Then Salter stops, eyes drawn to a cliff dwelling that he's not noticed before, far across the main canyon and halfway up the rock face. We decide to go off-trail to visit it and make our way carefully along the well-worn game trails and over the steep embankment of another washout, this time noting prominent landmarks. Carefully stepping over clumps of the living cryptobiotic soil, we reach the base of the cliff.

Nestled into a recess about halfway up the 100-foot cliff, the brick dwellings represent an ancient take on a familiar modern real estate maxim. They occupy a prime spot for observing, by sight and sound, the activity of the entire valley; they're highly defensible and to a great extent naturally warmed by the morning sun and cooled in the shade of the late afternoon. In other words: location, location, location.

Until recently, anthropologists paid little attention to the acoustics of ancient environments, overlooking the implications and impact of daily living conditions that were much quieter than today. But the very terms *acoustic anthropology* and *archaeoacoustics* show budding interest in a fascinating sonic spotlight on early civilizations, one that may illuminate our past and also point to present, unmet needs. Steven Waller, a biochemist by trade and by avocation, and an acoustic anthropologist, points to one common natural acoustic underpinning of many ancient civilizations: echoes. "Echo myths collected from around the world," he's written,

> attest to the spiritual significance attributed to echoes. . . . Systematic measurements at some of these sites have shown that sound reflections at decorated locations are at significantly greater decibel levels than at nearby non-decorated locations. The cultural significance of the phenomenon of echoing can be found in numerous ethnographically-recorded myths from around the world that attribute echoes to supernatural spirits. These ancient myths show that echoes were widely worshipped as divine gods, were considered to be the "earliest of all existence," and were systematically sought out.

Earliest existence. I like that. That is what I think America's national parks are about: rooting ourselves in our earliest existence in a quiet place.

It is evident by the solitude and the enormity of the view and listening horizon that the peoples who lived here—and there were more cultures in this valley than the ancient Puebloans—not only got the regional news 24/7 from Radio Valley thanks to their cliffside location, but could easily stay on top of local news, almost without trying, crowded as they were into such a confined and sonically reflective space. Under such quiet surroundings and close quarters, neighbors would hear their neighbors' lovemaking, gauge the difficulty or ease of childbirth and the severity of illness, and deduce how soon a family elder would die.

Scattered on the ground nearby lie shards of ancient pottery and jewelry. There's even a stone mortar used to grind maize; a pestle, the stone for the hand; and a bit of corncob. Time for a photo op. I place the OSI stone among these silent artifacts of those who preceded us here. *Click*. I've united One Square Inch with another sacred spot.

Modern Americans are largely a transient, rootless society that came from someplace else. My paternal grandfather came from Australia, my father from Nevada. I was born in California and lived in six more towns before I graduated from high school outside of Washington, D.C. Sitting quietly, beneath the cliff dwellings, I think my hunger for home may be nearly as great as my thirst for quiet. Everything about this place feeds me essential soul nutrients, nourishment absent from my childhood.

Salter and I head back to camp with hardly a word. Nothing is more important than what is being said by the place around us. "Don't speak until you can improve on the silence," I remember being told as a teenager by a wise man I considered a mentor, Dr. Henry Simmons, who worked for the Food and Drug Administration.

Salter is looking a little dried and haggard, but he's happy. If he's any kind of a mirror, I expect I appear much the same. At camp we lie back like hobos in the shade, just taking in the grandeur of this place and relishing today's exquisite accomplishment, connecting with the Earth exposed.

Then two female park rangers burst into camp—no greeting, barking questions, pointing fingers, and expecting answers. "Why did you put a question mark on the registration form for your vehicle license? Did you gather these?" One points to a collection of deer antlers that had been in the bushes for a long time. Both of the women have darting eyes, as if they suspect us of a crime. The bulldog of the two barks out the big question: "Why have you come here *ten years in a row* and stayed without moving on?"

They have us on this one. Salter doesn't know what to say. He holds out his hands, palms to the sky, as if an invocation to the silence. The less aggressive of the two rangers, scattering the deer antlers, explains that the mice need calcium in their diet and are too shy to come into camp to gnaw on the antlers.

When they settle down a bit, after they realize we're not the artifact poachers they had us pegged for, I ask them if they're planning on camping in the area. The bulldog says they haven't decided. Noticing that they are without backpacks, I ask, "Where's your gear?" The bulldog says, "At the trail, with someone watching it." I think, Someone watching it? Why does someone need to watch it? She comes from *someplace else*—a poisonous reminder of the world back on the canyon rim, the place where we must return.

Without so much as an apology for invading our privacy and sending our adrenaline racing, the rangers leave as quickly as they came. Their intrusion was temporary but spell breaking. No longer are we reflecting on the experiences of earlier in the day; we're replaying their words, trying to make sense of their actions. Salter figures that the only way they could have known about his 10 consecutive years at this campsite was through the computer. He'd been flagged, probably as a backpacking aberration, because he didn't hike campsite to campsite. Apparently, even here in the wilderness, where you least expect it, Big Brother is watching.

This sorry experience reflects a sad, illogical underpinning: the misguided notion that staying in one place is suspicious. It may not be the norm, but those who do, and do so with their senses alive to the subtle pageantry of place, are often much richer for the experience. Have you ever knelt beside a tidal pool? You'll need to wait a few minutes, but the still pool will come to life. All the little hermit crabs, small fish, and anemones will resume their normal activities. It is a tiny world unto its own. In the expanse of the desert you need days, not minutes, for your senses to adjust and new experiences and sensations to come your way. But come your way they will—unless the Rangerettes barge in.

I'm still pissed at the Rangerettes. And also at myself for not having the presence of mind to inform them of something that does need flagging: the main trail at the washout. Hikers could get lost or drop over the ledge in the dark, I wish I'd told them. I also wish I'd mentioned that, with the path not reestablished, the fragile desert floor we were warned to carefully avoid was being torn apart by an unnecessary maze of wandering alternatives.

Friday, around 3 a.m. I get up just to record the silence. The night-flying moths sound like those large rubber bands on balsa wood gliders when they unwind. I hear the small footsteps of rodents scampering all over the valley. Far off, a rock drops from high up on the canyon wall, another of millions of events that, added together, have formed and continue to shape this place.

I go back to sleep until, once more, the dawn chorus wakes me. I set up to record, but noise intrudes, this time from close at hand. A rip of Velcro.

A crinkle of cellophane. Salter's eager to record, too. He opens his recorder bag, grabs his microphone cables, and snaps them on. Then his footsteps also break the silence. Finally, I beg him to be quiet so that I can try to roll a few minutes of noise-free recording. He bows in apology and holds still. Inward comes the birdsong, immortalized in 1s and 0s.

Once morning passes we come to a sudden and mutual decision to hike out—now. Neither of us has ever cut short a trip like this. But we realize that, since last night, we've begun to get on each other's nerves. We've both gotten what we'd come for—and then had it snatched away. Our natural quietude had been shattered and, like Humpty Dumpty, couldn't be put back together again. At least not on this trip.

Hiking out, I fret about transitioning back to life above the rim. Just off the path I pluck a juniper berry and a small sprig of sagebrush. The world is full of symbols. Petroglyphs were man's first language. We raise flags. Slip on wedding rings. Place a small red stone on a mossy log in a temperate rain forest. I add these two Canyonlands keepsakes to the amulet around my neck and continue the climb.

We arrive, before sunset, at La Hacienda, a busy Mexican restaurant in Moab, and land in the bar, waiting for a table to be cleared. The cocktail waitress approaches. Pointing to Salter and holding up two fingers, I shout, "Heineken's!" The restaurant babble measures as high as 74 dBA. Salter looks as though he has gone feral. His green bandanna with little red flowers is knotted around his matted hair, and canyon dust is caked in the creases at his neckline. I don't even have time to imagine my own appearance, for leaning close to me, he says, "It's hard to believe, Gordon, we're coming from twenty decibels to this! A jet intrusion would be quieting."

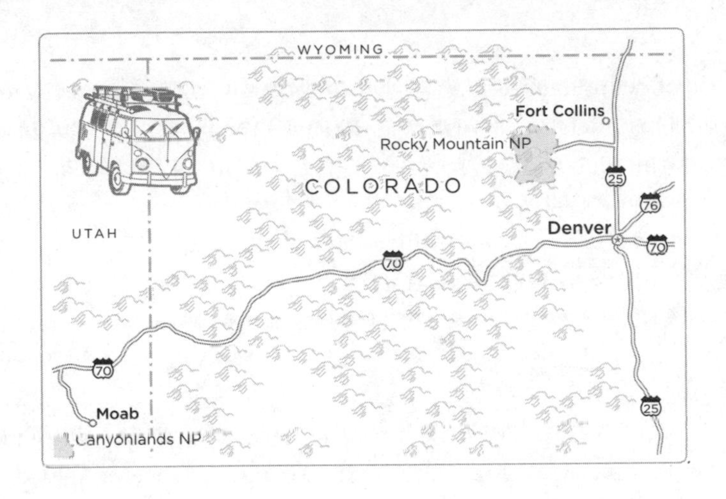

7 The Rocky Road to Quietude

One of the greatest sounds of them all—and to me it is a sound—is utter, complete silence.

—André Kostelanetz, orchestra conductor and arranger

I make good on my promise to stay at the Days Inn, Moab. The Vee-Dub looks untouched, except by envious stares. But I'm altered. Coming out of Canyonlands, just being in a room, any room, seems numbingly confining by comparison to the wild outdoors. Can society ever be truly environmentally conscious when our walls and windows are so thick and excluding? With few exceptions, our dwellings are isolation chambers compared to the ancient Puebloan cliff dwellings, and disconnect us, not connect us, with surrounding, sensual places. Camped in Canyonlands, sleeping in the open, feeling the breeze flow over me with a spiritual presence, or sleeping in a tent where the sound still comes through easily, I'm inseparable from place. I'm not proposing that we turn our motels or our homes into tents, but I do think we owe it to ourselves to spend more time in the wilderness to keep our senses alive and our moral compass properly oriented. As former Secretary of the Interior Bruce Babbitt said, "Our national parks are important because they are a gateway to the conservation ethic." Is it any wonder the Earth faces serious overlapping environmental crises when increasing numbers of the planet's peoples live so much of their lives out of earshot, and thus out of touch with the natural world?

In my motel room, Microsoft Outlook says I have several hundred unopened e-mails in my in-box. I have 20 phone messages. I have reentry problems. I'm going to have to ease back into the rat race.

As I head north out of Moab, across the Colorado River, still personally imbued with a bit of the luxurious, uplifting night silence of Canyonlands, this old Vee-Dub seems much louder to me. And, tooling along at 50 miles per hour, I miss the rewards of traveling at two miles an hour. I can't smell the plants or notice the subtle hues of the flower petals. I've lost the details of the landscape in the blur of highway, well, nearly highway speed.

I connect with I-70 and point east toward the Rockies. A downpour overwhelms my one-speed wipers, so I pull over to the side of the road. My nostrils are treated to the earth's scent, heightened by the freshly wetted desert soil, a combination of rock flour with a hint of herbs and musk.

Looking out over the immensity of the Utah desert, no one would think that solitude would be so hard to locate, and when found, so short in duration. But part of the reason is the very asphalt ribbon I've been driving. I-70 is part of the Interstate Highway System, what was once called the Dwight D. Eisenhower National System of Interstate and Defense Highways. Nowadays, some 46,837 miles of it—enough to circle the Earth twice at the equator—crisscross the nation. It's hard to imagine our country without these essential arteries. Modern highways gave Americans almost unbridled access to their land. They facilitated commuting, and thus the spread of suburbs. They made longer vacation drives possible, and thus helped build resort communities. Moreover, superhighways shifted the automobile age into high gear. Then came National Scenic Highways, National Parkways, and, finally, an increased stream of cars to America's most rural areas through the National Scenic Byways Program. All this has not been without cost, namely, the unintentional but nonetheless systematic assault on quiet places in much of rural America.

"Since 1992," one is told at the program's website, "the National Scenic

Byways Program has funded 2,181 projects for state and nationally designated byway routes in 50 states, Puerto Rico and the District of Columbia." The NSBP's website considers a scenic drive on America's byways a form of ecotourism because no footprint is left behind. There's even a quote from Muir: "Everybody needs beauty as well as bread, places to play and pray in where nature may heal and cheer and give strength to the body and soul." But the folks at the NSBP don't really know Muir, for they don't realize that he traveled virtually everywhere on foot, including from Indianapolis to the Gulf of Mexico, his famous 1,000-mile walk. And wearing the blinders built into the very name of their program, the NSBP fails to note Muir's skillfully cocked ear on America's natural wonders. The agency promotes scenic drives. There's no thought to the sonic beauty of these landscapes—or, for that matter, to the noise intrusions on the natural soundscape of all the combustion engines delivering these waves of ecotourists.

On my last trip back home I discovered that even Joyce's main street, Washington State Highway 112, a two-lane, curving road without stoplights or streetlights, has been recently designated Strait of Juan de Fuca Scenic Highway, one of the newest members of the National Scenic Byways Program. Gas prices notwithstanding, it may not be long until my backyard recording studio will be a thing of the past.

You would think that our national parks, with a mandate to preserve "America's secular cathedrals" in their original "unspoiled condition," would be immune from such development. Not so. From Great Sand Dunes National Park to Redwood to Tallgrass Prairie to Great Smoky and beyond, even our most precious natural wonders have been and will continue to be paved with asphalt. One of the great controversies at Yosemite was wilderness access for the automobile-touring public by constructing the Tioga Pass road by the National Park Service's first director, Stephen T. Mather, who in 1920 stated that the heartland of every national park should have a road through it for backcountry access. The "road problem" back then was that there weren't enough of them! Mather stated that this was "one of the most important issues before the Service" and predicted that tourists would soon use "every nook and corner of Yosemite."

Fortunately, Olympic Park escaped this madness. Franklin D. Roosevelt signed it into existence in 1938, nine years after Mather retired from the National Park Service. Today Olympic Park remains one of the very few national parks without a scenic highway cutting through its heartland. Its status as possibly the quietest park in the system is, however, *not* the result of planning. In fact, the NPS knows so little about Olympic Park's natural soundscape that its Denver Service Center—the Office of Park Planning, which recently drafted Olympic National Park's General Management Plan—contacted me to get the most basic information: "two or three" ambient sound level averages for backcountry areas. I gave them a lot more than that and was happy to do so. But just as a mother and father know and can interpret the slightest nuances in their children's voices, so should the Park Service know what its parks are saying.

Noisy car, noisy interstate. It seemed time to make a little noise myself. But rain over, rant over, I'm back on the road. Next stop: Fort Collins, Colorado, home of the National Park Service's Natural Sounds Program. Established in 2000 to "protect and restore soundscape resources whenever possible and to prevent unacceptable noise," the program was founded on four key principles:

1. Soundscapes, both natural and cultural, are integral to park visitor experience.
2. Each park has sounds that are appropriate to the purpose and the values for which that park was established.
3. The park soundscape is a resource that is necessary for the enjoyment of present and future visitors.
4. Appropriate soundscapes are essential for the overall health of park ecosystems, including the vitality of specific wildlife communities.

For its first three years, the Natural Sounds Program was housed in Washington, D.C., but then was relocated to Fort Collins. I've spoken with officials there and, in fact, was asked by Frank Turina, one of its planners, to write a commentary for the January–February 2007 issue of *Legacy,* the publication of the National Association for Interpretation. NAI is a professional organization dedicated to heritage interpretation, both cultural and natural, serving about 5,000 professionals, such as park rangers, who assist park visitors with gaining knowledge about and making emotional connections to the land. The entire issue was devoted to "the importance of sound in the interpretive experience." But I've never been to the Natural Sounds Program's offices, which, in a nutshell, provide soundscape planning and management assistance to any of the nearly 400 national park units that request it. If anybody knows what's getting done to clean up the noise in our parks, it will be these folks.

Frank Turina and Karen Trevino, both with the Natural Sounds Program, coauthored the lead article, "Nature's Symphony: Protecting Soundscapes in America's National Parks," in that special issue of *Legacy*. They make some telling points. They note that noise is the number one quality-of-life complaint in New York City, but also that the ambient noise baselines found in New York were comparable to those in developed areas of the national parks. They cite an Environmental Protection Agency finding that in 1974 more than 100 million Americans lived in areas with unsafe levels of noise. And they stress that a generation later, the impact of that din was clearly echoed in a 1998 survey by the NPS, which found that "72 percent of the American public says that one of the most important reasons for preserving national parks is to provide opportunities to experience natural peace and the sounds of nature." Citing ever-increasing noise pollution in the parks, they worry that opportunities to experience nature's symphony are disappearing at an alarming rate.

How can this be?

The near extinction of natural quiet in our national parks represents a clear violation of the very Act of Congress that created them, the National Park System Organic Act of August 25, 1916. Here is part of that legislation, with its "herebys" and "thereins," but noteworthy especially for the last few words, a mission statement now missing in action:

> There is hereby created in the Department of the Interior a service to be called the National Park Service. . . . This service thus established shall promote and regulate the use of the Federal areas known as national parks, monuments, and reservations hereinafter specified by such means and measures as conform to the fundamental purpose of the said parks, monuments, and reservations, which purpose is to conserve the scenery and the natural and historic objects and the wild life therein and to provide for the enjoyment of the same in such manner and by such means as will leave them *unimpaired for the enjoyment of future generations.* (Emphasis added)

The National Park Service seems unwilling or incapable of living up to its mission. The year before the Natural Sounds Program was established,

I was invited by Wes Henry, then the head of wilderness programs for the National Park Service, to attend a working group at Shepherdstown, West Virginia. In one week in May 1999 this group of acoustic experts and NPS resource managers from all over the United States roughed out Reference Manual 47: Soundscape Preservation and Noise Management. I particularly liked Section 4.2: "The National Park Service will preserve the natural ambient soundscapes, a natural resource of the parks, which exist in the absence of human-caused sound." And "The NPS will restore degraded soundscapes to the natural ambient condition wherever possible and will protect natural soundscapes from degradation due to human-caused noise." The Level III draft copy that I was sent for review is a well-thought-out, decisive document. But it was never adopted. Ignored, it was later called "obsolete" by those who followed. One year later another document took its place. Director's Order No. 47: Soundscape Preservation and Noise Management was signed by the director of the Park Service, Robert Stanton, on December 1, 2000. The watering-down is evident in this wording in its Soundscape Preservation Objectives section:

> The fundamental principle underlying the establishment of soundscape preservation objectives is the obligation to protect or restore the natural soundscape to the level consistent with park purposes, taking into consideration other applicable laws. Where natural soundscape conditions are currently not impacted by inappropriate noise sources, the objective must be to maintain those conditions. Where the soundscape is found to be degraded, the objective is to facilitate and promote progress toward the restoration of the natural soundscape.

"Progress toward the restoration of the natural soundscape"—nothing about the goal of a natural park ambience where human noise is absent.

Guess my rant isn't over, only heating up, the closer I get to Fort Collins. Fifty miles outside of Grand Junction I hit another squall and pull over again. The heavy rain registers 81 dBA. Might mean heavy snow in the pass. Just to the south of I-70 a couple of gas pipelines emerge out of the ground for about 15 feet, permitting control valves, then duck back into the earth again.

Yikes. The rain is dripping through the roof of the Vee-Dub, forcing me to put on my hat and grab a towel from under the dash to mop the seat. Just a few miles later my towel is still dripping, but I'm basking in the sunshine and the road is bone dry. If you don't like the weather in Colorado, wait five minutes.

Sixty miles per hour on the downhill—Wheee!

At the Colorado state line a sign says that engine brake mufflers are required and warns of a $500 fine, plus surcharge. Maybe Colorado has some pretty tough noise laws? Outside Rifle, I take in six new oil derricks in one glance. I reach Glenwood Springs, which I vaguely remember is the turnoff for Aspen. I pull over in a stretch of motels and fast-food restaurants. I see a pair of signs by a Comfort Inn: "Quiet zone 7 p.m. to 7 a.m." and "Noise ordinance strictly enforced. Violators will be prosecuted." I open my laptop, with "View network connections" on the lookout for unsecured networks. Got one: Best Western Motel. Time to play Priceline! How about a two-star moderate, $45, in Aspen? Do I think that's possible? Better go for $55 for a $140 room. I can taste the granola with heavy cream now! That's what I had the last time I stayed in Aspen, at the Mouse House Youth Hostel, in 1975 with my sister Holly after we hiked from Crested Butte over Conundrum Pass. She ate the orange begonia blossoms off the restaurant plants. Priceline interrupts my reverie: "We're sorry . . ."

I close the laptop and search for a hotel the old fashioned way, spotting the Motor Inn Motel right across the street. Hot tub, sauna, Wi-Fi, $69—all good. Time to do laundry and check e-mail. At the front desk I ask where I can buy some laundry soap. The manager, an attractive Eastern European woman (must be her elderly mother looking on from the back room), hands me not only a box of detergent, but also the key to the motel's private laundry room. Something tells me I might not encounter such goodwill and unalloyed trust a couple of time zones to the east.

Checking my e-mail I see that I have about $750 of new CD orders—orders I need to fill to help erase the $15,000 balance on my credit card but won't be able to tend to for another three weeks, when I'll break from the journey to fly back home to attend my son's graduation from the University of Washington. One e-mail in particular catches my eye. It's from a longtime customer named Gloria Fox, who shares why she bought the first of more than three dozen of my CDs. Listening to nature sounds at bedtime helped calm and relax her, silencing an as yet undiagnosed storm in her brain that had grown to banging sounds in her head. The problem was finally diagnosed as AVM, an arteriovenous malformation of blood vessels in her brain, and corrected by surgery. Fox still buys my recordings and often gives them as gifts.

"One of the hardest things to do is quiet the mind," she says in a phone conversation months later. "Like my friend, who has breast cancer, it's very hard to quiet that internal, frightened voice, which is really not who you are. You are something deeper and richer and much more profound. I gave her one of your CDs a year ago. She loves it. It helps her to reconnect with experiences in her life when she was outdoors, feeling robust and healthy—experiences in her life where nature played a major part."

Fox lives about two hours north of Manhattan "in the middle of the woods, with a stream behind the house and a view of the Hudson River out front. It's very quiet." Quiet enough that she can hear car traffic a couple of miles away, but ultimately not so quiet that she can satisfy her yen for nature listening by throwing open her windows or sitting in her yard. Which explains why she keeps buying and listening to my recordings, even playing them unannounced for surprised and delighted dinner guests. "They provide the gift of hearing nature, which helps us reset our bodies back to zero, back to a natural state. These are sounds we should be hearing, sounds that calm the mind."

A craving for Vitamin Q, let's call it, this hunger for quiet when you need to heal. Gloria Fox self-administered quiet in a time of physical crisis—and still takes her "meds." Is it coincidence that virtually everyone I've spoken with so far has sought quiet to heal? Bill Worf healed from Iwo Jima. Judy sought quiet in Twodot, Montana, when her hormones flipped. Doug Peacock healed from the horrors of Vietnam. Quiet heals me, too. Whenever I get overwhelmed and discouraged by ever-increasing levels of noise pollution, I head for the Hoh to recharge my soul.

Fort Collins looks more Rockville, Maryland, than Rocky Mountains. High-tech business parks, well-manicured lawns, names like Intel, AMD, Wolf Robotics, Hilton. I swing the Vee-Dub into the parking lot of what could pass for a modern community college pleasantly landscaped with spreading maple trees for shade, a few spruce trees for vertical accents, and a pond. I pull out my sound-level meter and measure 45 dBA: mostly distant traffic and the drone of building ventilators.

Two large, swirling bronze statues with petroglyph-like engravings tower more than 20 feet flanking the entrance to the building. Inside it's quiet, as if unoccupied. I have to search to find a practically unmarked solid door that reads "Suite 100, Natural Sounds Program Center." Frank Turina, a policy person with the title of planner and the salt-and-pepper goatee and easy demeanor of a folk singer, greets me and leads me down a narrow hallway past a couple of small offices directly to the large corner office of his boss, Program Director Karen Trevino. She knows all about One Square Inch and my cross-country journey to Washington and greets me warmly: "Gordon, thanks for all your work. We need someone out there smacking us around."

"Am I smacking you around?" I laugh.

"Trust me," Trevino continues, "it's helpful. It's a tough environment to work in. It's getting better. There's a new secretary and a new director, but the last six years have been trying. Anything and everything that anyone can do to help elevate these issues is really welcome."

Trevino has unfussed-over brown hair that frames an angular face. She's wearing a brown blazer over a long dress and brown boots. She's been described to me by a former high-level Park Service employee as "one smart lady, and tenacious." She's a lawyer by training, first serving in the World Wildlife Fund's office in the nation's capital in the early 1990s, next with the D.C. office of an Alaskan law firm specializing in natural resources, then as a senior counsel in the Department of the Interior, working mostly on Park Service issues. When she took this position three years ago, her husband told her she couldn't have found a better job because she's blessed with very sensitive hearing. I figure she'll be pleased to know exactly how quiet her office is and reach for my sound-level meter.

"You could be a street cop in New York," she says as I take it out of its protective case. "They carry them now, as part of Mayor Bloomberg's noise initiative. Did you know that?"

I get a reading of 35 dBA. That's with the lights and the AC on. Lights off, it's down to 30 dBA. That's excellent. It's reassuring to know that the Park Service officials responsible for soundscape studies and recommendations regarding natural quiet and man-made noise in our nation's parks spend their working hours in a quiet workplace. The office, however, serves in an advisory capacity. Frank Turina says that last year the program received

12 requests for assistance from Park Service properties and estimates that there have been some three dozen requests in his three-plus years on the job. Clearly, most national parks are not diligently working on soundscape management plans. "For many parks, the issue is not a priority for them," he says. "Other issues are competing for their time and resources. Usually it's precipitated by some kind of trigger, for example, Zion. They have a navigation beacon for commercial aircraft right outside the park and they get a tremendous amount of aircraft overhead."

Trevino mentions some good news. There is support for the Natural Sounds Program from directly above, the Park Service's Division of Natural Resources Washington office, which has had a threefold increase in its budget. But in virtually the next breath she cautions me to prepare me for my arrival in Washington two months from now.

"Take people's words and reflect upon their actions," she says. "I find people are very willing to say all the correct things, but their actions don't necessarily reflect their words. Take for example the FAA. They say they want to protect national parks and then turn around and do things like this—I just found out there are legislative proposals they're trying to cram in under the radar screen that would significantly hinder our ability to protect quiet in the parks.

"Publicly, they say they're going to do everything they can to work with the Park Service, but more often than not, it appears they're trying to write us out of the whole process. If the National Parks Air Tour Management Act is about protecting national parks, and I am pretty sure it is, then the relationship between the FAA and NPS must be fixed."

"What is the current standard for the National Park Service?" I ask a few minutes later. "Is there one noise standard?"

"It's funny you ask," Trevino says. "It's come to my attention, painfully and awkwardly in the past two years, that we are in dire need of a comprehensive noise regulatory scheme. It's difficult. We have national recreation areas. We have national parks, national seashores, battlefields. Within them there are developed areas, backcountry areas, wilderness areas, deserts and mountains and rivers, and all of them attenuate sound differently. It will be really complex, but I don't think that should deter us from doing it, because the need is really there. Because in the absence, we are all over the place. We have completely conflicting regulations at the moment."

I mention that I was in Bryce with *USA Today* for an article on quiet in that national park when a squadron of Harleys blew in on their way to Sturgis, the big motorcycle rally.

"We do a lot of work in Mt. Rushmore [National Memorial]," says Turina. "Last year during Sturgis, they got 119,000 motorcycle visits in one week."

"And Secretary of the Interior Kempthorne was one of them," I point out.

"Yes, he was," they both tell me. We talk about appropriate and inappropriate places for riding Harleys, with their patented sound.

"The same noise control in Yellowstone for snowmobiles should also apply for any vehicle, is that right?" I ask.

Trevino says that often these matters are left to the discretion of the individual park superintendent, but there is an automotive noise regulation standard, which, like all park service noise regulations, is decibel-based. "That's my other reason for wanting a comprehensive approach to noise," she says. "All of our regs are decibel levels. Like a boating reg that was done last year. First of all, it was decibel-level, and second of all, they got the level from the manufacturers, an OSHA kind of number, like how loud before it blasted your eardrums out. I kind of threw down a yellow flag at the eleventh hour, actually the twelfth or the thirteenth hour. This guy had been working on this reg for like eight years and it was at the final review at the director's office. I said this is unacceptable. Our mission, our mandate, is not about protecting somebody's ears from being blasted away. If we're going to be measuring something or making a regulation, it should be based on what we're supposed to do, which is protect wildlife and . . ."

I confess I interrupt Trevino, so eager am I to make a point. I tell her my training is in botany, and if I were charged with managing a botanical reserve, I wouldn't study the weeds, I'd study the native plants. Then, based on the study of the plants, I could implement whatever programs might take care of the weeds. "But what I've seen all along with the National Park Service, nobody is talking about the natural resource, it's the noise that's being studied. To me that seems totally backwards."

She agrees, and stresses what she terms an inherent problem of decibel-based regulations. "Nobody understands decibel levels, including me. They're just not logical. They grow exponentially. And the FAA loves it, because no one understands decibels. So we're trying now to come up with metrics that are more related to providing an enjoyable visitor experience, protecting wildlife or cultural resources, and putting it in terms that people understand, like how much of your opportunity to hear wildlife is lost because of a certain sound."

I mention what I see as one of the great benefits of One Square Inch: its simplicity. No man-made noise at that designated spot. Quiet there and in a far-flung circle of silence radiating out from the stone. It's immediately graspable.

"Can you make One Square Inch in all of our parks?" she asks, knowing full well that that is not practical.

I tell her that if One Square Inch is adopted in Olympic National Park, I hope the concept might be replicated in other carefully chosen and protected places in a couple of additional crown jewel national parks.

I thank Trevino for seeing me, and before my next appointment a few offices away with a staff bioacoustician, I talk a bit more with Frank Turina, who continues my education about the shotgun marriage between the Park Service and the FAA regarding the air tour overflights issue. It occurs to me that if there's a silver lining to the controversial issue, it's this: without the air tours, the Park Service would have no Natural Sounds Program. The program was created in 2000 in the wake of the National Parks Air Tour Management Act to represent the interests of the national parks in concert with the FAA in dealing with the issue of air tour overflights.

"It's a really complicated process," says Turina. "Since the FAA has jurisdiction over airspace, they're the lead agency on this, but the legislation also said the National Park Service is required to sign and agree with the plan. We're a cooperating agency, but we also have signatory authority, so it almost puts us at the same level as the FAA as far as being co-leads, because if we don't agree, we won't sign it."

I recall Skip Ambrose cluing me in to some of this ideological and cultural butting of heads (literally ground up versus cockpit down) and ask, "Is the FAA still treating it as if it's a noise study at an airport?"

"That's a lot of it. The way they collect and analyze data, the types of metrics they use are very different. One example: whenever the FAA looks at the impacts that an air tour has on the natural resources, they use as a baseline ambient all the noise that's there now. For example, at Mt. Rushmore, if you're standing on the view terrace, you can hear cars and buses and motorcycles on the highway and people around you talking and jets in the sky."

"They're saying, 'What is the impact of raw sewage on Lake Erie if the lake is already dead'?"

"Exactly. They're saying, since it's already noisy, the air tours aren't really adding much, so it's a very small impact. The Park Service policies require the use of the natural ambient as the baseline for determining impacts. So we get completely different levels of impact. We've talked with the FAA about this for a long time."

"Do they get it?"

"No. We couldn't come to an agreement on that."

Consequently, both ambients are currently being used in the slowly evolving air tour management plans—a telling reality, and symptomatic of the lack of progress. The Mt. Rushmore plan, Turina says, is furthest along, but far from completion. It was begun in 2003.

Turina hands me a two-page summary called "Laws, Regulations and Policies Related to Soundscapes." At a glance, it's easy to see there is a long legal history to protecting natural quiet in our parks. Plenty of federal laws have been passed and management policies codified, but very little progress has been made.

More than 30 years have elapsed, roughly one-third the existence of the National Park System itself, since the Grand Canyon National Park Enlargement Act recognized "natural quiet as a value or resource in its own right to be protected from significant adverse effect." Notably, it specifically mentions helicopter noise as an adverse effect on natural quiet. Three decades later, some 90,000 air tours, grandfathered in by the 2000 act, still fly over the Grand Canyon each year, until a mandated management plan decides their fate. In January 2006 the U.S. Government Accountability Office produced Report GAO-06–263, which assessed how well the National Parks Air Tour Management Act goals were being met. The report states, "Almost 6 years after its passage, the required air tour management plans have not been completed," and "the implementation of the act has so far had little effect on the 112 national parks we surveyed."

One glimmer in all this darkness is a few lines in Section 806 of the National Air Tour Management Act of 2000: "Effective beginning on the date of the enactment of this Act, no commercial air tour operation may be conducted in the airspace over the Rocky Mountain National Park notwithstanding any other provision of this Act or section 40126 of title 49, United States Code."

I ask Turina if the Park Service has ever suggested that air tours were incompatible with our national parks. If so, there'd be no need to create air tour management plans.

"I'm not aware of that [official] argument, but several superintendents have made that argument. Glacier National Park, in their general management plan, their preferred alternative is to phase out air tours."

Turina introduces me to Kurt Fristrup, who rises from the chair in his cluttered office and offers the hand that's not holding a banana. Fristrup is wearing a dark blue dress shirt, khaki pants, white socks, and white sneakers. His demeanor is equally unassuming. I mention the title by his name on the office's web page, stumbling on the word *bioacoustician*. "I think we're still struggling what to call me. I think I prefer 'scientist,'" he says, explaining that he arrived here about a year and a half ago from the ornithology lab at Cornell University to help improve the Natural Sounds Program's data-collecting and analysis methods.

"Historically, we used very expensive equipment," he says. "One monitoring set might cost twenty thousand dollars and have three or four solar panels and three or four lead acid batteries and take two or three people to pack in. And because of limited storage capacity we were recording ten seconds of digital audio every two minutes, enough to capture representative samples of some noise events, but for interpretive representations, it's terrible. You're cutting pieces out of things."

Newer, lighter, less expensive equipment now enables continuous recordings that last for days and have begun to produce 24-hour sound pictures, visual representations rather like EKGs, that can depict at a glance the sonic events at that location. "Instead of doing a full linear scale, we'll use one-third octave sound-level measurements, so now there are only thirty-four measurements that span full range and we can stack twelve lines of that on a graph. So you get two hours per line, twenty-four hours on a page," says Fristrup. "You probably can't see one-second events, but we can certainly see five-second events clearly and really get a sense of what the average diurnal patterns are, but also pick up the anomalous situations."

He leads me down the hall to a nearby room with several of these spectrogram posters on the wall. One, labeled Upper Kipahulu Valley, Haleakala National Park, identifies natural and man-made sounds on various lime-green lines on the graph. Arrows labeled "insect" and "birdsong" point to flatlining and only slightly elevated sound levels. Other arrows, denoting a helicopter overflight and that of a high-altitude jet, point to veritable mountain peaks on the graph.

Another poster, for a recording in Separation Canyon in Grand Canyon National Park, shows the beginning of a rainstorm, then a flash flood. "It's loud," Fristrup says, directing my attention to the next spectrogram. "But the interesting thing for me is, look how noisy Yosemite Village is in the middle of the night—the generator, heating and air conditioning. And there's traffic noise up into the fifties and sixties decibel levels. It's noisier than my neighborhood. I think this is something where the Park Service could look at the infrastructure that we support. Because wouldn't it be wonderful if, when you visited Yosemite Village, the dominant sounds you heard were the waterfalls? Backcountry Yosemite looks like this. It's deep blue, very near the threshold of hearing."

Fristrup believes these images could be very powerful in, say, a congressional hearing room. "Particularly," he says, "since we're moving toward more free-flight rules, an era of decentralizing the passenger air transport system, where traffic is not going to be channeled by ground based navigational beacons. That means the noise, instead of being concentrated under high-traffic corridors, is going to be spread out all over the United States."

He shares an interesting and potentially deadly discovery from these spectrograms. Before the onset of the flash flood, he explains, there's a few minutes of low-level sound, in effect, an early warning of the advancing deluge. But the warning is badly compromised by the jet intrusion, which falls in exactly the same portion of the spectrum and even sounds very similar. "I've twice heard oncoming flash floods and once heard an avalanche, and all three times my initial reaction was, it was either thunder or a jet," Fristrup says. "The cue was present for much longer before I realized, no, that can't be a jet. So I think we've got an issue here, particularly in some of the canyonlands in the Southwest. It could be a safety issue for people on the ground. I'm not sure how often this has made a life-and-death difference, but it could. I think most park visitors habituate to jet sounds. And the presence of that high-altitude jet masks your ability to hear the approach of the flood, so it's got to be much closer for you to hear it. The same is true of all other natural sounds in the park. When a jet is flying overhead, your aural world has shrunk."

"You're saying the noise impact of aircraft on the wilderness reduces a person's range of hearing?"

"Yes. Our auditory horizon shrinks, and it's affecting the very frequencies that travel the farthest. And the same is true of snowmobile sound. It's ironic to me that the Park Service has regulations that are weakly or inadequately enforced about how loud roadway vehicles and watercraft and snowmobiles can be, and we allow snowmobiles and boats to radiate more sound than roadway vehicles—even though over snow and water sound carries better than anyplace else."

"It's totally backwards."

"It's just an artifact of history. I think in many areas, noise control in the United States has not been driven by the values of acoustic resources so much as by the cost of annoyance or by what's easily achievable by industry in terms of controlling it."

I tell him that one of my favorite questions is asking people "How far can you listen?"

Fristrup, too, sees value in expansive listening experiences. "There is something remarkable when you realize that a bird is four hundred meters away, or you realize that frog is a half-mile away. It's a very powerful and dramatic experience when you realize the vastness of your capacity to hear and how silent it has been. I think that's a rare thing, an endangered resource.

"I think humans can hear some kinds of natural sounds at distances of several miles," he says, adding that low-frequency sounds can travel extraordinary distances. "You can set up an instrument here in the Rocky Mountains and hear the infrasound from waves crashing on both coasts. Humans can't hear that. We have very inefficient hearing down at those low frequencies, but pigeons can sense those very low frequencies."

Fristrup is speaking about infrasound, frequencies below 20 Hz, the lower limit of human hearing, which some animals, such as whales and elephants, use for long-distance communications. By comparison, frequencies above 20 kHz, which are beyond the upper limit of human hearing, are used by some creatures, such as bats and dolphins, for short-range communication, echolocation, and sometimes to stun prey.

Late in our talk, Fristrup expands the importance of silence beyond nature listening in national parks to other Park Service properties. "Deep silence is also valuable for providing the best backdrop against which you are hearing audible sounds. If you're visiting a military memorial, a contemplative mood is desirable, and the quieter it is, the more powerful the impression can be. If you're attending a ranger's interpretive session, it's going to be more compelling if the talk takes place in an otherwise quiet environment. It's why we focus so much attention on keeping school classrooms quiet. It's much easier to learn when there aren't a lot of competing sound sources."

He asks where I'm headed the next few days, and when I mention Rocky Mountain National Park, he suggests I visit one of his favorite spots, Wild Basin. On my way out of his office, my recorder stowed in my bag, Fristrup shares one final thought, which I immediately write down before I leave the building: "The loss of quiet is literally the loss of awareness. Quiet is being lost without people even becoming aware of what they're losing. It's tragic."

Next stop: Estes Park, on the eastern edge of Rocky Mountain National Park, where I've arranged to meet a member of the League of Women Voters who is going to tell me how they managed to make Rocky Mountain the only national park where air tours are expressly prohibited. The drive is spectacular. At my top speed of 35 miles per hour, making periodic pullovers to let traffic pass, I get time to enjoy the rugged, snow-capped mountains and their backlit spring leaves of aspen and jet-black shadows. Compared to the diffuse light of the Northwest, the views here are high-def sharp.

Eight elk, standing right in the middle of the road, greet me as I pull into Estes Park. Though they look similar to the Roosevelt elk of the Northwest, these are Rocky Mountain elk. Too bad it's spring and not fall. I would love to hear the Rocky Mountain elk bugle and judge if the two subspecies sound more different than they look. I'm almost certain that they would, since their performance halls are so different.

When I check into my Motel 8, the manager takes a long, interested look at my Vee-Dub and exclaims, "I used to own a VW bugger. Going over the Rockies was so slow."

In the morning it's downright frosty outside. Deer and elk are moseying around downtown doing God's gardening; 40 dBA, few cars, no aircraft, only a few birdsongs braving the brisk dawn. On my way to Starbucks (of course there's a Starbucks), I fill up at Phillips 66. As at all modern pumps that serve only unleaded fuel for my retro wide-mouth gas opening, I can't count on automatic shutoff. I need to bend my ear close to listen to the gurgling sounds to tell me when it's full or the fuel will come splashing out all over the place. But here at Phillips 66, the 66 dBA makes this a risky business. Up around $20 or so, I slow the flow and, happily, manage to stay dry.

I pull up at the house of Irene Little and see a new Toyota Prius in the driveway. Irene greets me warmly and introduces me to her husband, Steve, wasting no time informing me that she was just one of many at the League of Women Voters and other local organizations behind the ban of air tours over Rocky Mountain National Park.

German-born, from the town of Kassel north of Frankfurt, Irene still speaks with an accent. The Littles are a charming couple, friendly, full of energy and a love of the outdoors, the kind of people you'd want as neighbors. They have matching gray hair and wear unmatched vests; hers is knitted, a Christmasy, unbuttoned button-up with snowflakes and sleds and maybe reindeer; his is red fleece. Like good hosts, they ask about me, and I'm soon telling them about One Square Inch and pulling out the OSI stone.

"So this is the One Square Inch rock." They both laugh appreciatively.

"Is that pipestone?" Irene asks.

"It feels like pipestone," says Steve, accepting the stone from his wife. "There's a Pipestone National Monument in Minnesota where Indians used to mine stone they could use to make the bowls of pipes."

"I assume this was found on the beach, but David Four Lines may have swapped for this."

A second key reason for the success of the campaign to ban air tours over Rocky Mountain National Park—beyond the efforts of dedicated locals like the Littles—was the timing. In 1994, when the anti–air tour crusade was just starting, before folks started wearing "Ban the Buzz" T-shirts bearing the image of a helicopter with a big diagonal red slash through it and a moose with his front hooves over his ears, there were no air tours, just word getting around that would-be operators were asking for permission. "I think the only reason we managed to do it," says Irene, "is there wasn't an established industry. Once it's there, it's very hard to get rid of."

"You're taking people's jobs away," says Steve.

Indeed, without an established industry in the crosshairs of the League, momentum quickly grew. "The support was very strong," Irene says. "The park, the town, and, most important, the county commissioners, who eventually came down with a ruling that the only way you could have an overflight was if it started at an FAA-approved airport, and the nearest one is down in the valley, so it became economically infeasible."

"Larimer County," explains Steve, "basically shortcut the process and said, 'Let's put up a little barrier here.' Whether that would still be a barrier today is another question. I think it was primarily helicopters then, and helicopters are so much more expensive to run per mile. The biggest hang-up was clearly the FAA. The FAA was kicking and screaming the whole way. They didn't want to do anything that limited their power to decide this."

The more enduring solution, the League discovered, was to bypass the FAA with an act of Congress. Irene spreads a time line out on the kitchen table in front of us. In May 1996 Secretary of Transportation Federico Peña issued a temporary ban on air tours while the issue was being studied. In 1998 a provision banning air tours over Rocky Mountain National Park made it into an FAA authorization bill passed by Congress. Two years later Congress passed the 2000 National Parks Air Tour Management Act, which, in part, made the ban of air tours over Rocky Mountain National Park permanent.

And yet there's a Pyrrhic quality to this victory, which Steve explains by telling me how he and Irene like to relax. "We're both astronomers, both pretty sensitive to what's going on in the sky. We have a hot tub out on the porch where we like to sit in the evening. We can sometimes see four or five planes at a time. The park is due west of us, and this is east-west traffic, meaning a tremendous amount of air traffic is routed over the park." There's the rub. The 2000 bill speaks only about air tours, not higher flying commercial traffic, which also degrade the natural soundscape.

Before saying goodbye, I ask if they'd let me take a picture of them with the OSI stone. We step back onto the back deck and into the bright sunlight and I snap a couple of shots. Only when I look at these photos later do I realize that Irene and Steve each grip an end of the stone, and seem pleased to do so.

I fire up the Vee-Dub and head south on Highway 7 toward the part of Rocky Mountain National Park recommended by Kurt Fristrup, my interest primed by road names like Eagle Plume and Big Owl. The dirt road into Wild Basin shows only one set of tire tracks, and the fee station is closed—both good signs for solitude. I stop partway and record pine wind, 45 dBA, a soft whispering voice. At 8,500-feet elevation, the trail follows a stream that leads up to Ouzel Falls and then to Ouzel Lake.

What an invitation! The ouzel, also known as the dipper, is a fascinating bird that inhabits cold mountain streams, lakes, and rivers, hunting insects at the edge and by walking in rushing streams, grabbing onto rocks with its feet.

One spring morning, spying a six-inch oval, a dark gray ouzel doing deep knee bends at the edge of the Hoko River Canyon in Olympic Park, I hoped to learn more. Making my way down the wet, moss-covered banks of the canyon, I caused the bird to fly off, giving me time to set up my microphones close enough to hear it clearly with the white noise of the Hoko River in the background. Then I waited. After about 10 minutes, flying straight as an arrow and low to the water, my ouzel returned to the same rock at the water's edge and erupted into song—a song I'd heard before in Muir's description. It had "the deep booming notes of the falls" in it and "the trills of rapids, the gurgling of margin eddies, the low whispering of level reaches, and the sweet tinkle of separate drops oozing from the ends of mosses and falling into tranquil pools." The ouzel shows that a species can adapt to broad-spectrum noise in its environment, given enough time to evolve. But in the modern world, noise has come on so fast and so loud that it amounts to an ongoing explosion with no rest stops; there's no time to adapt.

St. Vrain Creek, the main stream that drains Wild Basin, is full of energy. Ten feet wide and approximately eight inches deep, at least today, it rushes over and around granite boulders the size of bowling balls. This is mass transit to winged insects hitching a ride in the cold air that drains over the stream, and many shoot by like rockets down into the valley. Patches of snow remain in the shaded areas underneath the pine trees. The groves of aspen have smooth, pale green skin into which a few hikers have scratched initials or words. I'm reminded of the petroglyphs in Canyonlands, but here, time and the healing of the bark have begun to blur many of the words. The trail leads me out of sight from the stream but not out of ear-shot. The higher frequencies attenuate more quickly than the low frequencies as the ever-expanding sound waves from the stream travel through the pine forest, causing the white noise to become deeper as a low throbbing beat of a more complex flow pattern becomes audible.

Hiking trails often follow the margins of streams and rivers, not only because the flow of water follows the path of least resistance, but also for purely visual reasons. Who doesn't like watching water? But those seeking a quieter path through the wilderness need to look elsewhere and follow those who use all their senses, for instance, deer and elk, which rely on their ears as an early warning system for possible threats and are ill served by traveling alongside rushing waters. I look for possible signs: hoofprints or scat. I see none, but I do spot five deer halfway up the mountainside browsing on tender shoots that have already emerged at that sunnier location.

A few other ouzels make bobbing knee bends, one after the other at the water's edge. But it is getting cold in the shadows in the valley and I return early to the Vee-Dub to warm up with a cup of hot tea.

Next I head to the northern part of the park to explore Upper Beaver Meadows because it is not a throughway and should be mostly free of traffic noise. But on my way I see that there is some kind of construction going on, and the brakes on the gravel-hauling trucks emit jarringly loud screeches, more than 90 dBA. Several of these trucks have collected, diesel engines idling, making a freight yard of what should be a pristine sonic environment: 54 dBA measured from 100 feet away.

So once again on this journey, I'm stopped dead by irony. Somewhere around here in Upper Beaver Meadows, I know, is the sign that Irene Little showed me in a photograph, a sign funded from an award bestowed on the Estes Park League of Women Voters by the National Parks Conservation Association. The sign bears the imprimatur of the National Park Service, color photos of flora and fauna and stunning landscapes of Rocky Mountain National Park, and these words under the heading "Protecting Natural Sounds":

> The call of an owl
> The music of a flowing stream
> The rustle of leaves in the wind
> The hush of a winter forest
>
> Natural sounds and natural quiet are a protected resource in Rocky Mountain National Park
>
> Nature's sounds and natural quiet are just as rare and precious as the native plants and animals of this park.
>
> Thanks to the dedicated work of concerned citizens, led by the Estes Park League of Women Voters, in 1998 the U.S. Congress recognized the value of natural sound and created a permanent ban on commercial tour overflights above this national park.

The sign also features this quote from the orchestra conductor and arranger André Kostelanetz: "One of the greatest sounds of them all—and to me it is a sound—is utter, complete silence."

But there is no silence here, not yet, not now. But someday.

To escape the noise I drive a few miles up a dead-end road to the End of Valley picnic area and observe a herd of Rocky Mountain elk and two coyotes. But the noise of the trucks travels, even here. I relocate again, to a mountaintop scenic overlook, hoping to listen to the fading sunlight, day's bookend to the dawn chorus, as the falling temperature gradually hushes the cricket-frog chorus below: 30 dBA. But a fixed-wing plane intrudes and takes its time crossing the park: 65 dBA, then 70, then 78, apparently a private sightseer. While I wait several minutes for the last, echoing vibrations of plane noise to dissipate, a car pulls up. A photographer hops out, apparently pleased to have made it just in time to catch the last traces of fading light. He leaves his car running: 46 dBA at 20 feet.

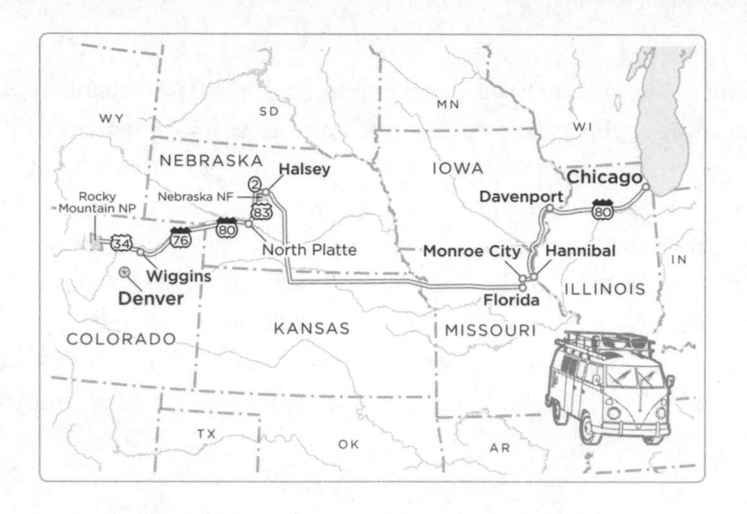

8 Nature's Symphony in Decline

> The echo is, to some extent, an original sound, and therein is the magic and charm of it. It is not merely a repetition of what was worth repeating in the bell, but partly the voice of the wood.
>
> —Henry David Thoreau, *Walden*

Breakfast comes early, coffee and honey buns at Stubbs Sinclair next to the Wiggins, Colorado, rest area just off I-76 at the junction with U.S. Route 34, where I spent the night, snug inside the Worm on the roof deck of the Vee-Dub. In the next booth there's a bunch of oil field workers wearing baseball caps so oil-soaked I can't read the logos. I hear a father and son discussing how $4-a-gallon prices in California will enable them to tap into new, previously unprofitable well sites. Amazing how something as simple as the price of gasoline affects the soundscape in places like Colorado, and how nobody ever factors this ripple effect into energy costs. Before you can say "wind power," let me blow a bit of cold air on wind energy as a quiet alternative. Wind farms I've encountered in California, Washington, and Hawaii make an incredible ruckus and seem to sprout in places where no people live (NIMBY at work?), many of them previously quiet places. No, indeed, preserving quiet isn't easy.

After breakfast I grab a cardboard box from behind Stubbs and tear it apart on the walk back to the Vee-Dub, which is still parked, stone cold, at the rest area. I slide a sheet of cardboard underneath the rear of the vehicle, then myself. Every 1,000 miles or less I try to check the valve clearances on all four cylinders, particularly on the number 3 cylinder. Improperly adjusted valves could be the number one killer of Vee-Dubs, something easily preventable with just the right touch. As Siddhartha learned while overhearing a sitar teacher instruct his student, "Not too tight or the string will break; not too loose or the string will not play. It is the middle way." I check my notes on my Treo's memo pad: "Slide a .006-inch feeler gauge through the valve clearance." Now there's a gross understatement. The thin strip of stainless steel must pass through the gap with the same sensation as a freshly stropped straight razor through butter. The correct adjustment is barely perceptible, yet utterly smooth. It is like listening at the threshold of human hearing. Remarkably, even after the long, slow climb through the Rockies, the valves are right-on perfect. The valve cover gaskets look a bit rough, but I wipe them clean and decide they'll do for a good bit longer. The covers snap back into place easily. Fifteen minutes total, and I'm on the road.

I am immediately rewarded with a postcard view: green hay fields beneath a perfect blue sky and cotton ball clouds. It is hard to go more than a couple of miles without finding something to photograph. After about five stops in less than an hour the dashboard idiot light comes on, indicating low oil pressure. I assume this must be a malfunction, but as a routine safety precaution I pull over, walk back, and lift the engine door with the help of the latch key. Smells like burnt oil and sounds like popcorn kernels just before they burst. Tilting my head below the underbody, I see that the right-side valve cover has leaked, completely soaking the heat exchanger. The dipstick confirms no oil. For the air-cooled VW, which, like a lawn mower engine, lacks a radiator, this is usually time for a rebuild, but with no other options I reseat the valve cover, rip up part of my bath towel, and jam little bits of terry cloth into and under the spring, hoping it will hold;

then I add oil and wait. Twenty minutes later, off I go. Tentatively. I stop every few miles to look for drops of oil on the asphalt under the engine and check the dipstick. Each time I'm amazed by the holding power of my makeshift solution. Moose and Dave would be proud of me.

I-76 takes me past Fort Morgan and follows the South Platte River past Sterling. The short grass prairie, like all grasslands, was among the first of our native lands to go into agriculture. Thick, fertile soils and the absence of trees facilitated rapid expansion of livestock and croplands. I'm making a beeline for Nebraska, a prairie state, but oddly, the home of Arbor Day and our only national forest made up entirely of planted trees.

Not long after a billboard tells me "Beef. It's what's for dinner" I decide to stretch my legs and take the exit for a sleepy-looking town in the northeast corner of Colorado about a dozen miles from the Nebraska border. I enjoy dropping in on such oxbow communities bypassed by the flow of modern events, hoping to find some new, previously unheard sound—at least to my ear. Or something other than the din of man's machinery, which has destroyed the pleasant subtleties of quintessential rural towns throughout America, disappearing sounds like the *whack* of a bat against a baseball, birds conversing amid rustling elm leaves, a flagpole clip clanging, the *chuch-chuch-chuch* of a rainbird sprinkler, all united by a passing breeze, as if the town itself took breaths between each of its sonic sentences. Soundscapes like these, once common in America but now rare amid constant motor noise, can tell a listener much about a town and its people long before you offer your opening hello.

At first glance through the Vee-Dub's split windshield, Sedgwick's main street appears to have no moving parts. Everything seems closed and boarded up. But the town is not deserted. A man with neat hair and attire, looking like an off-duty shoe salesman, strolls over and introduces himself: Patrick Woltemath, Sedgwick's mayor. Mayor Woltemath does about everything to welcome me but hand over the key to the city. He offers a tour of town hall, which nowadays doubles as a museum. He shows me a bona fide ballot box, a 1929 Bible used for swearing in duly elected officials, and an old hand-painted (but now empty) bank vault. Woltemath wants Sedgwick added to Colorado's list of endangered towns, seeking some financial assistance with the overwhelming costs of meeting new standards for waste water treatment and other problems of small town modernization. In 1957 Sedgwick's population was 504. A half-century later it's 182. What will Sedgwick sound like, I wonder.

The only place to spend the night is the town's old hotel, now operated as a B&B called the Sedgwick Antique Inn. The inn's owner, Loopy, seems pleased to see me, and though it's Saturday night, tells me I can have my choice of rooms. "In the morning there's breakfast downstairs. There's pizza across the bar," she tells me. "The rates are thirty or thirty-five dollars, whatever you care to pay. Or if you make the bed, twenty-five." I check out a few rooms and settle into the forward double suite above the boarded up Farmer's Bank entrance.

Having left my windows open, I'm awakened by nature's alarm clock—here, the predawn chirps of swallows darting about for the last of the night-flying insects. Bank swallows nest in colonies, and although to our eyes and ears they seem cloned, they are definitely not birds of a feather, but as different, one to the next, as a beach full of sunbathers. According to Michael Beecher, a professor of psychology and biology at the University of Washington and an expert on the sociobiology of bank swallows, they're able to recognize each other by individual sound signatures encoded in a simple chirp. Males and females form monogamous pair bonds and share parental duties of nest building, incubation, and feeding of the young. Notably, the flying young give a two-note "lost" call, to which only their parents respond. Back in the mid-1980s, when I was a bicycle messenger, Professor Beecher invited me to his lab and showed me what amounts to photo albums of facial characteristics paired with sonograms—visual printouts of very minute changes in emitted calls. There's no mistaking that even the simplest chirp carries an untold wealth of information unique to each individual.

A white stripe wriggles down the sidewalk on the opposite side of the wide and deserted Main Avenue. A skunk. Enough at the windowsill. I gather up my recording gear, slip on my headlamp, and hit the streets, stalking morning quiet in Sedgwick, hoping to add to my library of sounds.

The nearly flat landscape offers no natural barriers, and the dominant sound is surprising, something akin to the rumble of a distant train coming closer and closer, then fading. I anticipate a train whistle at any moment and its echo through the town, defining the broad streets and large yards. But I'm wrong. There's no train whistle. The rumble is I-76 truck traffic.

The dawn chorus of songbirds is far sparser and less diverse than what I remember and recorded across this stretch of America nearly two decades ago. The first two weeks of May are prime time for territorial and mating songs. When I passed near here in the spring of 1990, the migratory birds seemed to follow the river valleys, dispersing naturally into the landscape. I used that observation to intercept the northbound migratory wave of songsters and, in effect, limit the overall density of birdsong events in my sound portraits. Back then, there were so many songbirds that the spring dawn chorus was often too busy, actually overwhelming as a sound portrait. That's not a problem today. Where have all the songbirds gone?

Many people have asked this question in the face of alarming avian census data. Audubon's Watchlist 2000 reported that a quarter of U.S. birds are in decline. The painted bunting has been in a population free-fall of 50 percent or more during the past three decades. Cerulean warbler populations have plummeted by as much as 70 percent. Audubon's Watchlist 2007 lists 59 continental bird species and 39 Hawaiian bird species as imperiled. Another 119 bird species are listed as declining or rare. The 2007 bad news continues. Twenty common birds are believed to be in serious decline:

Northern bobwhite, down 82 percent
Evening grosbeak, down 78 percent
Northern pintail, down 77 percent
Greater scaup, down 75 percent
Boreal chickadee, down 73 percent
Eastern meadowlark, down 72 percent
Common tern, down 71 percent
Loggerhead shrike, down 71 percent
Field sparrow, down 68 percent
Grasshopper sparrow, down 65 percent
Snow bunting, down 64 percent
Black-throated sparrow, down 63 percent
Lark sparrow, down 63 percent
Common grackle, down 61 percent
American bittern, down 59 percent
Rufous hummingbird, down 58 percent
Whip-poor-will, down 57 percent

Horned lark, down 56 percent
Little blue heron, down 54 percent
Ruffed grouse, down 54 percent

Reciting each name evokes a lucid memory of some moment over the past quarter-century when the song or call of that species defined the moment for me. The bobwhite, for instance, calling its name through the eastern hardwood forests of my childhood outside of Washington, D.C. To think that current and coming generations may never hear nature's full repertoire is heartbreaking. The landscape of America is sick and losing its voice.

The avian choir is not just shrinking, but forgetting its repertoire. In 1999, while recording several dozen native bird species for the Smithsonian Institution in the captive breeding program at the Peregrine Fund facilities outside of Hawaii Volcanoes National Park on the Big Island, I recorded puaiohi, elepaio, and the Hawaiian crow, which at the time numbered only about 30 in the wild. When I was done, I sent these recordings for final review and species confirmation to Jack Jeffries, a field ornithologist working at Hakalau Forest National Wildlife Refuge. Jeffries reported back that none of the sounds from those individuals held in captive breeding was characteristic of their wild counterparts. Captive native bird species were reproducing, but the native language was not.

In Europe, where natural quiet no longer exists, except possibly in some far northern regions of countries such as Finland and Norway, entire populations of birds are adapting their songs to be heard above the din of noise pollution. The *New Scientist* (December 2006) reported that in urban areas birdsongs were losing lower notes and shifting to higher pitches that were less masked by traffic noise. Other studies note other noise impacts. In an article in *Molecular Ecology* titled "Birdsong and Anthropogenic Noise: Implications and Applications for Conservation," Hans Slabbekoorn and Erwin A. P. Ripmeester, working out of Leiden University in the Netherlands, write:

> The dramatic increase in human activities all over the world has caused, on an evolutionary time scale, a sudden rise in especially low-pitched noise levels. Ambient noise may be detrimental to birds through direct stress,

> masking of predator arrival or associated alarm calls, and by interference of acoustic signals in general. Two of the most important functions of avian acoustic signals are territory defense and mate attraction. Both of these functions are hampered when signal efficiency is reduced through rising noise levels, resulting in direct negative fitness consequences. Many bird species are less abundant near highways and studies are becoming available on reduced reproductive success in noisy territories.

Nor is the noise solely airborne. The oceans, too, have become increasingly noisy with the oil industry's seismic exploration and drilling, the low-frequency rumble of ever more commercial ships, and, possibly most harmful of all, military sonar, which scientists believe responsible for numerous whale strandings. "I've spent much of my life in the sea," writes Jean-Michel Cousteau, founder and president of Ocean Futures Society.

> A long time ago, my father said this was "a silent world." We now know it is far from silent. In fact, this world is home to whales and dolphins that depend on sound to communicate, to find food, to find mates, and to navigate. I'm very concerned that sound is being used for industrial, scientific, and military purposes at such high intensities that it may be harming whales and dolphins. The oceans are becoming more and more polluted by sound from many sources. Each additional insult further undermines the quality of the ocean environment for its residents.

For one marine species, the Yangtze River dolphin known as the baiji, the din may have been too much. This creature, called the "Goddess of the Yangtze," grew up to six feet long, weighed up to 220 pounds, and was practically sightless, relying on a sonar-based sensory system to navigate and feed. The extinction of the white Yangtze River dolphin, whose last reported sighting was by a fisherman in 2004, marked the first mammal extinction in 50 years. Although overfishing, dam building, and environmental degradation were cited, researchers also hypothesized that noise generated by shipping traffic may have overwhelmed the animal's sonar. That makes sense because dolphins use sound both to find food and to stun their prey with a jolt of piercing noise emitted at high frequencies. I have recorded this dolphin sound off the Kona coast of Hawaii and hope it doesn't go the way of the "Goddess of the Yangtze."

During the writing of this book the battle over ocean noise made it all the way to the White House, with President Bush overriding widespread scientific and public concern that the U.S. Navy's use of loud, midfrequency sonar can be harmful to whales and dolphins. Declaring the training "essential to national security" and of "paramount interest to the United States," the president exempted the Navy from two major environmental laws aimed to protect marine mammals and a federal court decision that also limited the Navy's use of sonar.

Sunday, May 13, Mother's Day. I pull into Lucy's Place Café on my way out of Sedgwick. Only nine cars in the parking lot, but the place is packed. Lots of moms and grandmoms, most of them probably spared a day at the stove.

"Sorry, it's going to be a bit of a wait," announces a woman I soon realize must be Lucy herself. In line by the door, pegged as a nonlocal, I'm told that when you order Lucy's special, "you get your dessert, you get a drink, you get a whole meal." Today's specials are halibut ($9.95), pan-fried chicken ($7.95), chicken-fried steak ($7.95), and roast beef ($6.25). On the wall beside a sliding-door drink refrigerator, a board with partially pounded nails holds a couple dozen coffee mugs, brought from home and identified with markered names of customers beside photos of animals and sayings ("Genius at Work"). I chat with the man in line ahead of me while planning my assault on the salad bar and assessing the various dessert choices. I don't have my sound meter with me, but this is easily the quietest busy restaurant of the trip. While I eat, I think it's easy to feel welcome in a place where the proprietor greets you, the portions are abundant, the prices rock bottom, and the conversations at nearby tables, distinguishable through a low-level babble, drift to your ears like a forest brook, adding to the sense of place, almost as a fourth course to a fine meal. (I had the pan-fried chicken.)

Full, and ready for Nebraska, I'm back on the road, eastbound on I-80. Within an hour I am engulfed by an unbelievable stench, a cattle feedlot outside of Ogallala. Sound and smell go hand in hand. The word *noise* can be traced back to the Latin word *nausea*. Noxious sounds and noxious smells are almost impossible to ignore. We evolved a sense of smell in part to protect us from toxins and actually have an inborn reflex to flee the stench of decay. Shouldn't we respect the wisdom imparted by our evolution?

Modern odor laws were first pioneered in the Netherlands during the 1960s and now exist widely in Europe under EN 13725:2003, a measurable standard whereby no "reasonable cause for annoyance" exists. But one of the first odor laws followed London's Big Stink of 1858, when the polluted Thames River caused members of Parliament to abandon their duties. Soon afterward, London's first sewer was built, in a city then home to a million residents. Mankind has caught on, after the fact, to the health implications of water pollution and air pollution, but has been much slower to face up to the far-reaching consequences of noise pollution.

Sound and smell played decisive roles in the success of early mammalian evolution, as William Stebbins explains in his book *The Acoustic Sense of Animals* (1982, Harvard University Press):

> There is no doubt that the mammals as a class have capitalized on their acoustic sense more than any other vertebrate or invertebrate group in the course of evolution. Their extensive use of hearing in all imaginable walks of life and the enormous diversity in their acoustic capabilities have increasingly become the subject of scientific research, and only recently have we begun to comprehend just what successful listeners the mammals are. How did all this come about?
>
> The picture of the early mammals that emerges is one of a progressive series of many complex adaptations that were to carry these animals successfully through the reign of the great reptiles and well beyond. They hunted and foraged primarily at night and were small—in fact, shrew-sized—which enabled them to fit snugly into trees or underground burrows during the day when the large reptiles were afoot. Although their visual sensitivity at night was probably fair, they lacked color vision. They developed an almost uncanny sense of smell, and that, together with improvements in their hearing, guided their movements at night.

Think of that: mammals make great listeners. We make great listeners! The range of human hearing far surpasses our ability to speak and make music. If you graph our hearing by frequency range and decibel level, you quickly see that speech encompasses the center of our range. Our music takes our ears further. The lowest note on a piano is C (27 Hz) and the highest note is C8 (4,186 Hz), well beyond our normal vocal range. And beyond the piccolos and kettledrums, there is more, much more. That more is nature's voice.

Nature often whispers, as in the faint tap of a Sequoia seed onto thinly crusted snow. And sometimes nature shouts. The farthest I have heard a sound travel was 172 miles, roughly the width of Washington State. This occurred on May 18, 1980, when I was fly-fishing in North Cascades National Park near the Canadian border. Dynamite, I thought. But Sunday morning? Minutes later, other blasts arrived from different directions. Hours later, ash fell from the sky. Mt. St. Helens had blown! Given that sound travels at roughly 1,130 feet per second and there are 5,280 feet per mile, a sound wave travels one mile in 4. 7 seconds. It took 13.5 minutes for the Mt. St. Helens sound wave to reach me, long enough to have caught and cleaned a rainbow trout, lit a fire, and had Sunday brunch starting to sizzle. The subsequent blasts were not, in fact, true echoes, but rather different sound paths traveling different paths at different speeds due to variations in temperature and atmospheric pressure.

Outside North Platte, Nebraska, I take U.S. 83 North through the Sandhill prairie. It is hard to imagine a forest in this grass covered, rolling hill country divided by the occasional river lined with a few trees. But back in the 1880s Dr. Charles E. Bessey, a botany professor at the University of Nebraska, proposed planting an entire forest by hand to provide wood for settlers and markets back east. Today Nebraska National Forest contains 90,000 acres under management, some 22,000 acres in trees, all originally planted by hand from seedlings grown at its nursery. In other words, a national forest not indigenous to the land. But still, I hope, a good listening opportunity.

Arbor Day began in Nebraska, started by one man, J. Sterling Morton, a journalist from Detroit who had moved to a treeless prairie and written of the virtues of trees for windbreaks, soil stabilization, and sources of shade and lumber. He proposed the first tree-planting holiday for Nebraska take place on April 10, 1872. It is estimated that on that day, more than one million trees were planted in the Nebraska soil. Within a decade the idea had caught on and grown to a national event. Today Arbor Day is celebrated in 31 countries, including Iceland and Tunisia.

I take the Thedford turnoff onto Route 2, head east toward Bessey, and continue on to Halsey, where I hope to load up on groceries. But Halsey has changed since I was last here in the spring of 1998 and appears to have met the same fate as Ingomar and Sedgwick. The only motel, where railway workers used to stay, is closed. There's no grocery store, no gas station, not even a convenience store to buy a snack. Whether chicken or egg, the railroad doesn't stop here anymore. The Burlington Northern still rumbles through town, engines roaring, whistle blaring at road crossings day and night, but no trains grind to a halt. A young mother laments, "We're a small town and we're dying."

Looks like I'll be reaching into my cache of canned food tonight. I head back to Bessey, where the Nebraska National Forest houses its administrative offices, visitors center, and a new recreational complex, complete with swimming pool, volleyball, basketball, and tennis courts, and even a baseball diamond. I've come here to listen in slightly elevated grassland areas called leks.

Each spring for the past few hundred years, if not thousands of years, chicken-like birds called the sharp-tail grouse have gathered at these leks to perform their annual mating ritual through song and dance. Not 100 yards that way or this way, but always at a lek. These leks have been mapped and blinds have been built nearby for photographers and bird watchers.

I buy a plasticized map of the forest for $10 from a woman who tells me the blinds are "first come, first served" and helps me understand the map. "Wherever you see a little tiny number with a windmill sign—there are over 200 windmills out here in the forest. We have permitees who raise cattle," she says, explaining that the windmills are for tapping into the water table. "People should now be bringing cattle on. So you might see cattle out there. So if you open any gates, be sure and shut them." She tells me to best be in the blinds an hour and a half before sunrise, which should be about 6 a.m. None of the brochures mentions anything about the wonderful concert at the lek.

Because I'll be arriving in the dark tomorrow morning, I want to stake out my listening spot ahead of time. I start driving on blacktop through a pine forest mixed with a few hardwoods, then follow a gravel road through patchy forest grassland. Finally, I'm on a packed-sand two-track through extensive grasslands. I stop near one of the windmills, a 50-foot-high steel structure that spins a wheel fashioned of 18 metal blades and measure the metal-clang sound it makes, which is similar to an out-of-tune bell buoy against a choppy sea, even turning slowly. The one-minute average of 50 dBA is significant noise intrusion in a grassland area, where sound travels farther than in forested areas. Another problem is that the noise is intermittent, meaning that it is more recognizable at low sound levels than a continuous noise. Fortunately, I expect to be listening well before the wind kicks up after sunrise.

I negotiate the loose roadbed of a few sandhills and see a rattlesnake make a couple of quick esses winding itself off the road well in advance of my wheels. No doubt the snake felt me coming. Even though snakes have no outer ears, their inner ears sense vibrations. Past another windmill and beyond the dip to a grassy knoll, I see the bare outline of a blocky structure low to the ground and park about a half-mile away to check out the blind, which turns out to have been designed for viewers, not listeners. Several large peepholes are big enough for a camera lens, but they're cut out of plywood, which excludes sound, rather than fabric, which would allow sound to pass through. A slight afternoon breeze catches the loosened fiberglass skylight, resulting in a clattering sound. Also blown by the wind, untied strings from fabric flaps tap the side of the blind, as welcome as a dripping faucet.

Hard to believe this lek will host a Woodstock-like sunrise rock concert, but studying the ground closely I can see the chicken-like footprints on the ground and some well-worn bare places about 30 feet directly in front of the blind. Lots of them. This is an active lek, all right. The sharp-tail males, I know, will do a Mick Jagger strut, wings outstretched, heads low, then high, prancing with uncontrolled excitement at the prospects of mating, making loud peeps and hoots, then rotating their hindquarters with a popping, twitching action. In theory my microphones could be anywhere, but I want this performance to happen center stage. I envision two spots low to the ground and in the lee of some grasses, figuring the grasses should add a nice touch; I want the possibility of a breeze to register a soft whisper and expand my listener's appreciation for the spaciousness of the vast prairie. Before walking back to the Vee-Dub, I make a mental note to pack in some extra weight to immobilize the skylight.

Dinner's a can of Dinty Moore beef stew. Afterward I retire to the top of the Vee-Dub and fall quickly asleep. I wake on my own to a sky full of stars, vast and uncountable. The prairie is silent, 20 dBA or lower. There's a gentle rustle through the grasses, 23 dBA, peaking at 30 dBA before the zephyr passes in less than 20 seconds. I gather my gear and hike back to the blind by headlamp. The illuminated cone in the darkness is intimate, each nod stroking the grassy hillside. It's cold. Too cold for rattlesnakes, I hope.

At the blind, I set a couple of battery packs as weights on top of the skylights, then place the microphones in position, peel out the cables to the blind, and listen through the recording system. Only a light wind passes in 15 to 30 seconds, well-defined breaths through the grasses. After about 10 minutes I hear the very distant twittering of horned larks. Then a deep rumbling of something, barely audible. Fifteen minutes later a western meadowlark sings from a barren stalk of yucca about 300 yards away, no louder than the subtle breeze. Then at close range I hear *Ou, ou, ou-ah.*

A sharp-tail has walked in undetected by my supersensitive microphones and now stands 30 feet away.

Wop, wop, wop. The rapid drumming of its wings as it flutters away into a choice position. *Hmmmmm.* Very distant drone of a plane.

Bop, bop. A tail twitch from a second male.

Arar-ei. Arar-ei. Ah-oo. Ah-oo. A third sharp-tail arrives.

Soon the lek is alive with about a dozen male sharp-tail grouse squaring off, each trying to outdo the others with ever more outrageous antics to attract the attention of any one of the four females that have been lured to the sidelines.

That barely noticeable rumble has been growing steadily for nearly 45 minutes. A whistle blast, probably at the Thedford crossing, almost surely identifies the rumble as a train. The train horn triggers a faraway coyote chorus but doesn't bother the sharp-tails, which continue their competitive swaggering and vocalizing, through three more horn blasts at other road crossings. At last, a female struts between two males and presents herself to the chosen favorite. Copulation happens so fast I fail to see it. But there's no missing the characteristic slow descent of several downy feathers after the birds untangle and part.

I sit motionless in the blind for several hours looking through the largest of the peepholes, headphones on, listening as if I'm right on stage. The sharp-tail show ends when an antelope ambles right through the center of the lek, scattering the birds upward; they sail off in a long glide. With the sharp-tails gone, I can now hear a faraway *Kaa-ooooo, kaa-ooooo, kaa-oooo* of a different member of the grouse family, the greater prairie chicken. That sound, coming from another lek, perhaps signals another wild dance party. What a morning.

One fascinating member of the grouse family that did not perform today is the greater sage grouse, a B-52 bomber compared with its close cousin, the sharp-tail. I have lain on the frost-covered high prairie in complete darkness outside of Walden, Colorado, waiting for the first rays of sun, when a *Whirrrrrrrrrrrrr* and a *Sssssssssssssss* combine to announce this *Spruce Goose* of a grouse gliding overhead to begin its age-old mating ritual. Loud pops and guttural growls ensue as the males seek the favor of the females. You must remain absolutely still and hidden because these animals are extremely shy and will flee at the first sign of an uninvited guest. With my recording equipment in place, I've stationed myself a football field away.

But far more disruptive to sage grouse mating than hovering bird watchers and listeners is round-the-clock noise from the recent proliferation of huge natural gas drilling rigs and associated truck traffic. One study has documented sage grouse population declines of as much as 50 percent downwind from deep natural gas drilling rigs in Wyoming. Though factors such as chemical pollution may also be at work, the prime suspect is noise: 70 dBA at a quarter-mile away.

Seeking scientific evidence, a University of California–Davis assistant professor of evolution and ecology, Gail Patricelli, and a team of grad students played recordings of drilling site noise and associated truck traffic through speakers hidden in stones at four different leks. Even though they used small speakers and a localized source of noise, early results point to a significant effect of noise. Says Patricelli, "We found we did cause a decrease in birds on the leks with noise. The decrease was about twenty-five percent on leks with drilling noise, so drilling noise is pretty disturbing stuff. It's a lot of low-frequency generator kind of noise and some high-frequency clanging and grinding sounds—and it's continuous. They drill twenty-four–seven for a few months."

Morning birdsong, even without such spectacular mating rituals, always fills me with joy. The sparkling sound reminds me of children waking up with innocent delight and enthusiasm. I guess it's become my *Weltanschauung,* the way I see the world. No matter what happened the previous day and no matter how bad things might have been, I'm always invigorated and filled with newfound enthusiasm whenever I listen to a vibrant dawn chorus. Einstein said that at some point each person decides, consciously or not, whether life is essentially good or essentially bad. This *Weltanschauung* affects everything that they do. I believe that life is essentially good, and I've got all the proof I need in the joyous chorus of birds at daybreak. Especially today on the prairie.

The Vee-Dub points east into the morning sun. The train noise during this morning's recording really didn't bother me because I have a soft spot for trains. I find them evocative of the landscape as they sound a horn at each road crossing, and also because at that distance, even a blaring horn is sweetened by the land, echoing across the hillsides in a many-layered note.

Like nature, trains have rhythm and make music. I have recorded famous locomotives in Europe, the United Kingdom, Asia, and the United States. Each link in a locomotive's powerful engine emits a sound and therefore connects us rhythmically to the experience of train travel. Even the rails have music in them, from the tingling sounds emitted when a heavy load rolls over a rail, to the *Clickity-clack* of the railroad track as the wheels of each car or wagon pass over the gaps between rail sections. These gaps alternate from one side to the other on older railways, creating a wonderful stereo effect while riding in a passenger or sleeper car. Or in a boxcar. I first learned to appreciate train sounds when I hopped my first freight in 1981 to record the stories of hobos in their jungle camps.

I'm always looking to record the *Clickity-clack* of the rails, for this, too, is a disappearing sound, as sectional track has given way to continuous rail, with its modern *Zzzzzzzsh* throughout much of the nation. So across much of Kansas and Missouri I drive the Vee-Dub on roads skirting railroad tracks, looking and listening for sectioned track. I find a few short, "unimproved" sections of track on bridges and turns, but nothing long enough to produce that sweet rhythm of the rails as the one-ton wheels rotate slowly out of the yard with a *Click-thump-thud, Bing-thud-clang* resounding over the rail gaps, ever quickening. Then the sudden explosion of *Bang-aclatter-cling-thud* across the switch and out onto the main line, where the rhythms really sing, leading to a crystal glass—like *Eeeeeeeeeee,* the sound of wheel flange around the first bend. Glenn Miller's "Chattanooga Choo Choo," John Denver's album *All Aboard,* and Johnny Cash's "I Hear the Train a Comin'" all pay homage to the contributions of trains to the American music scene.

Somewhere along the way, between Nebraska and Missouri, I notice fewer and fewer waves, those reassuring raised hands that seem inversely proportional to the local population. In sparsely populated rural areas, up goes the hand, either out the window or above the steering wheel, when you drive by another vehicle. You lift your hand in a friendly gesture, acknowledging respect for each other and the unspoken truth that if one of you broke down the other would stop to help. More cars equals fewer waves. Sure enough, the traffic's been steadily increasing.

Reaching the Mississippi River, I'm now halfway, psychologically, if not geographically, on my journey to Washington. I head north to take care of some unfinished business outside of Hannibal, Missouri, a town made famous by Mark Twain, America's most celebrated author. Twain's *Adventures of Huckleberry Finn* is perhaps the most widely read American novel, although I doubt many have listened to it. I didn't, the first time I read it in school. It wasn't until my cross-country sound safari in 1990, when struggling to find a noise-free place to record in the Mississippi Valley, that I picked up a copy of *Huckleberry Finn* to fill the time between recording opportunities. Sitting in the very landscape of the book, my own adventure changed when I came upon this passage at the start of chapter 19:

> Two or three days and nights went by; I reckon I might say they swum by, they slid along so quiet and smooth and lovely. Here is the way we put in the time. It was a monstrous big river down there—sometimes a mile and a half wide; we run nights, and laid up and hid daytimes; soon as night was most gone we stopped navigating and tied up—nearly always in the dead water under a towhead; and then cut young cottonwoods and willows, and hid the raft with them. Then we set out the lines. Next we slid into the river and had a swim, so as to freshen up and cool off; then we set down on the sandy bottom where the water was about knee deep, and watched the daylight come. Not a sound anywheres—perfectly still—just like the whole world was asleep, only sometimes the bullfrogs a-cluttering, maybe. The first thing to see, looking away over the water, was a kind of dull line—that

> was the woods on t'other side; you couldn't make nothing else out; then a pale place in the sky; then more paleness spreading around; then the river softened up away off, and warn't black any more, but gray; you could see little dark spots drifting along ever so far away—trading scows, and such things; and long black streaks—rafts; sometimes you could hear a sweep screaking; or jumbled up voices, it was so still, and sounds come so far; and by and by you could see a streak on the water which you know by the look of the streak that there's a snag there in a swift current which breaks on it and makes that streak look that way; and you see the mist curl up off of the water, and the east reddens up, and the river, and you make out a log-cabin in the edge of the woods, away on the bank on t'other side of the river, being a wood-yard, likely, and piled by them cheats so you can throw a dog through it anywheres; then the nice breeze springs up, and comes fanning you from over there, so cool and fresh and sweet to smell on account of the woods and the flowers; but sometimes not that way, because they've left dead fish laying around, gars and such, and they do get pretty rank; and next you've got the full day, and everything smiling in the sun, and the song-birds just going it!

I sped on to the end of the book and read it again, paying special attention to Twain's descriptions of sounds. Again and again, Mark Twain proves himself to be an extraordinary listener. He employs songbirds to foretell weather changes. And he correctly introduces thunderstorms at their proper time during the day up and down the Mississippi Valley, knowing that storms typically move south to north, occurring later in the day the farther upriver you are. Studying Twain I was able to escape the noisy modern world of the Mississippi Valley, if only for short stretches, and find a wonderful place to record, a place he surely would have enjoyed, a place with "everything smiling in the sun, and the song-birds just going it!"

Later I read his autobiography and learned more about Twain's life in the towns of Hannibal and Florida, Missouri. In search of his childhood haunts, I listened deeply, as Twain must have, for he was a stickler for quiet. A ticking clock would drive him from a room. He would not give a reading unless the programs were printed on paper with high cloth content—noiseless paper. In both of his greatest works, *Huckleberry Finn* and *The*

Adventures of Tom Sawyer, Twain transforms his heroes from boys into men, free-thinking individuals in society, in a strikingly silent setting. Huck receives his transformation alone on the Mississippi River and decides that he will help Jim even if this means going to hell. Tom is transformed in the silence of a cave.

One of Twain's autobiographical passages particularly caught my ear, this description of the farm belonging to his Uncle Quarles outside the town of Florida:

> Down a piece, abreast the house, stood a little log cabin against the rail fence; and there the woody hill fell sharply away, past the barns, the corn-crib, the stables and the tobacco curing house, to a limpid brook which sang along over its gravelly bed and curved and frisked in and out and here and there and yonder in the deep shade of overhanging foliage and vines—a divine place for wading, and it had swimming pools, too, which were forbidden to us and therefore much frequented by us. For we were little Christian children and had early been taught the value of forbidden fruit.

The "swimming pool" at Uncle Quarles's farm is where young Samuel Clemens had a near-death experience. He was pulled lifeless from the water and revived by one of his uncle's slaves. It is this slave who Twain immortalized as Huck's raft companion, Jim. That near drowning intrigued me. What if he had never written a single word? What if young Sam Clemens had died that day? So on a 1992 Mississippi River trek, first down the river from Lake Itasca to New Orleans, then back up, I decided to try to find that brook, whose music Twain knew so well. At the suggestion of the curator at the Mark Twain Museum, I finally wound up going door-to-door, knocking for leads. My search ended at the home of a man named Reynolds who pointed to the nearby treed hilltop. That, he told me, was where the Quarles farmhouse stood. Below, I'd find the creek.

But, even though it was May, the creek, whose banks had clearly fallen victim to the overgrazing of cattle and the silting of its pools, was dry. Finding none of the old farm buildings still standing and no water flowing in the creek, I had to return to the music of Twain's words in my mind. Staring at the stony creek bed I could hear Twain's "limpid brook which sang along over its gravelly bed and curved and frisked in and out and here and there and yonder in the deep shade of overhanging foliage and vines." I realized: the stones are the notes! I could gather a few of the stones and find a nearby foster stream in which to play them.

And that is exactly what I did, except not nearby. Every local creek and stream I came upon had met the same fate, or if they still flowed nicely, they were now surrounded by all manner of man-made noise. I had to travel all the way to New Albin, Iowa, some 270 miles. But it was well worth it. "Limpid Brook," one of my sound portraits available at my website and from iTunes, is a delightful piece that I believe re-creates a seminal setting in the life of one of America's greatest listeners.

And now, 15 years later, I'm back, to revisit what has become a personal shrine, to remember young Sam Clemens. I want to see how the site of Uncle Quarles's farm is faring, and having called ahead, again pull up at the Reynolds place, a two-story white house with red shutters and a red roof and a white rocking chair on the small, ground-level front porch. I'm barely on the walkway to the door before a small cocker spaniel mix races up to greet me, yipping at my legs, and is then joined by a terrier of sorts and a good old hound. Then Barbara Reynolds appears and says hello.

"We've lived here since 1971," she tells me outside on the lawn. "Bill's grandparents owned the land before us. Bill's grandfather knew this was the site of Mark Twain's childhood, though I don't think that's why they bought it. They bought it from an insurance company. I think the owners had lost it."

She points north. "That's the Quarles place. Where that clump of trees is." She identifies the owner of the property as a woman named Karen Hunt. Hunt wrote her master's thesis in 1981 on cultural influences on early farmsteads here in the Salt River Valley and, like me, learned about the historic importance of the Quarles farm. Eager to protect it from future development and someday see it as a state historic site, she bought the 28-acre property from Bill Reynolds's father in 1991. She's about to conduct an archaeological dig in search of building foundations and perhaps turn up artifacts.

"And the low draw down here," I begin, pointing below the old home site.

"Yes, that's the creek."

I ask permission to collect more stones to add to my growing limpid brook collection and am soon heading down the Reynoldses' driveway, past a tire swing in the shade of a huge spreading oak, then downhill to the nearly flat, mostly dry creek, which today retains shallow pools of water from an earlier runoff. The slope of the land, which helps determine the energy of the flow, is like a volume knob. And the limestone rocks in the creek bed are the notes that the water plays as it flows down the gentle slope. Walking the bank of this creek that belongs to the Reynoldses, I fret that, without recognition, this site may never be preserved. Some of the more fragile limestone notes, I see, have already been shattered by the heavy hooves of cattle.

Which stones, which notes shall I gather up? I see many with weathered faces pocked with cavities like craters on the moon and select a few dozen in various sizes, imagining these irregularities will bend and articulate the sound of the fingers of water flowing over them. I make numerous trips back to the Vee-Dub, wrapping each stone in a separate rag, then fitting them into hard plastic coolers brought along for the occasion. Nothing would make me happier than to return each of these stones back to its place under the legal protection of an acoustic historical site, but barring that, they will remain with me. Looking to fill another section of the orchestra, I collect a good many smoother, water-polished stones without craters. These will afford silent spacing between the delicate rills. Still not satisfied, I gather another several armloads because, well, because they appeared special to the eye and felt special when held in my hand. As a last gesture, I remove the OSI stone from around my neck and place it on a rock in the center of the deepest pool where Sam Clemens must have swum and take a photograph. Then I'm off, soon cruising right by Injun Joe's Campground and Waterslide to a Fed-Ex pickup point in Hannibal, where I ring up a whopper of a bill. I reserve a symbolic hatful of freshly gathered stones to put on the Vee-Dub's shelf next to the wood stove to keep me company.

I've made it to the outskirts of Chicago and to a second hiatus in the journey. I need to fly back to Seattle for Oogie's graduation from the University of Washington. My flight is tomorrow, May 25, and needing to clean up before I share a seat next to a fellow passenger, I check into a hotel not far from O'Hare International Airport. This time I've not surfed the Internet looking to pinch pennies. I've selected my hotel on the basis of its marketing promise. AmericInn, a chain with hotels in half the states in the union, guarantees a "perfect end to your day: a quiet night's rest." Inside its elevators, I've been told, where other hotels entice guests to their restaurant with close-up shots of juicy steaks, AmericInn displays a photo of its SoundGuard masonry block filled with its patented, sound-deadening foam.

This location will put AmericInn's quiet claims to the test. It's but an energetic baseball toss from the hotel's entrance to six busy lanes of I-55. The traffic noise measures 69.5 dBA outside the door to the hotel lobby, then 50 dBA between the double sets of doors. The sound level rises to 55 dBA in the lobby, where a TV drones on. Here's more of my sonic scorecard:

45 dBA at elevator.
55 dBA elevator ride.
35 dBA in the hallway outside my room on the second floor.
37 dBA at the window of my room, with a clear view of midafternoon highway traffic.
53 dBA at the window after I turn on the air conditioning on this hot day.

Later that night, at 10:55, I measure at my window once again and read 30 dBA when the AC isn't running. The hotel has also taken care with the windows, which are triple-pane and do open. With the window opened, the sound of traffic boosts the reading to 55 dBA. That makes 25 dBA of noise attenuation merely because of the triple-pane window. I realize I have heard absolutely nothing from my neighbors or the TV sets in their rooms. When inside the bathroom with the door closed and the lights off, I measure 28 dBA—truly remarkable, much lower than I would have predicted. I sleep soundly. Bravo AmericInn. Thanks for a quiet night.

Before checking out I go online and read some of my accumulated e-mail. A Seattle resident named Robin Brooks has written with praise for One Square Inch, but also with a gripe:

> Your One Square Inch project is an amazing and inspired idea, and I've been looking forward to visiting it for months. I learned about it through my Triple A membership this past fall, and have been waiting until warmer weather to take a trip out there. I'm wondering, however, why you would choose to send the original OSI stone on a "publicity tour." For me, that defeats the spiritual and symbolic purpose of the stone. I'm disappointed that I've waited out a long and rainy winter to journey to this special place, and the original stone won't even be there. Worse, it won't just be missing for a week or a month. It'll be gone all summer. It certainly has less appeal now, and I'm not sure I'll make the trip.

I e-mail her back, saying that, as it happens, I'm returning home for a short time and will be bringing the stone with me. I ask, "Would you be interested in joining me on a hike with the OSI stone up the Hoh Valley?"

Then I'm off to O'Hare Airport, first dropping the Vee-Dub at a Marriott Suites self-park, then taking the hotel's shuttle to the check-in level for Alaska Air. I'm early, so after checking in, I wander around for a bit.

Is there a decent-size airport in America that is ever free of construction? Here, the work is on the building itself, and loud noises resound at peak levels of 87 dBA from the other side of a plain white wall, where two men work on a beam with a welder and grinder. Back on the interior side of this temporary wall, I notice a posted sign, here, of all places:

> NOTICE: The City of Chicago, Department of Aviation, has designated this area as an appropriate place for individuals and groups to distribute literature, solicit donations, and engage in other activities protected by the FIRST AMENDMENT TO THE UNITED STATES CONSTITUTION. No City endorsement of the ideas or opinions expressed by those who utilize this area should be inferred from the City's permitting them to do so.

Since this spot is designed for free speech, let me exercise my right, at least here in print, for I could not have been heard at that specified spot. The natural human voice is 55 to 60 dBA; the construction noise of this free speech zone is 27 dBA louder. That is 400 times more powerful than the sound of my spoken words! Who gets the real say here? To post the sign in this vortex of construction noise is anything but an endorsement of the right to free speech. Let's remember that in 1787 the cobblestone streets in front of Independence Hall were quieted with gravel so that our forefathers could draft, unimpaired by noise, the Constitution of the United States: "We the People of the United States, in Order to form a more perfect Union, establish Justice, insure domestic Tranquility . . ." Each word entered into our Constitution was intentional. *Domestic:* adjective, "indigenous to or produced or made within one's own country; not foreign; native." *Tranquility:* noun, "quality or state of being tranquil; calmness; peacefulness; quiet; serenity." I'd like to argue that I have a Constitutional *right to quiet*—but would anybody hear me?

If a tree falls in the forest but there is too much noise to hear it, does it still make a sound?

Interlude

When we contemplate the whole globe as one great dewdrop, striped and dotted with continents and islands, flying through space with all other stars all singing and shining together as one, the whole universe appears as an infinite storm of beauty. The clearest way into the Universe is through a forest wilderness.

—John Muir, *Travels In Alaska*

My portal to the universe is the Hoh River Trail, the tall-treed and fern-lined path that leads to One Square Inch. It has been more than two months and many thousands of miles since my last visit. It's June 4, and my hunger for the Hoh borders on my need for food and water. Natural quiet is not a luxury; it is a human necessity. There is quiet, a stillness in all of us. But we require natural quiet to discover it.

Surprisingly, the visitors center parking lot is nearly deserted: one RV, one motorcycle, and four cars scattered on blacktop laid down for 10 times as many vehicles. Two of those vehicles have brought hikers who will join me. I spot a familiar face, Edward Readicker-Henderson, a journalist who has written about me and One Square Inch and who is now at work on a book about his own worldwide search for quiet. Two women approach. Robin Brooks introduces first herself and then her housemate,

Kate Parker. It was Brooks who e-mailed me, chiding me for taking the OSI stone with me on "a publicity tour," and who accepted my invitation to join me today.

We all reach for our backpacks and set off. The trailhead bulletin board provides a weather report: "Rain likely, cloudy, high near 59° F, SSE wind, 9–14 mph becoming WSW, chance of precipitation is 70%." This is the rain forest's version of a sunny day—perfect for aural solitude, subtle changes of light, and fragrant treats.

Brooks is colorfully attired in a peach-colored fleece jacket, white pants, and a light blue slouch hat. She smiles broadly when I hand her the One Square Inch stone. She describes herself as "an actor and playwright," explaining that she's currently a house manager at Seattle's Eve Alvord Theatre, a children's venue that opens its doors to busloads of schoolkids to expose them to live performances. "About fourteen hundred people come through there. I have to shout over the audience to get people's attention," she says. "I found out recently that if you have to shout over noise, you're exposing yourself to hearing damage."

Brooks recently moved to Seattle from Michigan, where she grew up. She's 32 years old, not even half Bill Worf's age. She, too, can recall profound quiet in her childhood home on the Upper Peninsula. "When I went to bed as a kid there was no light," she says. "It was completely dark, and you couldn't hear anything. The planes seemed to have stopped flying. And every place I've lived since, it's been bright and it's been loud." Brooks also hungers for quiet, the deep penetrating quiet of her youth. She seeks "a place to be at peace with my thoughts." Why? "I think that's who I am at my core," she says, mentioning that she's unsuccessfully searched for quiet in the usual city spots. "I've gone to the arboretum. You can still hear the traffic. And Discovery Park. You can still hear the boats." Which was why she got so excited about visiting One Square Inch after reading about it in a magazine.

"Can I get you to carry the One Square Inch stone today?"

"You bet!"

I put the stone back inside its leather pouch and, holding its cord in both hands, place it lei-like over her head.

My sound-level meter measures 35 dBA, then peaks at 38 dBA as my ears pick up a distant, familiar *Dud-dud-dud-dud-dud*. From a couple of football fields away, through giant Douglas fir, Sitka spruce, and western hemlock, a woodpecker drums from the Hall of Mosses with a sound signature all its own. But as often happens with groups of Hoh hikers, it takes a good many footsteps for everyone to begin to appreciate the natural quiet. For the first mile we are our greatest noise source. Overheated minds from other places need time to cool and quiet. And they do. Today, right on cue, chatter ceases at the footbridge with the view of the waterfall. The rushing creek defeats any attempt to speak, thrusts nature center stage, and silences human voices.

We continue on to One Square Inch, approaching the mossy log single file, and assume separate places in the forest clearing. Brooks chooses the top of a fallen hemlock tree and faces the quiet hush coming from the distant Hoh River, her back to us. Parker rests among wood sorrel, adopting Rodin's *Thinker* pose, facing the light coming through the spruce boughs. Readicker-Henderson melts into the ferns and rests comfortably in the deepest of the moss beds, sorting through the Jar of Quiet Thoughts. I have chosen to stand by my favorite huckleberry bush, savoring the red, salmon-egg-size morsels, sniffing for any recent evidence of Roosevelt elk. A two-toned faint murmur of various winged insects is barely audible. We all remain silent, immersed in our own thoughts and heightened awareness.

I am the first to leave. I wait back at the main trail. Everyone knows to stay as long as he or she wishes. On the walk back, muted conversations resume gradually. Brooks shares her frustration that the solace she'd sought proved elusive, because less than five minutes after we'd arrived at OSI a plane flew by, the first of two overflights that disrupted our quiet vigil. She says it seemed extremely loud, maybe the loudest plane she'd ever experienced. I tell her it wasn't actually so loud; it only seemed loud in comparison to the naturally quiet ambience of the Hoh Valley. Though disappointed, she still feels the journey has been rewarding. A bit farther down the trail she asks, "Do you ever feel like you're one person trying to do the impossible?"

"I am one person," I answer.

She looks puzzled, as though she expected me to answer differently. Frankly, my answer might have come out differently had we been in Seattle. That's what our national parks are for: to provide us places where we can step out of our modern world and be fully realistic, get the big picture. Yes, I'm only one person, but I do not feel my quest is impossible. Each side of this very trail we're walking is lined with miracles, from these giant spruces and firs to the equally stunning butterscotch slug that Brooks spotted on the way up. Saving One Square Inch of Silence seems like a modest task compared to creations such as these!

Our conversation on the way back to the parking lot continues in silence, that underrated third voice that never interrupts and always has something worthwhile to say.

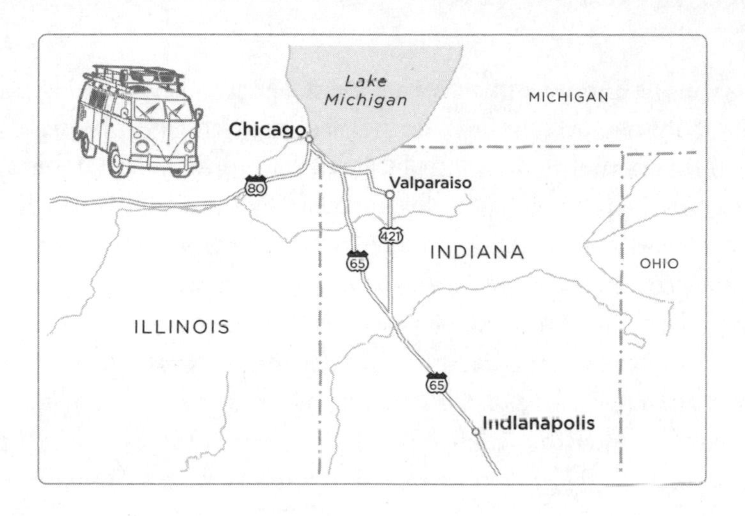

9 Toxic Noise

Calling noise a nuisance is like calling smog an inconvenience.

—Dr. William H. Stewart, former surgeon general of the United States

Heading back to Chicago to retrieve the Vee-Dub and resume my journey to the nation's capital, I can see checkerboard fields from my window seat on Alaska Air flight 22, probably Iowa or southern Wisconsin. It looks like I'm going to see more of the prairie than called for by the original flight plan because the pilot has just announced that we have been diverted around a thunderstorm. To the north a collection of deep, dark clouds with white tops serves as a dramatic backdrop for occasional flashes of lightning. Rough weather, all right, and roughly the size of Olympic National Park. Clearly, the FAA can, in the interest of safety, reroute planes at a moment's notice.

So why can't planes be rerouted around the real Olympic National Park to preserve the silence at One Square Inch? Though other airlines fly over the park, many coming from or headed to Asian destinations, Alaska Air has the most flights in and out of Sea-Tac Airport south of Seattle. In 2005 I wrote the airline to tell them about One Square Inch of Silence and ask them to fly around the park. Kevin Finan, vice president of flight operations, wrote back: "Normal departure routings from Seattle to Alaska do overfly the park. This is the FAA preferred routing from an Air Traffic Control perspective. Deviations to this traffic pattern would increase delays, causing higher fuel burn and increased emissions." I prepared myself for a polite brush-off. But he continued: "Alaska Airlines is an environmentally conscious company. To assist in your effort, for all non-routine flight operations (e.g. maintenance flights, test flights, etc.), we will enact a policy encouraging our flight crews to avoid flying over the Olympic National Park."

Every step toward progress is worth celebrating. Maybe the Seattle-based Boeing Airplane Group, which, I learned using WebTrak, tests its jets over the remote Olympic Peninsula, will follow suit and step up for quiet. So far, however, Boeing has not responded to my pleas. A 40-year veteran of the skies, 30 of them flying for commercial airlines, recently retired Captain Ron Nielsen provides this view from the cockpit: "If I'm flying from Phoenix to San Francisco, I'm going to go right over Yosemite. That's where the flight plan, the highway goes. I could reroute if I was motivated—or if somebody was motivated—but you can imagine now, the original reason that you flew airplanes is 'cause you could go from point A to point B directly, and because you didn't have to follow curving roads. That saves time and it saves fuel. That gets corrupted sometimes in today's environment because there are so many airplanes going from point A to point B and from point C to point D, and there are conflicts, so they may make the highways bend a little bit, if you will.

"We as pilots never involve ourselves in what we're going to fly over that day. And the airlines, it's not in their interest to do anything like you suggest [flying around One Square Inch] because it means they're going to have to put in a special request for clearances and burn more fuel."

But Captain Nielsen does admit that special interests on the ground do affect flight patterns, as at Orange County Airport in southern California, where wealthy Newport Beach residents situated near the departure runway pressed for and won changes in the ways jets take off. "You're going to climb very steeply to one thousand feet, then level off and go virtually into slow flight, where we pull the thrust back on the engine to just barely keep us flying at a slow speed until we get six miles off the coast and then we'll resume a normal climb. We do that totally for noise abatement," he says, terming the flight maneuver an approved procedure. "They [locals] have influence. 'If you don't abide by the noise standards that we've set in our community we won't let you fly out of the airport anymore.' It takes more attentiveness on the part of the pilot. In a pure sense it would be absolutely safer if we didn't do something like that, but because of the economics and the desire of the airlines to operate in and out of there, because ticket prices are lower out of Orange County than LAX [we make the adjustment].

"Pilots aren't going to unilaterally take it upon themselves to say, 'Avoid Yosemite or Yellowstone,'" he continues. "They don't have enough people asking for that. What percentage of people buying tickets want this?"

He's right. Currently, most people don't give a moment's thought to the noise shadow of their flight on the ground below. But what if they were better informed?

"Hey, you're preaching to the choir," Nielsen says. "I would love to have areas that are pristine with no kind of distraction." He further heartened me with these words: "It's really quite simple [to change an aviation highway]. Someone who has the authority has to say 'We have a new priority. Now instead of going directly from point A to point B, we want you to divert course just a little bit to avoid an area on the ground that is noise-sensitive.'

"It's a matter of priority—a matter of saying 'From today, from now and forever more, we're going to fly this way because we don't want to upset the pristine nature of the national parks down below.'"

As for the fuel issue, Nielsen estimates that a minor rerouting around One Square Inch might add 30 seconds of extra flight, requiring "negligible extra fuel consumption." In 2006 the Air Transport Association estimated that every minute of flight delay added $66 to operating costs. Yes, fuel prices have shot up since then, but a slight change in course in support of quiet does not seem unduly expensive. Nielsen stresses, "It's not so much about fuel as it is about the energy and attentiveness it takes to change flight plans."

I can't wait to ask the FAA about this when I get to Washington.

Errrrrrrrrrrrmmmmmmm. I hear the comforting sound of the descending landing gear followed by a boost in the jet engine roar, then feel the jolt of touchdown. I grab my bags, then take a shuttle to the Hyatt-Regency O'Hare, a Priceline score, sitting shoulder-to-shoulder with chatty pilots and flight attendants. They babble at 70 dBA. That's 10 dBA above, or twice as loud as, normal conversation. But now, hours into my reimmersion in airport clamor, jet cabin noise as high as 108 dBA on take-off, and big city highway traffic, this sound level seems normal to me. So does joining the crowd at the hotel bar for a double Chivas.

Eeek, eek, eek, eek—click. My Radio Shack alarm clock does its job using a piezoelectric speaker common to many modern alarms, sirens, and beeps that say "Wake up," "Help is on the way," or "Your meal is ready." All are made more attention-grabbing by emitting sound waves unlike those commonly found in nature. Instead of smoothly flowing, up-and-down, sinusoidal waves, these mechanical noisemakers produce square and even saw-toothed sound waves that interact with the ear much differently. The difference between a morning birdsong and a saw-toothed piezoelectric alarm is the difference between a morning massage and a karate chop. This morning's quick blow to my ears propels me out of bed.

I insert both packets of grounds into the Hamilton Beach coffeemaker, which kicks into operation with a watery concerto of gurgles and sputters and hisses not unlike the Flying Scotsman steam locomotive on idle. A quick, strong burst of steam followed by a gradually decreasing hiss. Then another burst. Finally, a long sigh. My coffee is ready.

The strong aroma of coffee brings back memories of college days in the Midwest when I first started grinding my own beans, subjecting myself to noisy gnashing. I couldn't help but recall those noise exposures not many years later, when I read of a startling and still controversial discovery that certain strains of young rats and mice were prone to audiogenic seizures, sometimes fatal, when exposed to loud sounds. This was made more fascinating by the fact that the animals that succumbed to the effects of loud sounds were first exposed between the ages of 15 and 25 days and then reexposed as adults. Fortunately, most scientists agree that audiogenic seizures like this are restricted to these specific genetic strains of laboratory animals.

I'm gearing up for engine work. My plan is to recover the Vee-Dub from long-term parking and replace the oil-soaked rags that have been holding in the oil around the valve covers with parts I had mailed to the Hyatt-Regency. When I ask for my package, I find it enclosed in a heavy-duty plastic bag. At least one of the bottles of fuel additive I also ordered must have burst open in transit, making a mini–Exxon *Valdez* of my order. Omen?

I answer a call from a family friend, Amy Burk, who asks how the journey is going and recalls how she used to wake up as a child on the beach near Gig Harbor, Washington. Water lapped in the stillness, and she could hear the sounds of a dog walking along the beach inspecting what the tide had brought. By the sound of the dog's steps she could tell when it found something interesting.

A nice piece of quiet-induced deep listening. But it troubles rather than comforts me. Like a food lover about to go on a bread-and-water diet, I'm already missing my recent restorative indulgence in quiet back in the Hoh. Since Kansas, my cross-country journey has been without quiet. I'm edgy and irritable. I confess I'm not all that interested in resuming this trip.

"Is it better to know or not know?" I ask her. Better to experience true quiet and know what it offers, and then necessarily suffer in its absence? Or better never to know the powerful pull of true quiet, blithely settling for less because you don't know better, admiring a framed velvet Elvis instead of a Rembrandt?

"Oh, it is better to know," she answers immediately. "Much better to know. It is just like love. It is better to have loved and lost, than never to have loved at all. It is better to know quiet than never to have known quiet at all."

So from here on, I'm bracing myself, because I know quiet. And I don't expect to find it again until I return to the Hoh.

I pay a parking fee of $166 to a lot attendant who's obviously pleased to see me. "So that's your vehicle," she bubbles. "That got so much attention. People were stopping, old and young alike, and got their pictures taken."

Within minutes I'm ensnarled in Chicago's seemingly 24/7 highway crush of traffic. The city's tall buildings are visible in the distance. I-90 East is a slowly moving parking lot, bumper to bumper. The Vee-Dub's air-cooled engine is pouring heat at my feet on an already hot morning. A truck brakes just in time, with a loud screech. I stayed in bed an extra two hours this morning, lying there with my eyes open, dreading exactly this kind of metropolitan morass. Now I'm worried the VW will overheat because the engine is pretty much a glorified lawn mower; it just gets hotter and hotter until, if there's not enough air passing over it, it seizes. Just like I probably would without a break. Recent studies cited by the Noise Control Foundation point to "a clear correlation between [long term] exposure to high levels of road traffic noise and cardio-vascular diseases."

I'm doing all of 10 miles per hour as I pass the Addison Street exit, which heads to Wrigley Field. Truck to the right of me, truck to the left. I get an 80 dBA reading as the light rail train running down the highway median whooshes by. According to Lisa Goines, RN, and Louis Hagler, MD, in "Noise Pollution: A Modern Plague," published in the March 2007 *Southern Medical Journal,* "Noise levels above 80 dB are associated with both an increase in aggressive behavior and a decrease in behavior helpful to others."

Me? I'm down to four miles an hour. So I whistle. A passable, and certainly inspired, version of the Chicago blues.

An hour later, now looking at the skyline from the south, still crawling along, I haven't made it into Indiana. With time for out-the-window photos, I snap an interior documentary shot as well: my face in the rearview mirror. I look committable, the distress etched in deep lines across my brow, furrows plowed by urban congestion and noise.

In the same *Southern Medical Journal* article, Goines and Hagler compare noise pollution to secondhand smoke:

> There is growing evidence that noise pollution is not merely an annoyance; like other forms of pollution, it has wide-ranging adverse health, social, and economic effects. A recent search (September 2006) of the National Library of Medicine database for adverse health effects of noise revealed over 5,000 citations, many of recent vintage. As the population grows and as sources of noise become more numerous and more powerful, there is increasing exposure to noise pollution, which has profound public health implications. Noise, even at levels that are not harmful to hearing, is perceived subconsciously as a danger signal, even during sleep. The body reacts to noise with a fight-or-flight response, with resultant nervous, hormonal, and vascular changes that have far reaching consequences.

A U.S. Department of Transportation document, *General Health Effects of Transportation Noise* (June 2002), estimated from EPA data published in 1981 that at that time more than 200,000 Americans lived and worked in areas where the average daily noise level exceeded 80 dBA.

One of the demographic groups potentially most affected is school-age children, as Arline Bronzaft documented in a classic study on the Upper West Side of Manhattan. Bronzaft, a feisty people's advocate in the Bella Abzug mode, gravitated to noise issues while working as a transit consultant appointed in the Lindsay administration. Eager to quiet New York's subway trains, to lessen the impact on those living near the elevated tracks, Bronzaft conceived of a groundbreaking, pragmatic study.

"I had the idea if I could demonstrate that children can't learn because of the noise, maybe we can get those trains quieted," she recalls. Now a professor emeritus of environmental psychology at Lehman College and the chair of the New York City Council on the Environment, the street-smart New Yorker realized that victimized schoolchildren made far bigger headlines than low-income residents living with transit noise. Of the nearly six dozen city public schools located near subway tracks, Bronzaft chose PS 98 for her study, which was published in 1975 in the journal *Environment and Behavior*. Not only did the school's activist principal grant her access to four years of test scores for grades 2 through 6, but the building itself helped frame her experiment. "Half the classes faced the tracks. The other half were on the other side of the building," she says, describing the built-in control groups.

"When a train went by, the noise level [in the trackside classrooms] jumped from fifty-nine dBA to eighty-nine dBA. Trains came by every four and a half minutes, disrupting the class. The teacher had to stop teaching for thirty seconds," Bronzaft explains. Teachers lost their train of thought, if you will. Students' attention wavered. Effective teaching time plummeted. Bronzaft also interviewed the teachers and found that those with classrooms on the noisy side of the school more often reported being exhausted at the end of the school day. But more tellingly, she documented that by the sixth grade, children assigned to the noisier classrooms were a grade level behind in reading ability.

The Transit Authority had been willing to mitigate noise in the stations, but Bronzaft's study helped spread the noise abatement work farther along the tracks. Rubber cushioning pads on the rails helped achieve classroom noise reductions of 6 to 8 dBA, she says, noting that a follow-up study at PS 98 showed the reading gap had disappeared. "I wish it was quieter [in the classrooms facing the tracks]," she says, "but it was enough to get both sides reading at the same level. Enough for the T.A. to say they'd go out and do the same for all the schools in New York. But to me the second study was more important. It showed if you did something, it made a difference."

Even with Chicago behind me, the traffic is worse. Much worse. Cars idle in both directions. Drivers have started pulling illegal U-turns across the median. I think of a wonderful little book called *Honku: The Zen Antidote to Road Rage* by Aaron Naparstek, who neatly distills the downside of America's love affair with the automobile in haiku like these:

Though impressive
your vehicle's sound system
triggered my migraine

A forest of high-tension lines announces Gary, Indiana. Decades ago, when fresh out of high school, I worked as a deckhand onboard the SS *Sparrow's Point,* an ore boat that carried taconite ore from Thunder Bay on Lake Superior through the locks at Sioux St. Marie and into Lake Michigan, bound for modern-day Dickensian industrial centers like Gary and Burn's Harbor. Back then, in the late 1970s, the U.S. Environmental Protection Agency's Office of Noise Abatement and Control called "unwanted sound America's most widespread nuisance" and termed noise a real and present danger, with more than 20 million Americans exposed each day to noise that is permanently damaging to their hearing. And that was before the din of leaf blowers, boom cars with state-of-the-art stereos, earbuds sprouting on millions of MP3 player listeners, and a national census topping 300 million. Noise has become ubiquitous, inescapable even in space. Astronauts and cosmonauts aboard the International Space Station live around the clock, month after month, exposed to equipment-generated sound levels as high as 75 dBA in work areas and 50 dBA in sleep areas. Surely they feel the irony: the silence of the universe so close at hand but unreachable.

Finally the highway unjams and, like an unclogged drain, everything begins to flow. A cicada lands on my windshield, breaking the road trance and prompting me to leave the interstate in favor of back roads to let my ears do the navigating.

What a strange, eerie combination of cicadas and frogs I've discovered—58 dBA, climbing in long pulses to 67 dBA—at the Talltree Arboretum Gardens outside of Valparaiso, Indiana. I grab my recording gear. The mature hardwood forest offers a lush relief from the highway asphalt. I find cooling streams and a pond with footpaths. But even at this relatively remote place with a fairly prominent natural ambience as loud as a human conversation, the thrum of distant motor noise still comes through loud and clear, littering the soundscape, wanton aural graffiti over nature's art. After an hour of recording and changing microphone positions trying to avoid the noise, I give up in frustration. I realize that many who know this scenic spot will find it impossible to believe, but during my stay, from 3 to 4 p.m. on a Monday in mid-June, not a single second passes without the sound of a motor somewhere, sky or land. Lovely name, Talltree Arboretum Gardens. But my equipment doesn't lie. Much as I long for it, there's not even the merest quiet moment.

Room 102, Motel 8, Valparaiso, Tuesday, June 12, 3:25 a.m. Even at the back of the building, away from the interstate and highway, the sound of traffic still comes through my open window, along with the fragrant smell of the early-morning dew of the forest. The higher-frequency detail of the traffic noise is reflecting off the forest leaves and into my bedroom (48 dBA). masking the more subtle sounds of nature.

Even after the mind has habituated to noise, the body still listens: "Cardiovascular disturbances are independent of sleep disturbances; noise that does not interfere with the sleep of subjects may still provoke autonomic responses and secretion of epinephrine, norepinephrine, and cortisol" (*Southern Medical Journal,* March 2007).

How does one tally the true impact of noise? You can start with the health effects: hearing loss, loss of sleep, possible harm to the unborn fetus, increased risk of heart disease and therefore probably a shortened life span. But don't stop there. There are also detrimental effects on learning, decreased productivity, more sick days. Factor in misunderstandings and miscommunication: the wisdom of a teacher or a subtle warming in a first-date dialogue, words uttered but unheard above the din. Doug Peacock would remind us that significant man-made noise, a phenomenon of just the past few centuries, strikes a major blow against humanity. By overriding natural quiet, noise isolates us from the natural world, muting its infinite beauties and obscuring its essential interconnectedness, cutting us off from our very origins as a species.

Awake in my Motel 8 bed, I'm missing sleep, but I'm also missing something more subtle. I can hear car after car pass over the individual seams in a roadway miles away, but my animal ears have not heard a twig drop to the forest floor. How different from Amy Burk's recollection of hearing a dog walking the tide line. Are we among the last generation who will remember listening to true quiet?

Daylight. Time to finally repair the Vee-Dub before the heat of the day builds into the 90s. Sliding underneath the rear, I yank out the oil-soaked rags jammed into the cracks around the engine and pop the springs that hold on the bent valve covers. I see my new valve covers are chrome. Yeah. Looking good. I give the engine a quick wipe, then set the cork-composite gaskets onto the covers, and *click-click*. The two units snap snuggly into place. I ease out from beneath the van and fire up my wheels.

The Vee-Dub is running fine, even if I'm not. Self-diagnosis is easy: an acute case of QDD, Quiet Deficit Disorder, for lack of a better term. Yesterday I simply wasn't myself. My attitude was nasty! Got to work on that.

I've pulled up near the emergency department entrance at Porter Memorial Hospital in Valparaiso. I'm still of sound mind and body; I've just come to pay a visit to Dr. Samara Kester, chief of emergency care and an e-mail-writing customer of mine since she started listening to my CDs to relax after a stressful day in the ER. This one-time rock band flutist turned medical doctor has started cocking an ear to the issue of hospital noise. In her domain, the often frenetic emergency room, hard surfaces reverberate with the movements of gurneys, the clatter of instruments, the staccato commands, the tensions surrounding lives in the balance. The din, which averaged a "concern raising" 61 to 69 dBA in a study in the emergency department of Johns Hopkins Hospital, makes communication more difficult in a setting where a misunderstood word can be fatal, and, Dr. Kester says, adds greatly to the stress of the job. She's also uncommonly aware of the inherent oxymoron of the noisy nature of her workplace. Hospitals are places of healing. Yet noise does not heal; noise injures. A study of ICU patients at the University of Texas Southwestern Medical Center found sleep patterns so superficial that patients "barely spend any time in the restorative stages of sleep that aid in healing."

"Two major things contribute to abnormal sleep in these patients," notes Dr. Randall Friese, assistant professor of burn, trauma, and critical care at UT Southwestern, "the pathophysiology of the disease process itself and the stressful environment of the ICU. If we can neutralize the stressful environment, maybe we can shorten the hospital stay, lower infection risks, and increase patient wound healing."

Even at night hospitals are not quiet places. In-room monitoring equipment makes noise. So does a portable X-ray machine rolled down a hallway. Ditto shift changes. Those were some of the noisemakers studied in 1999, pre- and postintervention, by a team of concerned nurses in the surgical thoracic intermediate care area at Saint Marys Hospital, a Mayo Clinic–affiliated hospital in Rochester, Minnesota. The study, which was published in the *American Journal of Nursing,* found that noise levels in occupied semiprivate rooms averaged 53 dBA, exceeding EPA recommendations for nighttime hospital noise levels of 35 dBA. The sound level in unoccupied rooms also exceeded that level, averaging 45 dBA. Shift changes typically coincided with the highest hospital noise levels, peaking as high as 113 dBA. In a hospital! At night!

A host of changes, from switching to quieter paper towel dispensers, to rescheduling 3 a.m. to 4 a.m. supply deliveries, to adding padding at the bottom of the pneumatic tube system for delivering document-filled canisters helped lower the peak noise level to 86 dBA and the average noise level in occupied patient rooms to 42 dBA. Awareness, concern, action: that's what it took at Saint Marys Hospital.

That message isn't new, however. It traces back more than 100 years to

Florence Nightingale, who in 1898 included this advice in her seminal *Notes on Nursing:* "Any sacrifice to secure silence for these [patients] is worthwhile, because no air, however good, no attendance, however careful, will do anything for such cases without quiet."

Kester isn't optimistic that Nightingale's call for quiet will be universally answered anytime soon. "Fifty years from now, what we're doing now will be primitive," she says. "They'll look back at our time and say, 'What?' But the human body will still be the same. The body needs solitude. I tell every single patient, 'Get rest, quiet, and drink plenty of liquids.'"

After saying goodbye to Kester, I've got a straight drop down Route 421 to I-65 and on to Indianapolis, where I'll spend the next couple of days. I can't wait to get off the road. The heat is unbearable. It feels like flames coming out of the floor duct at my feet, and my rear end is playing a game of slip-and-slide on my driver's seat, which is broken. All those rough roads in Montana rattled it loose from its moorings and no amount of jerry-rigging has steadied it for more than 500 miles at a time. I'm ready to scream. Happily, I'll be meeting tonight with a patient listener.

John Grossmann has kept an ear to the ground about vanishing natural quiet for two decades. My introduction to him began with a phone ring on the afternoon of September 13, 1988, when he was researching an article on quiet for *Creative Living* magazine. Since then, he's joined me on recording projects on both coasts and along the Mississippi River to write articles for *Omni, The Philadelphia Inquirer* magazine, *USA Weekend,* and recently *Sky,* the in-flight magazine of Delta Airlines. It was that article about One Square Inch of Silence that led to this book and our partnership. He's flown in from his home in New Jersey to meet me here in Indianapolis, which is an important stop on the journey, for moral support.

Tomorrow we will take a tour of Aearo Technologies, one of the world's largest manufacturers of hearing protection, now owned by 3M. Elliott Berger, a senior scientist at Aearo, is a good friend whom I trust will give us straight answers. Elliott has made plans, too. He wants to have my ears tested (so do I), then fitted for high-end custom earplugs. After a quick dinner, we'll listen to an outdoor concert by the Indianapolis Symphony Orchestra performing Led Zeppelin hits. The next day, the three of us will head to the Indianapolis Speedway for time trials. Then we'll seek out a downtown historical marker commemorating what

I consider the birthplace of our national parks: the site of an industrial accident in 1867. I'm cheered by that schedule and the prospect of reuniting with friends.

Aearo's relationship to decibels and hearing loss is rather like that of Weight Watchers to calories and obesity, though Aearo enjoys a good bit less name recognition, even with the very nature of its concern in the heart of its name. The company has nearly a dozen plants and some 2,500 employees around the world, but its main offices and production facilities are here in Indianapolis, in three buildings at a sprawling industrial park.

Berger greets us in the lobby of one of those buildings. He has the mien of someone with investigative medical curiosity: with a calm expression, his eyes look directly into yours (but not piercing); he wears large, wire-rim glasses, and his hair and beard are closely cropped. Instead of a stethoscope around his neck, two earplugs dangle on each side of an elastic cord. We are about to go on tour.

"This building deals primarily with noise control materials and vibration isolators—damping materials," he says, stressing that the proper term is *noise damping,* not the commonly misused term *noise dampening.* "Nothing gets wet." He has arranged for us to tour the building with Greg Simon, a materials development engineer, who begins by taking us through the plastic materials development lab, which tests and develops different kinds of foam and elastic materials the company uses to help to bring a modicum of quiet to our noisy world. Manufacturers of just about anything with moving parts, from laptops' spinning disk drives to planes climbing to cruising altitude, elicit Aearo's help to make their products quieter. Simon points to a stack of pink, yellow, and green foam samples, each color denoting a different stiffness that can be selected for a particular application. "This is for aircraft. It's very lightweight. It's a skin damping foam for the inside of the fuselage, to damp its resonant vibrations. This is actually a kit," he says. "We'll send a big box for one plane, with as many as one hundred fifty of these pieces stamped out. Each one is labeled. They have a little map." Installing them is not unlike putting together a jigsaw puzzle.

Aearo's work has sunk as well as soared. For the U.S. Navy the company has customized four vibration isolators to still and quiet the navigational gyroscopes aboard submarines. For others the challenges can be less momentous but still meaningful. In another room we see somebody bent over a typewriter-like device, a Braille machine, working to silence the clacking of its keys for the especially sensitive ears of its user.

"I'm trying to prevent noise," summarizes Simon. "Elliott's trying to protect people from noise."

Though Aearo sells a wide range of hearing conservation products, its workhorse product is an inexpensive, reusable, one-size-fits-nearly-all product that far too few people use when they should and far too many use improperly when they do: earplugs. The company's flagship design, its Classic E-A-R foam plugs, which it has been selling since 1972, are stamped by the millions from sheets of proprietary yellow foam in a nearby building, where we're headed next.

As with Post-it notes, nobody set out to invent foam earplugs. In fact, their creator, a young chemist named Ross Gardner Jr., was working on a new generation of joint sealants when his attention turned to acoustics. Employed in the mid-1960s for the National Research Corporation, the company responsible for mass-producing penicillin and commercializing instant coffee and frozen orange juice, Gardner noticed an unexpected property of some of the resins he was studying: they absorbed energy extremely well. These energy-absorbing resins (father of the acronym E-A-R) led first to vinyl damping materials to control noise in the foundry industry and then, almost counterintuitively, to effective vinyl foam earplugs.

Nobody expected significant attenuation, that is to say, reduction in strength of sound waves, from such a low-mass material as light-as-a-feather foam. But Gardner thought differently. In a paper cowritten with Berger and presented in 1994 at the 127th meeting of the Acoustical Society of America, Gardner reflected on his suspicion "that energy absorbing materials which act like a composite shock absorber/spring system on a molecular basis could be very soft statically but very stiff dynamically, especially to incoming sound waves." His hope: a major improvement on the preformed vinyl or cotton-and-wax earplugs then on the market.

Gardner finally produced optimum results with a short cylinder 0.61 inches in diameter that he rolled between his thumb and forefinger and inserted in each ear canal. He marveled as, second by second, the plugs expanded back to their off-the-job size, effectively making a seal inside the ear canal, isolating him from much of the noise in his laboratory. Even the low-frequency noise of his laboratory hood all but disappeared. Knowing there would be Doubting Thomases ("But they're only little pieces of foam . . ."), Gardner invited some of his company's key distributors for a demo, one he staged with all the bravura of a door-to-door vacuum cleaner salesman dropping a handful of dirt on a potential customer's carpet.

He gave everyone a set of his earplugs and demonstrated the proper way to insert them, which happens to involve a chimpanzee-like, over-the-top-of-the-head reach to tug an ear skyward to facilitate a deeper insertion. Then Gardner asked for confirmation that all could still hear him. Next he reached for a steel plate and a hammer and started pounding, asking again if everyone could hear the result of his action. While still hammering away, Gardner instructed them to remove their earplugs. The proof was in the pounding. The din, only then objectionable, had everyone reaching for his ears—even running for the door, in Gardner's account. The noise-blocking ability of the foam earplugs became instantly evident.

At the second Aearo building, Berger introduces us to Brian Myers, vice president of Hearing Conservation Products. Short of the factory floor we don safety glasses, then the obvious: earmuffs for me, earplugs for Grossmann. Myers leads away. Near the machine that punches out the stubby cylinders from the company's proprietary foam, I get a reading of 83 dBA. Soon we're watching a machine that's joining a different design of plugs with a blue plastic cord. These molded earplugs, which are preformed to fit the ear, look a bit like mushroom-shaped sidewalk lights. They feature a tapering, three-flanged, conically shaped head atop a post. The machine inserts the ends of the cord into the bottoms of the posts and applies an adhesive that marries the two polymers, connecting the plugs via the cord. Here the sound level is 84 dBA. I get readings of 88 near a packaging machine, and as close as I care to go to a production line with blasting air hoses the meter registers 105 dBA.

The lightweight earmuffs I've donned have a Noise Reduction Rating (NRR) of 20 dB. The earplugs protecting Grossmann's ears have an NRR of 25 dB. Those numbers point to an idealized decrease in the sound levels—protective drops of those amounts from the potentially harmful occupational decibel levels here in the factory in the high 80s, 90s, and low 100s. But those NRRs are based on optimized laboratory conditions and perfectly inserted earplugs—not commonly the case with everyday use. A company brochure written by Berger called *Life can be LOUD: know your hearing protection* explains that NRRs for various hearing protection devices "typically range from 15 to 35. In practice the protection that can normally be achieved is about 10–20 decibels." Hence the call for doubling up—plugs and muffs—in especially noisy environments.

Here, in such a noisy setting and with all of us wearing attenuating hearing protection, brief exchanges are possible, but it's not easy carrying on much of a conversation. We wait until we've left the din for the relative quiet of the hallway that offers our ears the visual equivalent of stepping from a cave into bright sunlight.

Government estimates of the numbers of Americans occupationally exposed to noise range from five million to 30 million. "There's no real data you can hang your hat on," says Mark Stephenson, PhD, a senior research audiologist and coordinator of hearing loss research at theNational Institute for Occupational Safety and Health (NIOSH). He's comfortable with the number Berger provides here in the hallway: probably 10 million Americans—20 million ears at risk on the job.

"And most of them, maybe ninety to ninety-five percent of them, are exposed to noise levels slightly elevated to what you heard in the packaging area," Myers adds.

An October 2006 report from the Federal Bureau of Labor Statistics called occupational noise-induced hearing loss (NIHL) the number one occupational illness, topping skin disorders and respiratory harm. The federal government does have an occupational noise-level standard, but it fashions an imperfect safety net that leaves many individuals still subject to harm. Moreover, in focusing on hearing as the ability to perceive human conversation, the standard tragically ignores other kinds of hearing loss. Two federal agencies are involved: NIOSH and the Occupational Safety and Health Administration (OSHA). NIOSH is charged by the Occupational Safety and Health Act of 1970 (Public Law 91–596) with studying workplace safety concerns and making recommendations "to assure so far as possible every working man and woman in the Nation safe and healthful working conditions and to preserve our human resources." OSHA is charged with regulation and enforcement.

NIOSH recommends 85 dBA as the upper limit for an eight-hour workday. That, however, is not the number that effectively triggers broad federal concern, because OSHA uses 90 dBA. The difference is significant. For an

85 dBA workplace exposure, the most recent NIOSH noise standard criteria studies forecast an 8 to 14 percent excess risk of developing occupational noise-induced hearing loss over a 40-year career. Increase the noise level to 90 dBA, the governing OSHA standard, and the excess risk shoots up to 25 to 32 percent of exposed workers: more than one in every four workers! And there's no way to know in advance if you're among the lucky or the unlucky. Thus the charge that the OSHA standard is "really just a way to document hearing loss."

Stresses Berger, "If you wanted to be absolutely safe for everybody, protecting individuals more susceptible to inner ear damage, you'd need to drop [the NIOSH number] about ten decibels [down to 75 dBA]." This, incidentally, was the recommendation of the Environmental Protection Agency back in the 1970s and the current recommendation of the World Health Organization.

The regulatory standard isn't the only problem. Entire categories of workers, agricultural workers, for instance, are not covered by the OSHA noise standard. Then there's OSHA's definition of hearing loss, which focuses narrowly on hearing loss between 2,000 and 4,000 Hz, frequencies pertaining to human speech. I lament our anthropocentric point of view, which is largely responsible for the global environmental crises. OSHA standards aim to protect our ability to hear what we tell each other, but not what Nature is saying. The song of a winter wren falls above 4,000 Hz, as does the music of many songbirds. Yosemite Falls, Rialto Beach, and Mississippi Valley thunder resound well below 2,000 Hz. Appreciation of these sounds, which fall outside the range of human speech, remain unvalued and unprotected. No wonder many Americans are becoming deaf to Nature.

In Aearo's noisy production areas every ear was protected. But that's clearly not the case in other factories and on other job sites. The construction industry, Myers points out, is notoriously lax in using protective earwear. Berger estimates that 30 to 50 percent of those exposed to a wide variety of on-the-job noise *do* don hearing protection—meaning, of course, that upward of five million do not and therefore leave themselves at risk of noise-induced hearing loss.

This partly helps to explain why Aearo produces some four dozen different models of earplugs. In addition to designs for specific applications, such as for those discharging firearms, the company also offers different colors and graphic enhancements, hoping to appeal to Gen X and Gen Y workers and possibly counterbalance some of those crank-up-the-noise ads in the stereo and motorcycle magazines. "We came out last year with a product we call the Skull Screw," says Berger. "It's an interesting-looking push-in product, a piece of foam with a soft stem so you can insert it in your ear without rolling it up first, and it performs about as well as the roll-down kind. The ad campaign shows an X-ray of a guy's head with an earplug going straight to the brain. We have another plug with what looks like barbed wire printed on the outside. It's called the Tattoo. You want to get all generations."

So add the earplug indicator as yet another sign of rising planetary clamor. Each year, worldwide, a half-dozen or so companies manufacture them by the billions. Nobody makes any more than right here at this plant in Indianapolis, where double shifts aren't sufficient to meet demand. "We're running twenty-four hours, seven days a week. The world is getting to be a noisier place," says Myers. "We need eight days in the week."

Humans, of course, can reach for earplugs. Animals cannot.

Noise-induced hearing loss doesn't disappear at the end of the workday. Recreational motor sports using Jet Skis, snowmobiles, dirt bikes, and 4x4s (to name but a few) also pose hazards. So does listening to loud music. "I'm on a music-induced hearing loss committee at the National Hearing Conservation Association," says Berger, when the hallway discussion turns to the issue of cranked-up MP3 players. I mention my frustration with the volume level on my daughter Abby's iPod.

"We're developing a product than can isolate you from ambient noise, and also limit the sound level you can put in your ear. And there are probably other products out there like that. But I'm wondering, who will buy this? Would your sixteen-year-old, Gordon?"

"Just the word 'limit' would scare them off," I agree.

"If you could convince them," Myers suggests, "that by sealing off the ear from the ambient noise—that even when they listen to it at a lower noise level, it's a better sound. It seems so obvious to us. But it's a tough nut."

Indeed it is, as becomes readily apparent in a later phone conversation with a hearing expert. Dr. Brian Fligor, the director of diagnostic audiology at Boston Children's Hospital, has delved deeply into the potential harm of personal music players. He's written for the *Journal of the American Medical Association* on the subject and a paper for the popular press titled "Hearing Loss and iPods: What Happens When You Turn Them to 11?" His question is a playful reference to the legendary past-the-max amplifier setting in the movie *This Is Spinal Tap*. Dr. Fligor, however, is dead serious about protecting millions, especially teenagers and young adults, from damaging their ears.

"We should be concerned," he says. "Plain and simple: abusive use of headphones and earbuds will cause hearing loss. The best estimate I can give you is a small but significant number of people who regularly use them will have some degree of hearing loss, probably on the order of half of a percent."

That small percentage becomes significant, appropriately enough, by volume. Fligor pegs Apple's sales of iPods and iPhones at around 140 million worldwide. Half of 1 percent of that is 700,000. "And that's just Apple," he adds. "With all the cell phones getting into music reproduction—Sony, Ericsson, Nokia, which is huge internationally—I think it's on the order of billions. Everyone's plugged in. Everyone's capable of being plugged in."

Many do so, he points out, to "take back control of their own soundscapes" in an increasingly noisy world. "Instead of trying to quiet everything down, people are just replacing one noise that they can't control for another one that they can control," Dr. Fligor says. "Except, in order to be able to hear at an enjoyable level, that level has to be quite significantly above that background noise. And unfortunately the ear doesn't know the difference between one noise and another, a pleasant musical passage you enjoy versus factory noise. The ear will treat both about the same, even though the brain treats them very differently. Both can damage the ear."

France has passed legislation limiting the sound level of personal music players to 100 dBA. Dr. Fligor says iPods can go as high as 102 dBA and stresses that a 2-dBA difference is not insignificant: it increases the sound level by more than 60 percent. "Here's why I don't like the French approach," says Les Blomberg, founder and head of the Noise Pollution Clearinghouse. "It implies, because they've set a limit on it, that it's safe, which is absolutely not true." How long one listens makes a big difference, and, Blomberg points out, so can the very nature of the music.

"If you have classical music with a wide dynamic range, hitting a hundred decibels is not that uncommon. If you're listening to a Wagner opera or Verdi's *Requiem,* where you're supposed to see hell, you might want to be there—for a second or two. But the problem is, rock and roll has a dynamic range of maybe five decibels, so you start getting into trouble. I kind of agree with the libertarians: you're in charge of your ears; you want to ruin them, go ahead. But the problem with that is, many of the users are children; they are not really capable choosers of things like this."

Moreover, few adults, let alone children, know much about the array of sound-level exposures recommended by government and health organizations, guidelines that help pinpoint the problem of cranked-up MP3 players. Take NIOSH's workplace recommendation of 85 dBA for an eight-hour exposure. It comes with a 3-decibel exchange rate to caution against higher noise exposures. This so-called exchange rate invokes an exposure half-life for each 3-decibel increase in sound level: no more than four hours at 88 dBA, two hours at 91 dBA, one hour at 94 dBA, a half-hour at 97 dBA. Accordingly, 15 minutes is the maximum exposure at 100 dBA—France's limit for personal music players, including iPods. Hence Blomberg's concern. Fifteen minutes barely begins much of a shuffle on an iPod.

Blomberg has come up with a different, educational approach to the problem: he has applied for a patent for a software addition to MP3 players that would enable users to measure a temporary threshold shift in their hearing. "My invention will tell you if you've listened to your iPod too loud," Blomberg says, explaining that his hearing test would present a series of tones at a single frequency, 4 kHz, one of the most commonly affected frequencies in hearing loss. Those using his test would count the number of audible test tones before an iPod listening session, then again afterward. Hearing fewer tones afterward would indicate a noise-induced temporary hearing loss, a cause for concern, and provide quantitative feedback on the need to lower the volume. "It's meant to play as you turn the machine on and again as you turn the machine off. It takes only about five seconds," Blomberg says. Apple, he claims, won't answer his calls.

Is Apple, a company that claims to think differently, concerned about the potential of hearing loss in those using its products? One would hope so, but it's hard to know. Apple is listed as the assignee on a U.S. patent application filed on December 7, 2005, for "a portable audio device providing automated control of audio volume parameters for hearing protection."

An attempt to ask about that, other iPod specifics, and total sales to date triggered the following phone conversation in response to a message left with Apple's office of corporate public relations:

"I understand you have some questions about the iPod."

"I do."

"So, I wanted to direct you to our website. Everything that is public is available there. It's at apple.com/pr."

"In other words, there's no way I can ask questions of a live human being?"

"Nope."

"Nope?"

"No. Everything you need for your book should be available on the website."

"So if I have a question about a patent application that you've made and don't see any reference to that on the website, then what do I do?"

"You can try and send an e-mail, but typically that's not a question we would answer."

"Would I send that to you?"

Christine Monaghan provided her e-mail address.

Apple's website proved of little help. *No match found. No match found.* That's what came of typing in "iPod and decibels" and "iPod patent application" in the search box. A follow up e-mail to Monaghan went unanswered.

Dr. Fligor stresses just what is at stake in the misuse of MP3 players: "The best way I can describe it is that it will prematurely age the ears. Imagine you're a twenty-five-year-old with fifty-year-old ears. At thirty-five, having the ears of a sixty-year-old. And you may get some ringing in your ears. And that's another issue, a huge quality-of-life issue: tinnitus. It's a very big deal, because it takes away any opportunity for quiet. The worst thing you can do for a tinnitus sufferer is put them somewhere quiet."

Back at Aearo, factory tour concluded, I am headed for somewhere quiet. Berger leads us to an acoustics lab that houses a reverberant chamber. This is about as soundproof as a facility can get without spending a million dollars. It has a base noise level of 17 dBA.

The quietest indoor place on earth, with a base noise level of -9 dBA, is an anechoic chamber at Orfield Laboratories in an outlying part of Minneapolis. Hemi-anechoic chambers ("hemi" for hemispheric) have hard floors. Orfield's Guinness-certified chamber has sound-absorbing wedges sticking out not just from the walls and ceiling but also beneath its wire floor. The lab's founder and principal, Steven Orfield, describes it as a six-sided chamber within a chamber, with an overchamber built around it to further isolate its innermost sonic sanctuary from noise—a kind of architectural Russian doll. He bought the whole works about two decades ago at a fire sale of sorts for less than the cost of a Mercedes. The Sunbeam Corporation, he learned through the acoustic world grapevine, was closing its Schaumburg, Illinois, research center outside Chicago and had no more use for its anechoic chamber. The word was out: a good price for anyone who would move it off the premises in two weeks. Two big companies inquired, but only Orfield could move that fast, wrangling help from the University of Chicago football team (his brother taught there) to dismantle it and load it into three semitrailers. Orfield shipped it to Minneapolis, where it sat in storage, in pieces, for seven years.

Anechoic chambers are measured point to point, not from the deepest recesses of walls and ceiling but from the tips of their sound-absorbing, orderly stalactites and stalagmites and horizontal projections. Orfield's chamber is 10 feet by 12 feet and 8 feet high. The chamber has hosted refrigerators (part of verifying Kenmore's claim of ranking quietest in consumer polls), sleep apnea machines, heart valves, and cell phones (not for anything to do with their ring, but to pinpoint the sound made by their lighted displays). Orfield has a standing bet of a case of beer to anyone who can stay in the chamber in the dark for 45 minutes. He admits he's hedged that bet by adding the lights-out requirement, a major second blow of a one-two sensory deprivation punch.

"You have a couple significant cues to your spatial orientation," he says. "One is acoustic; one is visual. If you take both away you much more easily become disoriented." He says that NASA does something similar with astronauts. "I'm told within about a half-hour they have visual and acoustical hallucinations." Sitting in a chair in his chamber with the lights on at -9 dBA would still be challenging. He guesses somebody might last a half-hour, maybe an hour.

Meantime, I'm headed into Aearo's chamber to have my hearing tested. Generally, however, this sealed room is used to test hearing protection products, the company's own and those of other companies, such as a parachutist's helmet. At about $25 a session, with sessions typically lasting two hours, Aearo has a go-to list of mostly its own employees, as well as friends of employees and the general public, that it hires to step into this room and be paid to listen. Like them, I'll be responding to a series of test tones, but my ears will be unimpeded, as they are in the wild.

Leading me inside and closing the door behind him, Berger explains, "This is a concrete-block room that is structurally isolated. There's a concrete plank ceiling and fiberglass on top of that. Then, inside this room is a steel wall room which sits on a separate isolated foundation on its own spring system with magnetic seal."

Boop! My sounding of the space delights my ears with the emptiness of it all. Finally, an air-conditioned room without a mechanical decibel side effect. In the center of the dimly lit space there's a plain metal chair with a seat cushion but no armrests. The word *interrogation* pops into my mind. But this will be painless. I'm fitted with headphones that cup my ears and further isolate me from any noise that might make it through all the walls. My hearing goes black. It's like a switch has been thrown. Instantly, I am wide open, searching for even a hint of vibration. None found. I'm handed a custom-built T-handle (ingeniously designed to be silent by dropping my finger across a photo-switch) by an Aearo senior acoustic technician named Ronald Kieper, who closes the door behind him, sealing me off from all sounds except those bound for my headphones.

A soft voice whispers in my ear, "Hello, Gordon."

Will today's measurements show a cost of my journey: temporary threshold shifts, or worse, some permanent hearing loss? Stop thinking, I tell myself. Just listen.

Fed through the headphones, the instructions are really quite simple. From the other side of the wall in the acoustics lab, Kieper tells me to drop my finger into the gap of the T-handle when I first hear a pulsating beat. The operative word is *first*.

We run through a couple of examples at midrange. *Ump-ump-ump-ump*. I'm a little slow on the draw, he tells me. I relax a bit more, wiggle my trigger finger, and empty my mind. Bring 'em on! In my mind, I'm still sheriff at Dodge City, a world-class listener. But there are a lot of bike messenger miles on these ears; before that, a fair amount of boyhood firecracker foolishness; and, no way around it, 54 years of 24/7 on the job for my hearing organs. In other words, my star is likely a bit tarnished, tarnished by *sociocusis* (hearing loss due to living amid the noise of society) and *presbycusis* (hearing loss due to aging). Separating the two can often be tricky.

When the test is over and I rejoin everybody out in the acoustics lab, Kieper hands over the results to Berger.

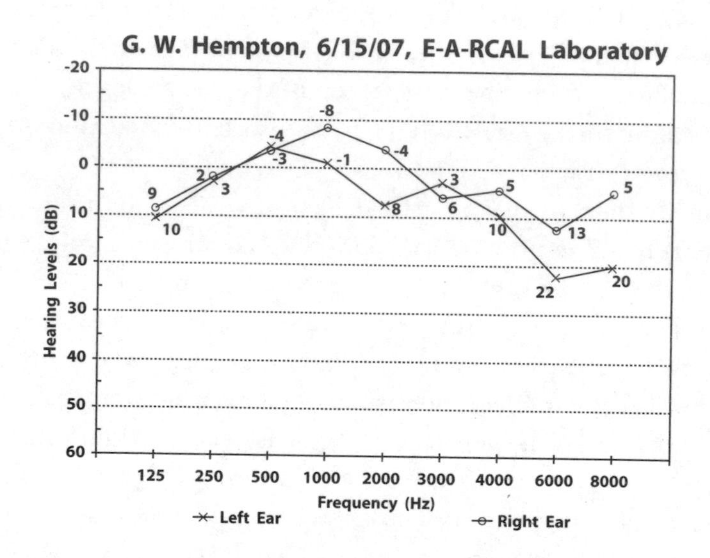

"I'm jealous," says Berger. "Gordon has great hearing. Normal is zero decibels plus or minus ten, and anything within fifteen decibels is considered just fine. The onset of detectable HL [hearing loss] is taken to be about twenty-five decibels. So Gordon is great, with incredible hearing at one kilohertz in the right ear and a tiny bit of what looks like the beginning of age-related hearing loss at six and eight kilohertz, primarily in the left ear." Berger is referring to what I have suspected for years: a decline of high-frequency sensitivity in my left ear between 4 and 6 kHz, typical of many insect sounds. I've noticed recently that I sometimes hear an insect through my right ear but not at all through my left, even when I rotate my head. To compensate for this hearing loss while recording, I sometimes have to glance at the meters on my digital recorder and rotate the microphones until both left and right channels are equal.

Now for the wet willie, epoxy-style. A cool substance with the consistency of thick syrup is injected first into one ear, then the next. This enables Aearo to make a replica of each ear canal that will produce earplugs that will fit me perfectly. In them, acoustic filters will be added to turn down the music. The whole process will take a few weeks, not in time for tonight's outdoor performance of the Indianapolis Symphony Orchestra. But how loud can they be?

As I thank Kieper I can't help but notice a magnetic sticker on an equipment rack that seems meant just for me, but no, he tells me, the sticker has been there for 15 years. It features a quotation from Chief Seattle: "The Earth does not belong to us. We belong to the Earth."

Next Berger takes us into his office for some serious show-and-tell, with my ears again on the receiving end He has me sit down and cock my head slightly to the side. Then he grabs a flashlight-enhanced otoscope and carefully inserts the cone-shaped tip into my left ear. Thanks to the tiny fiber-optic camera on its tip, the fantastic voyage down my auditory canal begins on the screen of his computer monitor.

"And now we're going to start going in," Berger begins to narrate. "This is the front wall of the canal. Those are the hairs. This is your ear canal."

"Those are hairs?" Grossmann wonders about the sparse forest growing on the inner wall of my ear canal. Flecks of road dust are scattered about, adhering to wax.

"We're looking at the front wall of your ear canal," Berger says, and delicately pushes the otoscope further, toward the hidden workings of auditory perception. "So this is your eardrum here." He traces the circular outline on the computer display with his finger. His finger skips over to a bulge on the membrane. "This is actually the first of the three middle ear bones that protrude through the center of the eardrum."

I am fascinated and filled with wonder at what must lie beyond that bulge on my eardrum. At Berger's suggestion we watch an incredible video animation by a Baltimore-based medical illustrator named Brandon Pletsch, whose videos help explain to medical professionals how the body works. His lifelike animation *Auditory Transduction* (available from Radius Medical Animation, www.radiusmedicalanimation.com) shows how sound waves travel through the ear and get converted to electrical impulses that the brain learns to interpret as sound. The video teaches that once sound waves pass down the narrow ear canal and reach the eardrum, or tympanic membrane, the wave energy causes this membrane to vibrate. Low bass sounds cause the membrane to vibrate slowly; higher pitched sounds excite the membrane to vibrate quickly. If the sound is very loud the membrane moves in and out a lot, and if the sound is faint, the membrane hardly moves at all. The membrane of the eardrum is almost always vibrating in complex patterns, similar to a wavy sea, because most sound events are made up of many sounds simultaneously.

A side view of the eardrum reveals that it is cone-shaped and leads to a very small bone, the malleus (that's the bulge that I saw on my eardrum), which then connects to another bone, the incus, in a tiny three-bone chain. The vibrations made on the tympanic membrane are passed along this tiny mechanical system, which amplifies the vibrations and passes them on to a fluid-filled spiral system called the cochlea. Inside the cochlea, thousands of microscopic hair cells convert sound-wave energy into electrical signals. Mammalian hair cells come in two varieties: inner and outer hair cells. The signals generated by the former produce electrical impulses in the auditory nerve that are sent to the brain. Researchers have tracked this process down to the molecular level and have, moreover, clocked the speed of this response at 1,000 times faster than the reaction of similar channels in the eye in response to light. One might win a carefully phrased bet about the speed of sound being faster than the speed of light. Remarkably, the brain also sends messages back to the ear. In fact, there are three times as many receptors listening for brain impulses (outer hair cells) as there are receptors that talk to the brain (inner hair cells). This begs the still unanswered question: What is the brain saying?

On the way to dinner, Grossmann asks Berger about what looks a bit like hearing aids in his ears.

"These," Berger says, with glum resignation, "are tinnitus maskers."

Berger is one of 12 million Americans who suffers from tinnitus, or ringing in the ears. Two in 12 suffer badly enough, according to the American Tinnitus Association, that their daily activities are impaired. Although noise exposure is the leading cause, tinnitus can also be triggered by vascular disorders, allergies, and an ear infection, and it can be a side effect of more than 200 ototoxic (ear-poisoning) substances listed in the *Physician's Desk Reference,* among them ibuprofen and streptomycin.

Berger has suffered with tinnitus for three months. The irony equals if not trumps my previous hearing problem.

"I've lectured on tinnitus," he says.

"I know," I say. "Your little red dot thing." To describe tinnitus, Berger uses a visual analogy of a blind spot, a red ball smack in the middle of your view. No matter which way you look, there's that damn ball!

"That doesn't begin to describe it. Every pain issue I've ever dealt with, whether it's been emotional or physical, I've addressed with meditation or relaxation. Now that doesn't work. Because the tinnitus substantially interferes with quiet relaxation."

Quiet, he explains, is no longer a friend, but an enemy, something to be avoided. For without the competing background conversations of the restaurant we're headed to, or without an equivalent, introduced soundtrack from his tinnitus masking devices, he's at wit's end because of the inescapable ringing.

"At first I couldn't sleep, couldn't concentrate. I couldn't relax. I lost weight. People have committed suicide over tinnitus," he says.

We've reached a perfect quick dinner stop before the concert, an award-winning gumbo restaurant, Yat's, where we follow Berger's lead at the counter and order red beans and rice with mushroom étouffée at the service counter. While we wait for our dishes and names to be called, he continues where we left off.

"The idea of the masker is to kind of rewire the brain, which is plastic and can relearn things. My brain is focused on this tinnitus. Undoubtedly, some of the problem is in my ear, which has been damaged, although my hearing has been tested, of course, and I don't have any additional hearing loss. But part of the problem is in my auditory cortex, so even if I had my ear removed I would probably still have the tinnitus.

"And that's one of the problems," he explains. "The auditory cortex, unlike the visual cortex, is wired to always get stimulation. Primitive humans wouldn't have wanted ear lids, because their ears would warn them if something was after them. The auditory cortex doesn't go to sleep. It always wants some input, so, especially if you have a hearing loss due to noise, there's an area where the ear isn't getting as much input, the brain turns up the gain, and says, 'I'm listening for something. I'm listening for something. Where is it?' That's one of the thoughts why tinnitus is so often an accompaniment to hearing loss."

While we eat, we talk about combat exposure to ear-damaging explosions. Weapons are among the loudest of the loud. Handgun, 22 caliber pistol: 140 dBA; M16 rifle: 155 dBA; machine gun, .05 caliber, M2: 153 dBA; shotgun: 160 dBA; howitzer, 155 mm, Model 198: 178 dBA, loud enough to potentially cause permanent hearing loss from a single event. The cost of hearing loss for all veterans from 1977 to 2005 was nearly $7.5 billion; another $418 million just for tinnitus, according to the Navy Medical Center in Portsmouth, Virginia, which states in the executive summary of *Noise and Military Services: Implications for Hearing Loss and Tinnitus* (2005):

> VA reported that the 2.5 million veterans receiving disability compensation at the end of fiscal year 2003 had approximately 6.8 million separate disabilities related to their military service. Disabilities of the auditory system, including tinnitus and hearing loss, were the third most common type, accounting for nearly 10 percent of the total number of disabilities among these veterans. For the roughly 158,000 veterans who began receiving compensation in 2003, auditory disabilities were the second most common type of disability. These veterans had approximately 75,300 disabilities of the auditory system out of a total of some 485,000 disabilities.

Aearo makes special earplugs for the military. This two-ended design is actually two earplugs in one. With the green end of the Combat Arms Earplug inserted, soldiers receive constant protection and a noise reduction rating of 22 dB. This application works best attenuating constant noise, typically from vehicles or machinery. For combat situations, soldiers switch to the yellow end, which has a noise reduction rating of 0. This permits the wearer to hear a sergeant's orders or the warning of a fellow soldier or enemy sounds, while also providing hearing protection when needed thanks to a patented nonlinear filter in the stem. This filter automatically activates at the discharge of a weapon or an unexpected explosion, suppressing the noise impact to a much safer level.

Berger checks his watch. We scrape our plates and hustle out the door, bound for the Lawn at White River State Park, a downtown venue with an uprolling lawn that forms a natural amphitheater. A middle-aged crowd pours in, buzzing with anticipation. Our tickets say "The Music of Led Zeppelin Performed by the I.S.O." All three of us remember Zeppelin's hard rock from our college days in the seventies, and we look forward to some familiar licks—maybe reinterpreted and a bit mellower. We'll see the performance up close. Berger snagged some choice seats: Row H, right up front, to the right of center stage.

"Rock on through to the other side . . ."

The preconcert taped music measures 79 dBA, a little loud for my ears, but if we lean close we can still converse. Berger has brought along a dosimeter, which measures our noise exposure and records the data for study later. It should give us a good idea if the public concert stays within safe limits.

We applaud as the orchestra members, dressed in white polo shirts and khakis, take the stage first. The conductor is introduced. We quickly realize that this will be a mixed marriage. A rock drummer takes his seat at a drum kit immediately behind the conductor as his band mates, electric guitarists and an appropriately longhaired lead singer, squeeze on stage.

We should have known better and been prepared. In hindsight, the massive tower of speakers at each side of the stage provided two big clues. Like the screech of brakes before an accident, we hear a *Hummmmmmm:* the soundman pushing the volume up, waaaaaay up. Instinctively my hands go up, like the figure in Edvard Munch's famous painting, and my index fingers block my ears, just as a great rock wall of sound crashes down on us like a monster wave. Berger, still wearing the dosimeter microphone clipped to his shoulder, is not so lucky. His ER15s (Musician's Earplugs by Etymotic Research, the same earplugs custom-molded for me earlier today) aren't up to the task. He gets blasted until he's got a set of E-A-R foam plugs in place. Grossmann, too. In a quieter moment, I reach for my own pair. Berger frowns and shakes his head. I agree. We don't even make it through the first song in our seats. We get up, exit our row, and quickly move a safer distance from that bank of speakers.

Berger searches for an usher or event official to see about claiming other seats. Earplugs and expertise on his side, he lands us seats in the corporate VIP section, a tiered platform with beer, wine, and cocktail service that's mercifully a good 50 yards from the stage. Our earplugs stay in, and rightly so. One guitar solo spikes past 100 dBA, one of several instances when the sound level exceeds 100 dBA at our distant location from the stage. Later, the printout from Berger's dosimeter pegs the initial assault on our eardrums at 115 dBA and shows several sound-level readings over 100 dBA—and we spent 99 percent of the concert far from the stage.

Returning after a chat with the sound crew at the mixing boards, located between our new seats and the stage, Berger tells us that they, too, are wearing earplugs. But the five of us may well be the only ones in the sea of spectators wearing them. Until it gets too dark to check, we look and look. We see no one else wearing hearing protection. The World Health Organization recommends that patrons to music events not be exposed to more than 100 dB LAeq, or A-weighted average, during a four-hour period. Our exposure, according to Berger's dosimeter, was 94 dB Leq, nearly all of it far from the stage (see appendix B). Had we remained down front, the exposure would have been much higher.

If a lawn is treated with pesticides, warning signs go up. If a restroom floor is newly mopped, orange cones and a caution about a potentially slippery floor mark the entrance. With no warnings posted here, no signs mentioning the need for hearing protection, no earplugs provided at the gate, there's seemingly no cause for concern, no reason to think that attending an outdoor function in a public place could damage one's health. Yet I have little doubt that along the banks of the White River in Indianapolis, Indiana, on June 15, 2007, on a balmy night with a rosy sunset, there was a taking, without warning, of public health.

What about noise ordinances? Yes, most cities and towns have them, but when it comes to loud music at outdoor venues like this, or inside bars and nightclubs or big arenas, noise ordinances don't even pay lip service to noise-induced hearing loss. They're really nuisance ordinances, written to shield nearby residents from unwanted noise. If concerned cities or club owners wanted to protect music listeners (and more problematically, occupationally exposed employees), a ready remedy awaits. Seven hundred fifty dollars will buy an SLC1 sound-level controller from a company called Gold Line of West Redding, Connecticut, one of many companies that sells equipment that provides a running sound-level record and flashes a warning light when the music exceeds a preset decibel level. If desired, the device can either dial down the volume or shut down the music when the limit is exceeded. Sound-level controllers, however, are orphan products in Gold Line's catalogue of offerings.

President Martin Miller explains that the very few units he does sell don't go to altruistic club owners. They're installed in venues where neighbors have called in the noise police (in the United Kingdom they're called noise wardens) and owners find themselves under a court order to maintain a record of sound levels and not exceed a level that will again bother those living nearby. Or, Miller says, "it's a deejay situation. Often some of the kids who are running these rigs get overenthusiastic, and they really overdrive the amplifiers and system and destroy a lot of the equipment [belonging to club owners], so very often the owners are more concerned about protecting their equipment than anybody's ears."

A longtime noise control official in a big city environmental protection department confirms that his crew's attention to loud music is entirely driven by the complaints of neighbors. Any citations issued pertain to excessive sound levels next door or across the street, not to much higher levels blasting the ears of patrons only a few feet from the club's speakers. "If you buy a ticket to a venue, what we call the common law, . . . you assume the risk of whatever the volume is."

"Really?"

"Yes," he replies. "Noise is defined as unwanted sound. So it's not noise when you go and buy a ticket to your favorite musical event."

Because the effects of noise exposure are cumulative, the real and lasting effects of tonight's orchestral-enhanced rock concert (and others) may not be known for years. How can the City of Indianapolis or any event sponsor or music venue absolve themselves of this responsibility? Hearing remains a blind spot, if you will, of public concern.

There's no question that for some, loud music is pleasurable. In my younger days, I turned up the rock on my dorm room stereo and even opened my windows so everyone else could enjoy it too. (So, Abby, I understand and sympathize with your iPod listening behavior, but still fear for you.) Since then, scientists have learned that an organ near the ear known as the sacculus, responsible for making the body aware of its position, is stimulated by loud music. It also turns on the pleasure center of the human brain—in some individuals, apparently quite powerfully. Habitual listeners to loud music, when deprived of their decibel fix, show withdrawal symptoms similar to those of addicts, according to Mary Florentine at the Institute for Hearing, Speech and Language at Northeastern University in Boston.

One person's pleasure might be another person's pain, but the fact remains that for both, high sound level experiences can produce a hearing loss, which, in fact, boils down to hair loss. According to James F. Battey Jr., MD, PhD, director of the National Institute on Deafness and Other Communication Disorders:

> Scientists once believed that NIHL [noise-induced hearing loss] damaged the hair cells [in the ear] by the pure force of the loud sound vibrations. In that case, the only means of prevention was to reduce the sound exposure and/or use ear protectors. Recent studies, however, have found that noise exposure triggers the formation of molecules (free radicals) known to cause hair cell death.
>
> —from a February 14, 2006 letter to Massachusetts Congressman Edward J. Markey, responding to concern about hearing loss caused by MP3 players

Just as excessive alcohol consumption has produced a number of hangover remedies that include apples, bananas, and Vitamins A and C to replenish a poisoned body, so-called *bangover* remedies are beginning to appear. At the website of a San Diego pharmaceutical company called American BioHealth Group it is now possible to order the Hearing Pill. For $44.95 you get 95 pills, which the company claims reduce the amount of destructive toxins in the cochlea and work to keep hair cells healthy. American BioHealth states openly that although the product's active ingredient, n-acetylcysteine, is FDA-approved for other medical uses, the FDA does not yet support the company's hearing protection claims. Still, the field has attracted the interest of researchers.

"On the horizon is a nutrient bar to fight off ear damage," Dr. Josef Miller, director of the Cochlear Signaling and Tissue Engineering Laboratory of the Kresge Hearing Research Institute at the University of Michigan, told *US News & World Report,* which wrote in its July 16, 2007 issue, "He and others have shown that a combination of the antioxidant vitamins A, C, and E and magnesium not only protects the inner ear when taken before noise exposure but can limit damage for up to 72 hours after the insult."

The next morning Berger takes us to a local institution and, for millions of race fans, a national shrine: the Indianapolis Motor Speedway, host on Memorial Weekend of the Indy 500 and crowds exceeding 300,000. Today there will be Formula One time trials and considerably smaller crowds. Neither Grossmann nor I have ever been to an auto race of any kind, so out of curiosity we agreed to go. We want to learn how loud it is in the stands and see how many people wear hearing protection. We've got earplugs—and sunscreen and hats. It's going to be a scorcher.

We feel the buzz about a mile from the racetrack. I haven't seen a crowd like this since 1969, when I was swept up in the giddy tide to Woodstock. RVs and party tents have staked claims in some of the larger lots. As we drive closer, I'm surprised to see that the Speedway is located in a residential neighborhood. You'd better be a race fan to live here, or at least enjoy the company of race fans. Homeowners fill their front lawn (and on prime corner lots, their side lawn as well) bumper-to-bumper and nearly door-to-door with cars. Twenty dollars seems to be the going rate. Berger turns in at one house a few blocks short of the racetrack, claiming a spot with curb egress to enable an early getaway. The guy who holds out a hand for our money stands next to a sign advertising earplugs. Berger takes the lead by asking about them.

"A lot of people don't have earplugs. I sell them for two bucks," he says. "I get them from a company in New York. They're OSHA-approved. They cover you up to one hundred twenty-five decibels. Formula One runs at one hundred seventy-five, so you've got three-quarters protection."

Berger asks to see a package, recognizing them as made by a competitor in California.

"One weekend I sold two hundred sets in about forty-five minutes," the man adds.

That may be accurate, but very little else of what he said is on the mark. Closer to the racetrack Berger sets us straight. On race day, with a full complement of cars, sound levels peak not at 175 decibels, but at around 125 to 128, he tells us. "The loudest place on earth where people work, the deck of an aircraft carrier, is about one hundred fifty decibels. As for the plugs, you don't talk about 'what plugs cover up to.' The rating level on those plugs is thirty-three, so nominally, if your safe level is eighty-five, they would protect to one hundred eighteen decibels. Again, that's for an eight-hour exposure, so they're certainly adequate for your time at the race." Bottom line: misinformation about decibels. No surprise there, but nothing wrong with offering hearing protection for sale. I see others selling earplugs on the walk to the Speedway. Maybe this bodes well for protected ears at trackside.

We buy general admission tickets and walk for blocks and blocks along the outside of the stands before entering the Speedway. Grossmann and I take seats in Row R, 18 rows off the track, in a bit of welcome shade beneath an upper deck of seating. Berger has called a friend on his cell phone and headed off to join him in the upper deck. It's 12:40 p.m. The Beatles' "Twist and Shout" plays on the sound system. My sound-level meter registers 85 dBA. Our earplugs are in.

A nearby spectator fires an air horn and the PA blares. The announcer says something, but with all the other noise, I can't detect much more than his British accent.

I hear a sudden burst of sound and a four-wheeled rocket emerges from the pits, turns sharply to the right, and shoots right past us. I have never seen a land object move so fast—or make so much noise! A glance at my sound-level meter shows 102 dBA. Moments later another car emerges. Then another. The three circle the track like angry hornets, passing by in a dizzy blur and a whine of noise (111, 112, and 115 dBA) that's apparently music to the ears of those gathered all around us. Well, almost everyone. I see a woman to my right with her fingers in her ears. A quick canvass of all those within eyeshot brings this earplug scorecard: 16 with earplugs or earmuffs (some of which offer protection and the ability to listen to special race channels); 31 barc-eared, clearly risking noise-induced hearing loss.

Unprotected ears are just as prevalent elsewhere in the stands. Berger sees about four of every 10 spectators without earplugs or muffs in the upper deck by the turn before the long straightaway. And the automotive roar is even louder there, under the roof. He gets readings as high as 126 dBA, approaching the threshold of pain. Later, when we're back together, Berger says above the noise that he wishes he'd brought along a camera to document the lack of hearing protection. I offer him mine. In one of his photos, 15 ears are visible—not one has ear protection. Other photos show folks with foam earplugs sticking out of their ears about as far as the knobs on Frankenstein's neck. In other words, they've not rolled them tight before inserting them, not properly sealed off their ear canals, and thus have achieved only a minimal reduction in decibels. But in the plus column, bravo for one set of parents (or grandparents): a baby in a carriage lies sleeping—amid all this hair-cell carnage—wearing the kind of headphone-style protection worn at shooting ranges.

By 2 p.m. we've made our way via a tunnel beneath the track to the infield and a popular vantage point by two of the hairpin turns of this grand prix track. When a helicopter passes overhead, adding to the din, I decide I've seen and heard enough. As soon as the helicopter is far enough away to make my words intelligible I say to my host, "You want to go find Muir?"

Berger knows of my "thing" for John Muir. So he's happy to go a bit out of the way on the drive back to his house so I can check out a historical marker I've read about, a marker erected in downtown Indianapolis by the Hoosier chapter of the Sierra Club. Yes, the Sierra Club. We discover the marker at the confluence of South Illinois and West Merrill Streets at Russell Avenue. It stands in a triangle of shabby grass near a parking garage and a major post office distribution center, and virtually in the shadow of the under-construction RCA Dome football stadium. Pretty far removed from a natural setting.

Nor was it pristine or quiet nearly a century and a half ago, when John Muir took a job at the Osgood, Smith & Company carriage factory that occupied this spot. He had come to Indianapolis in 1866 after studying botany at the University of Wisconsin and later wandering Canada and collecting plants, some say draft-dodging the Civil War. The carriage factory was powered by revolving belts, and these belts were joined by lacing. Muir was tightening some of these laces, using a file, when he lost his grip and the file pierced his right eye, draining the vitreous fluid into the cup of his hand. That proved one of the last images that he would see for some time. His left eye went blind by sympathetic reaction. In the ensuing darkness and relative quiet away from the noise of the factory, Muir made a promise to God, vowing, should his eyesight ever return, to devote his life to the inventions of God and not the inventions of mankind.

Muir did regain his sight in both eyes, and as it says here on the historical marker, he "left Indianapolis on September 1, 1867, to begin extensive travels, which ended in California in March 1868." During those travels, immersing himself in natural settings, he did, indeed, make good on his promise. He traveled first by rail to Jeffersonville and then by ferry to Louisville, Kentucky. After that he was on foot. "My plan was to push on in a general southward direction by the wildest, leafiest, and least trodden way I could find, promising the greatest extent of virgin forests," he wrote. His destination was the Gulf of Mexico, about 1,000 miles away. I suspect that his vision was still impaired, his eyes still healing, and that he was therefore especially aware of sounds and reliant on his ears—not just for his wilderness education, but for his very survival. The man who would found the Sierra Club and who would later become known as the father of our national parks wrote this in *A Thousand-Mile Walk to the Gulf,* his account of his inaugural ramble:

> *September 12.* Awoke drenched with mountain mist, which made a grand show, as it moved away before the hot sun. Passed Montgomery, a shabby village at the head of the east slope of the Cumberland Mountains. Obtained breakfast in a clean house and began the descent of the mountains. Obtained fine views of a wide, open country, and distant flanking ridges and spurs. Crossed a wide cool stream, a branch of the Clinch River. There is nothing more eloquent in Nature than a mountain stream, and this is the first I ever saw. Its banks are luxuriantly peopled with rare and lovely flowers and overarching trees, making one of Nature's coolest and most hospitable places. Every tree, every flower, every ripple and eddy of this lovely stream seemed solemnly to feel the presence of the great Creator. Lingered in this sanctuary a long time thanking the Lord with all my heart for his goodness in allowing me to enter and enjoy it.

"There is nothing more eloquent in Nature than a mountain stream, and this is the first I ever saw." Since I first read that line 15 years ago, it has haunted me. Like an FDR buff eager to tour the Hudson River home of the 32nd president for insights into his life and character, I've long wanted to find Muir's "first" mountain stream, to sit beside this cool waterway and listen as he did. Now, bound for Washington, D.C., I've got time for one last side trip.

Berger, whom I've known for years, shares my enthusiasm for that mission. But I detect a different tone in the pauses between the first two words of a question that's obviously been on his mind for the past two days: "So, Gordon, tell me honestly. After everything that you know, do you really think that you can make a difference in Washington? Do you really expect to be able to talk to the right people and have an impact?"

Berger and I first met when, unable to attend my Joy of Listening class at Olympic Park Institute, he hired me to give him a private tutorial at Rialto Beach, and we've been friends ever since. We've hiked a five-mile chunk of the Washington coastline together. He's been with me to One Square Inch. On many topics, our thoughts travel the same path. I hear the doubt, or at least the concern of a friend, in his voice. I feel my own doubts. They've been rattling in my brain the entire trip, like all the *clinks, clunks,* and *tings* inside the Vee-Dub.

"I don't know if I can make a difference," I reply. "But I do know that I can try."

I have learned, probably in the Hoh Valley more than anywhere else, that it is not a matter of choosing between what is doable and what is not when making a stand. One chooses between what is right and what is wrong. And when you're alone in nature the difference between right and wrong sharpens, often becomes blatantly obvious. Silence gives me the strength to be a good Earth citizen. I know that it is right to save silence. I do not know if silence can be saved. Saving silence will take many voices.

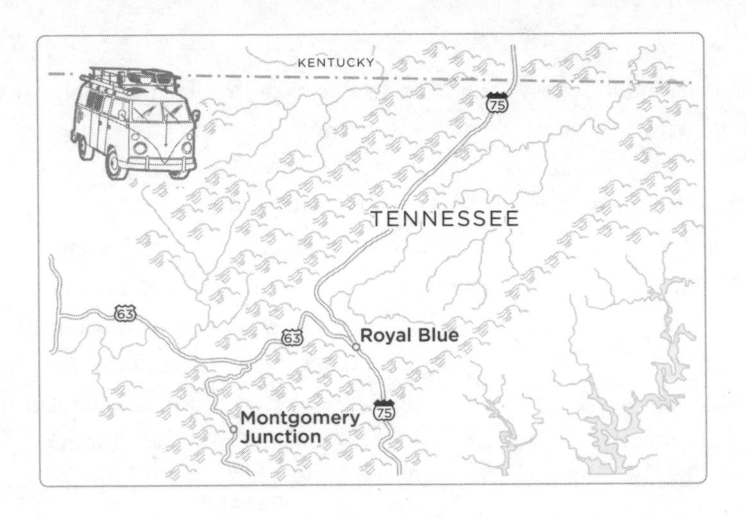

10 Seeking Muir's Music

> There is nothing more eloquent in nature than a mountain stream, and this is the first I ever saw.
>
> —John Muir, *A Thousand-Mile Walk to the Gulf*

Back on the road again, alone and happy: happy to be in Tennessee, enjoying the first mountains since Colorado, and happy to be on a long-overdue hunt. My old Vee-Dub is my Tennessee hound sniffing out a little peace and quiet. I can't wait to get out of this rig! Can't wait for my feet to touch the ground and travel as John Muir did through the state, perhaps walking the same ground as he did and listening to his cool mountain stream, "the most eloquent voice of Nature," which he wrote about in 1867.

Muir became my much-needed mentor after I cautiously approached his writings from the second half of the 1800s. I feared that this legendary figure, touted as the father of our national parks, might be something of a historical figurehead, an embellishment of Madison Avenue on behalf of the National Park Service. But quickly I fell under the spell of Muir's words:

> I drifted on through the midst of this passionate music and motion, across many a glen, from ridge to ridge; often halting in the lee of a rock for shelter or to gaze and listen. Even when the grand anthem had swelled to its highest pitch, I could distinctly hear the varying tones of individual trees—spruce, and fir, and pine, and leafless oak. . . . Each was expressing itself in its own way—singing its own song, and making its own peculiar textures. . . . The profound bass of the naked branches and boles booming like waterfalls; the quick, tense vibrations of the pine-needles, now rising to a shrill, whistling hiss, now falling to a silky murmur; the rustling of laurel groves in the dells, and the keen metallic click of leaf on leaf—all this was heard in easy analysis when the attention was calmly bent.

I was hooked. Muir, clearly, was a nature sound recordist using the available technology. His listening prowess and ability to capture the varied range of nature's symphony astounded me. I devoured his *Eight Wilderness Discovery Books* and systematically entered each nature sound description into a searchable database, then analyzed his perspective, hoping to identify his favorite subjects and listening haunts. Determined to close the gap between his words and my understanding, I set out to literally follow in his footsteps. I walked to Yosemite "on or about the first of April," grew a beard, even approximated Muir's wilderness vegetarian diet, trying to get as far into the spirit of Muir as I could, so he could teach me to become a better nature listener. He describes valleys and rivers as musical instruments and the sounds of nature as music, sometimes with the granite boulder notes rearranged by earthquakes. It was Muir who truly opened my ears to listening to nature as music.

> The air is music the wing forsakes. All things move in music and write it. The mouse, lizard, and the grasshopper sing together on the Turlock sands, sing with the morning stars.

The sound he most often describes is water. Here Muir describes the 2,425-foot-high cascade of Yosemite Falls:

> This noble fall has far the richest, as well as the most powerful, voice of all the falls of the Valley, its tones vary from the sharp hiss and rustle of the wind in the glossy leaves of the live-oaks and the soft, sifting, hushing tones of the pines, to the loudest rush and roar of storm winds and thunder among the crags of the summit peaks. The low bass, booming reverberating tones, heard under favorable circumstances five or six miles away, are formed by the dashing and exploding of heavy masses mixed with air upon two projecting ledges on the face of the cliff, the one on which we are standing and another about 200 feet above it. The torrent of massive comets is continuous at time of high water, while the explosive, booming notes are wildly intermittent, because, unless influenced by the wind, most of the heavier masses shoot out from the face of the precipice, and pass the ledges upon which at other times they are exploded. Occasionally the whole fall is swayed away from the front of the cliff, then suddenly dashed flat against it, or vibrated from side to side like a pendulum, giving rise to endless variety of forms and sounds.

Muir's water listening started at a cool mountain stream outside of Montgomery, Tennessee. As far as I'm aware, no one has knowingly listened there, as Muir did with attention calmly bent, or with Muir's words echoing inwardly since 1867. I'm going to find that stream, Sound Tracker–style.

The next morning in the orange glow of a beautiful sunrise, I detect a familiar, spicy, herbal scent in the air. It's . . . it's . . . just like the wild, deciduous tang of the woodlands I roamed as a boy, when I would set off ever farther from my home in Potomac, Maryland.

About 38 miles outside of Knoxville, I take the Route 63 exit and find myself disgorged into a commercial olio: Comfort Inn, Stuckey's, an As Seen On TV Outlet, and a huge fireworks store called Titan. State Route 63 West winds with the hilly terrain, offering scenic but discouraging valley views: almost every major valley has a road cutting through it; the hills will echo traffic. I realize that I missed my intended turnoff when I pass the Buffalo River crossing, so I pull a U-ey, and this time take the turn onto Norma Road. My heart is beating fast. There's a river far below and a set of railroad tracks that follow it. The tracks look shiny. Probably still in use. Eyes back on the narrow, potholed asphalt road. It is time for me to calm down, look, and listen to where I am. I pull onto the shoulder next to an empty, rusty, log-hauling trailer parked in the cool shade of an oak tree.

Just off the passenger side of the Vee-Dub a wooded hillside drops steeply toward a river. Stepping closer to the edge, I spy all manner of garbage that has been chucked from this very spot: plastic bottles, paint cans, Styrofoam cups, tires, artificial Christmas garlands. Then my eye stops on some books with tan covers and dark green and gold bands on the spines. Volumes of the World Book Encyclopedia, exactly like the ones that I used to have as a kid! In fifth and sixth grade at Lake Hills Elementary in Bellevue, Washington, I loved poring over our set at home, page by page, until I got my fill.

Easing my way down the steep bank, carefully avoiding the prevalent poison ivy, I open the first volume I come to. Volume 16, a fat one dedicated to the letter S. I note a copyright date of 1961 and start peeling apart the pages. *Saliva, salmon, Siberian husky . . . silence, Tower of Silence (see Tower of Silence)*. Darn! I can't find the T's! I keep turning until I come to *Sound*.

> *Kinds of Sound. Noise.* Noise can harm people in two ways. Intense noise can produce actual deafness. Boilermakers, steelworkers and others who are exposed to intense noise for long periods sometimes become deaf. Jet airliners often make such loud disturbing sounds that they are not allowed to use some airports. Continuous or periodic noise may cause people to become tired or irritable even if it is not exceedingly loud. The steady whine of a saw or the periodic ringing of a telephone can cut a worker's production in half. Builders often cover the inside walls of offices and factories with felt, cork, or other materials that absorb sound. This reduces the noise and improves the efficiency of the workers. *Musical Sounds.* Musical sounds are made by three types of instruments, stringed, wind, and percussion.

I think about Muir's music back in the late 1800s and recall his words:

> As long as I live, I'll hear waterfalls and birds and wind sing. I'll interpret the rocks, learn the language of flood, storm, avalanche. I'll acquaint myself with the glaciers and wild gardens, and get as near the heart of the world as I can.

I'm jolted back to my present location by a loud rumble on the road above me, a much louder roar than any I've heard so far. Here's the E volume. Let's see what the folks at World Book had to say about ecology in the sixties.

> *Ecology* is the branch of biology that deals with the relationships living things have to each other and to their environment (surroundings). Scientists who specialize in studying these relationships are called *ecologists*. No living thing, plant or animal, lives alone. Everything depends in some way upon certain other living and non-living things. Animals and plants that live in the same area or *community* depend on each other in some way. . . . The study of ecology increases man's understanding of the world and all its creatures. This is important because man's survival and well-being depend on relationships that exist on a worldwide basis.

Climbing back to the Vee-Dub, I can still hear the distant echo of that giant rumble. Before it dissipates, another roar starts to build, and then I see it pass by: a coal hauler. When I do find Muir's cool mountain stream, I'll have to wait for nightfall, past quitting time for the coal truck drivers, to have much hope of quiet.

Back behind the wheel, road signs like "Caution: Road Broken Off" provide a bit of encouragement, reminding me I'm heading deeper into backcountry hollows. Here's a turnoff for Swinging Bridge Road. Surely Muir would have made a right here to cool off. No swinging bridge these days, but I'm rewarded with a Norman Rockwell view of two children taking turns on a rope swing at a riverbank swimming hole. They up the ante on their dismounts as soon as they see my telephoto lens pointed their way. Big country grin, then *Ka-splash!*

Norma Road downgrades to gravel with asphalt covering the more precarious stretches and is now barely wide enough to accommodate the huge trucks carrying coal, let alone my Vee-Dub in the opposing "lane." Each truck heading toward me passes within inches of my door and kicks up billowing clouds of dust that adheres to my sweat and coats everything inside the van. A water truck comes along spraying down the dust, but the water quickly evaporates, and within a few minutes the dust is flying again. A modern-day coal country Sisyphus, that truck. This road might be a country road in name and locale, but it carries industrial traffic.

I know that I have gone too far when I see a sign announcing Cumberland Trail. I must have driven right through Montgomery without even noticing it. Well, now it's my turn to cool off in the river and rinse off the road dust. The wide, slow-moving New River runs clear over rock bordered by tall majestic walls of hardwood, making a wonderful natural amphitheater. My cool immersion into the deeper pools provides an eye-level view of the river highway used by insects, occasionally rippled by the splashes of feeding bass, while truck traffic echoes off the leafy canyons. I spot small, iridescent puddles of oil among the riverbank pebbles.

After drip-drying, I study my topographic map more closely and get a pretty good feel for where I need to go now: up a side valley that points east up the Montgomery Fork toward Roosting Ear Spring. Along the way a "Wildlife Management Area" sign offers encouragement.

This protected side valley spreads out into a number of hollows, nice little amphitheaters completely surrounded by a mix of oak, hickory, and maple. And coursing down the center of the valley is a little mountain stream. This must be it! If it's not Muir's cool Tennessee mountain stream, it's surely kin.

I park on high ground so the *Pops* and *Clicks* of my cooling engine won't interfere with listening, shoulder my recording gear, and set out on foot down the dust-covered road that leads toward the calling voice of a mountain stream. The overhanging foliage offers glimpses of pools and assorted stones, flickering the last rays of sunlight. Within minutes I come upon a nice little rill, and before I can settle into it, a wood thrush sings its harp-like stringed song as a final blessing for this moment. I freeze, afraid I might disturb this shy, robin-size bird, then quietly place my tripod into the shallow stream and push "Record."

But here, too, I'm thwarted. A truck operated by the National Coal Corporation drives in, parks about 100 yards upstream from where I'm crouched, and starts sucking water like a thirsty elephant. So this is where the road-spraying trucks get their water. The thirsty beast fills this little valley with a piercing, mechanical whine. My sound-level meter shoots up to 71 dBA—in a wildlife management area, no less!

I abandon any thought of recording and walk over to talk with the driver. The name on his shirt is Randy. He tells me he takes 4,000 gallons at a time. Works till 3 a.m., then somebody else takes over for the next 12-hour shift.

In addition to keeping down the road dust, the water's also taken to replenish the coal mine's pond.

"No water, no coal," he says. He eyes my clothes. Sweat-soaked tan T-shirt, sun-bleached khaki hunting vest, canvas shorts, and Tevas. "What y'all doin'? UPSin'?" Randy must think I'm a UPS driver in hill-country uniform.

I tell him I was trying to record the birds.

Each fill-up, he informs me, takes about 15 minutes. He makes seven to 12 runs a night. This one's bound for the intake pond at the coal mine, which has been running dry. "And they can't run coal without water. Little shars, one or two a week ain't helpin' any. The creek usually ain't this low until August or September." I deem that fortuitous recording news because this should make a better match with Muir's September 12 visit.

Between the noise and his accent, I'm often a sentence behind parsing the meaning of his words. Showers, I realize, just as Randy asks if I'm recording what birds come down the creek.

"No, I'm listening for sounds that John Muir wrote about in his journal back in the 1860s, so it's kind of a historical trip for me."

"I reckon."

"So," I ask, "how far is that National Coal mine from here?"

"I'd say about three and a half, four miles at the most."

Well, that's the last straw. Even if I were to try to record in the brief lull of Randy's shift change, that lull would only reveal another layer of noise: the deep, earth-hugging drone of man's dependence on energy. Same problem as at the old Worf homestead in Montana.

"Do you know where the town of Montgomery is?"

"Right here. You're at it. This is what's called Montgomery Creek. The little creek that runs right here under this trestle. That's Roach's Creek. It's a pretty creek. It's beautiful up there."

We see the water spilling over the top of the truck. "I'm over," Randy says and scoots over to shut off the pump motor.

I wave goodbye as his truck warns *beep, beep, beep, beep* in reverse before he turns to head back to the mine.

At Randy's suggestion, I'm walking Roach's Creek up the valley, imagining I'm in Muir's footsteps. Walking in the stream itself, I'm able to hear its changing voice. No two rocks are the same. No two flows are the same. At every step, a new combination, new notes. In a different time, Muir's time, for instance, I could have lost myself in these subtle variations, but with the distant din of National Coal Corporation Mine No. 11 in the background, it's impossible to bend a calm ear. "Nothing more eloquent in Nature than a mountain stream," Muir wrote in his journal very near to where I'm standing. What would he write today if he were charged with the Wikipedia entry on ecology?

I do know that Muir wrote this: "How infinitely superior to our physical senses are those of the mind!" I hope he's right, for we will need our wild intelligence, born of sensory input, to make it through the next millennium.

It is time to walk out before the day's down to starlight, which won't be enough to safely negotiate these slippery rocks. After reaching the Vee-Dub and waiting for my tea water to boil, I take encouragement from Muir's words to his wife, Louie, in July 1888: "The morning stars still sing together, and the world, not yet half made, becomes more beautiful every day."

Yes, that is what is in my heart, too. I'm now more certain than ever that I must continue on to Washington. The world is in a process of becoming; it is "not yet half made." The crescendo of noise pollution is a call to action.

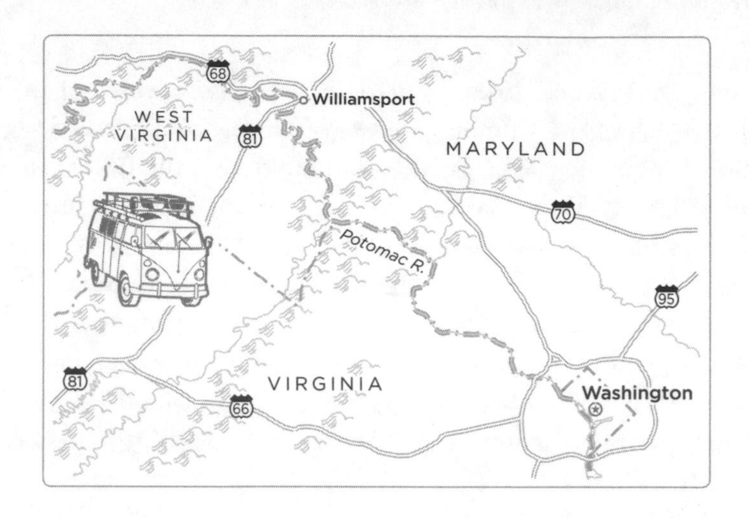

11 Hundred-Mile Walk to Washington

Only in quiet waters do things mirror themselves undistorted. Only in a quiet mind is adequate perception of the world.

—Hans Margolius, philosopher

Just outside the town of Williamsport, Maryland, alongside an eight-foot-wide gravel path, stands the 100-mile marker of the C&O Canal. Williamsport (established 1787, population 1,868) is a quaint, somewhat quiet town made up of colonial brick and stone homes with white picket fences and large spreading oak trees nestled on the shore of the Potomac River. In 1850 the town was flourishing along a new commercial artery. The Chesapeake and Ohio Canal, linking Washington, D.C., and Cumberland, Maryland, brought canal boats loaded with tobacco, grains, whiskey, furs, and timber towed by horse or mule along a well-maintained path that paralleled the canal. The C&O Canal shut down in 1924, and Williamsport slumbered.

Thirty-seven years later, in 1961, primarily through the efforts of

Supreme Court Justice William O. Douglas, the entire length of the C&O Canal, 184 miles from Cumberland to Washington, D.C., was designated a National Monument by legislative action. In 1971 the C&O Canal was renamed the C&O Canal National Historic Park, managed by the National Park Service.

Now that I'm within striking distance of the nation's capital and will soon, I hope, be meeting with federal officials, including the director of the Park Service, I want to walk the final 100 miles. Traveling in the mode of Muir, I'll attempt to gather my thoughts and "walk it off," as Doug Peacock might say. My 10,000-mile pinball route across country in the Vee Dub will switch to a shaded footpath that Justice Douglas extolled as "a refuge, a place of retreat, a long stretch of quiet and peace at the Capital's back door."

Nearer to Washington, I'll reach familiar ground. I spent my high school years in Potomac, Maryland, attending Winston Churchill High and running wild along the path and in the river with my buddies, John Jacobsen and Scooter James. We camped, too, sometimes for days, cooking pork and beans and brewing coffee over a campfire, or draining a six-pack. Some mornings we awoke with frost on our flannel sleeping bags. We got to know some of the largest sycamore trees by climbing high up in their branches. There, we'd see collections of pawpaw seeds and meet, if we were quiet enough, perhaps the very squirrel that gathered them. We imagined great things would come from these footloose times, those last precious moments of youth. Scooter's draft lottery number was 1. He joined the Army right away. John and I got higher numbers. After graduation I went off to college and John worked construction.

Now I'm back near the river of my youth, in the parking lot of the Red Roof Inn on East Potomac Street in Williamsport, a modern-day way station for canal travelers. Lots of folks mill about in outdoor gear. A few pour water over their heads to cool off on this hot, humid day. Everyone is burdened with backpacks or bike helmets. A dozen or more mountain bikes, loaded to the gills, lean against the motel.

The woman behind the reception desk hands me a set of room keys and offers to let me park the Vee-Dub here for as long as it takes for me to reach Washington on foot. At dinner I carbo-load—pizza and beer—for the miles ahead.

The next morning, preparing for my trek on the towpath, I spread my professional sound gear on the lawn next to the Vee-Dub beside my camouflaged backpack, water jug, food, and umbrella, piquing the curiosity of another early riser, who introduces himself as Bob. After I explain a bit about my profession, Bob tells me a bit about himself. He's 68, lives in Evansville, Indiana, calls himself a hunter and fisherman. Figuring he spends a good bit of time outdoors and quiet, I ask if he's noticed any changes in the environment. He mentions the country's recent swing to ethanol production. "The greatest change we're going through in the Midwest right now," he says, "is the ethanol program. The farmers are buying back their conservation reserve contracts. Ground that's been fallow for ten years, they're planting. Everywhere they can, which is going to basically destroy a lot of songbirds. Bobwhite quail are gone, practically. Out West, where I hunt, in Wyoming and Montana, global warming has caused the game birds to disappear the last five years. I think they're in the twelfth year of a drought now. Ranchers can't raise their silage crops. The environment is changing at such a fast rate it's scary."

"When you're out there hunting, have you heard any difference?"

"You've got seven species of game birds in that area," he says. "They're all gone."

"The sound has changed?"

"Yes, and it's changed completely, and it's scary. Like at home," Bob says. "You just don't hear what you used to hear years ago. There's no longer any frogs in our lake. You don't hear them croaking at night. Used to be thousands of them."

I say that sound is our warning signal. Bob mentions things changing "about the time they invented [the weed killer]Roundup." He's just getting started.

"It's a completely different world that's coming on fast," he continues, "and it's scary. It really is. Because once you see this happening, eventually it's going to reach man—and I think it has already with all the cancers. We need to make some changes, but unfortunately, money and industry speaks louder." He continues for a bit, then shakes the doom and gloom like a dog drying off. "It can change. There's no reason it can't change. Just the other day, Tyson announced they were no longer putting antibiotics in their chickens."

With the last gulp of my coffee and final door lock check of the Vee-Dub, I sling my pack onto my shoulders and head down the hill toward the canal. I'm wearing a lightweight long-sleeve shirt, a swimsuit, and Teva sandals. I expect that the 80-degree temperatures forecast for today and the unaccustomed miles will cause my feet to expand; Tevas give them plenty of room, a trick I learned (painfully) on my 255-mile walkabout from San Francisco to Yosemite about 15 years ago.

The C&O Canal National Historic Park interpretive sign offers this greeting:

> Look around you. The park you stand in exists because people cared. In January 1954, Justice William O. Douglas of the Supreme Court of the United States responded to a Washington Post editorial recommending that the C&O Canal be turned into a parkway. Writing in support of preserving the canal as a national park, Douglas wrote, "It is a sanctuary for everyone who loves woods—a sanctuary that would be utterly destroyed by a fine two lane highway."

The sign goes on to tell of a walk made by Douglas on March 22, 1954, with naturalists who interpreted the natural history of the canal for editors and journalists who also made the journey on foot. That walk is credited for generating the public support that favored a national park over a highway.

Doesn't that go to show! A place, and an idea born of a place, can be a powerful thing—if we take the time to listen. I think Douglas exercised supreme wisdom by informing and framing the key land-use decision by listening to the facts along what was then "a long stretch of quiet and peace at the Capital's back door."

After only minutes on the towpath, I'm all alone. I can hear the *Cheeseburger, cheeseburger* of a Carolina wren and *Thew-thew-thew* of a cardinal echo through the hardwood forests that arch over both sides of the canal. A black mulberry tree droops with sweet berries, similar to my northwestern blackberries in appearance only, for the fruit is sweet, not tart. The tree, I'm happy to discover, is thornless, making for easy pickins and a refreshing snack.

Rrrrrrrrrrrrrrrr. An outboard motor churns the Potomac. (They are banned from the canal itself.) And I hear a *Hummmm-boom-hummmm-boom-bop,* what sounds like traffic on a nearby highway bridge. I hope these 54 dBA and other noise intrusions will disappear as I put down the miles and leave the town farther behind me. But through a wide break in the trees I see a tower of power and noise, a coal-fired power plant: several buildings and piles of coal and conveyor belts. Here's the sound I took for bridge traffic. And now comes the *Beep, beep, beep* of a backup alarm. Then a loudspeaker announces someone's name. The echo resounds: 65 dBA at a distance of 400 feet.

At Lock 44 a gate across the towpath bears a sign posted in large block letters for bicyclists:

SOUND PRODUCING DEVICES (BELLS/HORN)
REQUIRED, MUST BE SOUNDED WHEN
APPROACHING WITHIN 100 FEET

Another impending noise. I hope bicyclists will instead deploy common sense and slow down when passing me. I could have been speeding along on two wheels myself. My mountain bike from my messenger days still has wire cage baskets and enough bungee cords to carry just about anything. But going it on foot is a fitting conclusion to this journey. I started on foot in the Hoh Valley, and I will end on foot. My parents named me well: my middle name is Walker.

An hour on the towpath finds me inside an arbor tunnel created by the lush forest. It's quiet enough to hear my footsteps and enjoy the birdsong. I get a sound-level reading of 43 dBA beside the three-foot-high reddish-brown concrete pillar that announces Mile 98.

Two hikers approach, holding a conversation. Four bikers pass, none of them obeying the warning alarm rule, thank God! The next two miles show no evidence of human activity except this towpath, at least not visually. Motor noise often dominates the background, but it's still quiet enough that my crunchy footsteps alarm a turtle, who slips into the canal at my approach. At 9:05 a.m. a fixed-wing plane, the third of the morning, passes with a mild drone of 48 dBA.

Rampant, glorious birdsong, echoing through this forested corridor, stops me. I get a reading of 39 dBA as I listen to the first nature-dominated soundscape since Tennessee. A great horned owl departs from its perch, wings flapping silently, apparently anxious at my presence. *Twip,* a twig falls to the forest floor.

"Morning." A duet greeting from a husband and wife I met at the Red Roof Inn. They ride past me on a tandem: 71 dBA from the gravel spit out from knobby tires. The smell of mosquito repellent (Cutter's) lingers long after the noise fades. That's a good sign: the air is stable enough to hold scents in place. Stable air makes for excellent listening.

The overarching hardwood forest displays cathedral-like acoustics, not unlike the National Cathedral, which I hope to visit once I get to Washington. I believe that these original, natural cathedrals—places of worship, awe, respect, submission to higher powers—still hold great significance.

Pausing at mile marker 96, I hear a faint *Tweeeeeeeee,* then a pause, then *Tweeeeeeeee:* an unseen insect hidden beneath a mitten-shaped sassafras leaf. *Chew-chew-chew.* A cardinal blasts his sonic laser down the shady tunnel with a *Star Wars*–like effect: 46 dBA.

Whew-o-wee. Who-o-who. The thrush! But before I can measure its song, a prop plane overrides this communion with 44 dBA of airborne graffiti. Through a gap in the trees I see a fisherman floating in his anchored johnboat on the Potomac, apparently content just being still, watching the river pass by.

At age eight I held my first fishing rod and learned the art of still fishing—no small challenge for a kid reared on Sugar Pops. My parents were amazed. For long stretches of time I would focus on nothing but the tip of my fishing rod, a metallic blue True Value six-foot spin-cast rod made out of solid fiberglass with chrome-plated line guides and ferrules. I sat on the stonework at Violet's Lock, near mile marker 22. The mantra-like effect of shear anticipation at catching something as yet unseen was enough. So few experiences in this world seem like enough anymore, but still fishing is one of them.

I stop to read the Falling Waters historical marker: "Retreating after Gettysburg the confederate army was trapped for seven days by the swollen Potomac River."

I know how fast and high the Potomac can rise. My father took two days off from work as chief of aviation and maritime telecommunications at the Federal Communications Commission to search for me. As a high school kid I'd boated out to one of the islands in the Potomac just after a torrential rainstorm. I mistakenly assumed the worst was over, not knowing that much more rain had fallen upriver and a huge bulge was coming. I rowed out and pitched camp, throwing my anchor into a tree as a last-minute lark. This bit of whimsy proved to be a stroke of luck, because the next morning my johnboat was afloat and the island of more than 100 acres had shrunk to just a few. I thought this all great adventure! Never for a moment did I consider my father frantically beating the towpath for his river-smitten son.

The first mountain biker under the age of 30 approaches, iPod earbuds in place. Apparently lost in his workout or his own little world, he stays in the middle of the path, forcing me to step aside. In his wake, I take note of tulip poplar, ash, basswood, hickory, ironwood, sumac, black cherry, black walnut—trees of every kind and sound.

Euell Gibbons became my high school hero with *Stalking the Blue Eyed Scallop*. I devoured all his books, including *Stalking the Wild Asparagus*. My goal was to be entirely self-sufficient, and nearly all my weekends were spent here, along the Potomac River, camping out, foraging the best I could from the wild, building smoky fires to beat back the mosquitoes. Usually I ventured out alone, but sometimes I went with friends. We'd push large forked logs into the river, forked so they wouldn't rotate and toss us into the drink, and after a joy-filled downstream ride, slide off before we reached Great Falls, usually around Swains Lock. Then we'd jog back, singing as we air-dried on the towpath to Violet's Lock. We had an enchanted life. Away from society's laws and pressures, we invented our own happiness.

Our biggest entertainment at night was beating rhythms. We'd search for particularly resonant logs, gleefully auditioning every horizontal candidate we could find. The good (and movable) ones we'd carry back to camp, position them off the ground around the fire, and then go into wild beating rhythms—Scooter, John, Wayne, Mark, David, myself. Many wild nights. Never did we think, even for a moment, that we might be disturbing the peace. Hey, we were teenagers.

I foraged in the wild as often as I dared. I feasted on wild asparagus from hedgerows and soon learned to dig for the nut-flavored corms, or stem base, of succulent Spring Beauties. Roughly the size of hazelnuts, they lie only about two inches deep in soft soil below the beautiful white flowers.

I also gathered mustard leaves for salad greens and even dug through the snow in winter to probe for day lily tubers. This last bold exercise taught me the value of a botany degree. I lay in agony all night in the woods, unable to move until daylight. Apparently I had mistakenly also harvested a poisonous tuber or two in my hasty dinner foraging.

As it happened, my botany degree enriched much more than my wilderness diet. It has proved most valuable in my career as a nature listener. Plants not only form much of the architecture of forest concert halls, but have such a close association with wildlife, providing both food and shelter, that I've come to think of vegetation as a musical score. By carefully studying a photograph of a forest it is possible to correctly identify the season, time of day, and what kinds of bioacoustic events were at play when the photograph was taken. With this knowledge, I've provided audio content to clients who were designing museum exhibits or enhancing online encyclopedia entries.

Thinking back, I realize I listened intuitively in my youthful ramblings along the C&O Canal. Just downriver from Violet's Lock a small stream enters the canal. That was my marker. About 50 paces farther I would stow my gear under a tangled mass of honeysuckle vines and then venture off—to swim, to forage, to explore. Those nights that I returned in the dark, I would locate my gear by listening for the particular murmur of this stream, which sounded brisk, even joyful, as if spring-fed, compared to the others that also fed the canal but would ebb and flow in volume, registering the history of recent rainfall.

The birds have quieted down now. Midday sound-level reading: 32 dBA. A lone cardinal sings out. *Pur-ep, pur-ep, tu-tu-tu-tu-tu:* 58 dBA peak at 60 feet at the top of a bare tree. Then the entrance of a gentle breeze at 38 dBA. A motorboat 100 yards away in the river registers 58 dBA, as loud as the cardinal, but not nearly as pleasant. I'm alone on the towpath, and my legs feel great, like two happy dogs that finally escaped the confines of the Vee-Dub.

A mother raccoon runs across the towpath, but her three young kittens opt to climb quickly up a tree rather than risk following. *Click.* I take a photograph of them gazing curiously downward at me, as a low *Grrrrrrrr* registers Mom's disapproval from inside another tree.

I'm just in time for wild strawberries and spot some growing in the shade of the forest, veer off the towpath, and pick and pop a good many in my mouth. *Mmmm, mmmmm*. The wood strawberry is not as sweet as the field strawberries we all eat on cereal and shortcake, but it is crunchy, offering snappy mouth bursts like flying-fish roe. I'm too early, though, for the pawpaw, although I see its immature fruit. I shake a small but full-grown tree, but fail to dislodge any. But I do startle a deer, which leaps out of seclusion and dashes farther into the shaded forest. A few steps later, I find plenty more pantry items: wild allspice; greenbrier, identifiable by its corkscrew-shaped tendrils; and black cherries, so ripe on a two-foot-diameter tree just off the towpath that they're already falling to the ground. Cyclists may unknowingly zip right past this feast, but this is my kind of slow food.

A helicopter passes somewhere nearby: 45 dBA, 53 dBA, 63 dBA. A detour around a washed-out section of the towpath forces me onto a sweltering, unshaded, shoulderless local road for six miles. My feet start to blister and the water in my gallon jug dwindles. I knock on the door of the first house I see. The woman who answers seems perplexed at the sight of me, backpack on, baking in the sun. But she's thrilled at my simple request and the opportunity to be a good neighbor. In a flash, she returns with my jug full of clear, cool well water.

Regaining the towpath I soak my burning feet in the river near the spillway of Dam No. 4: 56 dBA at 400 yards. Oh, man, this feels *goooooood*. On the dark horizon I can see that a thunderstorm is moving in, and with it, I hope, relief with cooling temperatures. Wind gusts and soft rain: 58 dBA. Moderate thunder (distant): 68 dBA. Then the peak thunder as the storm passes, still at a distance: 77 dBA with a resounding roll.

I press on to the Horseshoe Bend Campground, a distance of approximately 20 miles from Williamsport, not including the five-mile detour. The calm forest under fading light is ideal for listening. It offers the same preferred reverberation time (approximately two seconds) of Benaroya Hall. But distant traffic noise hums like an ungrounded wire in a sound system.

Exhausted, and well fed from my trailside nibbling, I feel no need for dinner. I set up my tent and slide inside. Too hot for even a T-shirt, I lie in only boxer shorts. The mere closing of my eyes ushers me toward silence.

I've been awake for about a half an hour listening to the soft tapping of rain on my tent roof when I first glance at my watch: 3:30 a.m., 30 dBA. It is remarkably quiet and relaxing, even though, with my ear close to the ground, I can hear somebody's music coming from a long way away, possibly from one of the homes that line the C&O Canal or perhaps from a boat anchored somewhere out on the river. There's a *Whoo* of an owl. Then another.

I remain still, transfixed by the subtle awakening of a place: 4:07 a.m., a distant train horn from the other side of the river; 4:55 a.m., the distant beating of traffic; and now 5:00 a.m., the first call of a bird across the river, so distant that I cannot discern what kind, only that the sound is bird-like: 29 dBA, just two decibels above the silence of Benaroya Hall. This must be about as quiet as it gets around here.

I enjoy two cups of tea and a Balance bar, break camp, and reshoulder my backpack. Tevas on the towpath once more, but my double-blistered feet have me walking in an awkward stride for the first half-hour as the pain dims. No way I'll be able to keep my determined pace of yesterday. The distance to Washington ticks off in descending mile marker numbers: 79, 78, 77. I pass my second set of backpackers, a man and a woman, also in their 50s. We exchange "Good morning"s. Later, two fishermen carry on a conversation in the middle of the river, silhouetted in the morning light. Faces unrecognizable but voices clearly heard.

An overhead jet at 45 dBA. Then another at 56 dBA. I must be getting closer to Dulles International, one of three airports serving the nation's capital.

A young woman riding alone whooshes by on a mountain bike with iPod and earbuds in place. Near mile marker 75, a cooling breeze passes through a creaking sycamore tree (58 dBA), sounding more vocal than mechanical, more fauna-like than floral. A large bass breaks the surface of the canal with a resounding slap: 54 dBA at 45 feet. A jogger passes with iPod and earbuds in place.

Past mile marker 70, I arrive at Antietam Creek, which shows on the Park Service map as a walk-in campground. Dog tired, I quickly pitch my tent and crawl inside, eager to grab a nap before I can enjoy the evening light on the river.

I wake about an hour later to grunts and hollers. Emerging from my tent, I observe more than a dozen Boy Scouts and a handful of Scout masters setting up camp. This campsite can be entered by bicycle, which is how these Scouts have traveled. But just across the canal there's a very large parking area, where they parked a support vehicle attached to a large trailer that identifies their troop number and the volunteer fire company that sponsors them. The back of the trailer is open, and from it the dads and Scouts unload their gear: lawn chairs, stoves, full-size mattress pads, coolers.

"Hey, are we having any bowl meals?" a Scout shouts to a scoutmaster at 55 dBA measured from 75 yards away.

On through dinner, the hollering continues.

"He almost lit his lip on fire."

"Bring me one!"

A couple of Scouts shove each other playfully, like bear cubs. A conversation between a visiting mother and daughter hits 70 dBA from 30 to 40 yards away. Nobody's still. Nobody's quiet. It is as if a sign asked "Please talk loudly." Compared to the quiet voices of nature around me (the heron squawk, the crickets, the splashes of fish), these campers seem excessively loud.

A nearby field is coming alive with the flash dance of fireflies. A soft medley of crickets provides a lovely soundtrack for this luminescent ballet, which I haven't seen in years. But no one else seems to be looking or listening. All seem to suffer from chronic sonic urbanitis, CSU, let's call it. Accustomed to speaking louder than prevailing, man-made, city noises, they fail to readjust their conversational volume when in quieter places and continue to talk in their city voices. CSU-afflicted individuals fail to quiet themselves. Thus they miss the very quiet and quieting experiences they seemingly seek.

"It's boys' night out!" yells a Scout, leaping out of his lawn chair with outstretched arms.

The park map shows at least five miles to the next campsite. I crawl inside my tent. When the Scouts are not disturbing the peace (I propose a merit badge for quiet), it's possible to hear a fish splash or the *Crawk* of a night-fishing heron. At 11:45 p.m. and 27 dBA, I can discern distant jets, distant car passes, snoring campers, a barking dog, and a muted conversation from the far end of the Antietam Creek Campground. Silence is not the absence of something, but the presence of everything.

Early morning finds me back on the towpath, easing into my stride because of the blisters. A lone fisherman motors upriver at 7:10 a.m., standing midship in his 15-foot johnboat, casting a wake in the perfectly serene Potomac. I heard him coming five minutes ago. Now, about even with me, about 100 feet away, he casts a sonic wake, too: 78 dBA. It will take another five minutes for the noise to fade.

The relatively cool morning temperatures quickly soar into the 90s by midday, unbearably muggy until an afternoon breeze fans through a grove of giant sycamores, a sound similar to the fine-grained splashes of alpine waterfall. *Brack-brack-brack-brack-brack.* Frogs argue from seclusion somewhere in this nearly drained section of the canal. *Konk, konk, knok –Konk, konk.* A woodpecker searches for insects buried deep inside a dead tree. *Twee-del-eee.* A thrush begins its tone poem. Even the *Eeeeeeeeee* of a mosquito is clearly discernible, concurrent with, of course, the drone of a distant plane. A sun-baked broad leaf drops to the forest floor. Even this is audible. All in all, 43 dBA.

Swim call in the Potomac never sounded so good. I drape my clothes on a clump of bare branches, allowing them to air, wade to the shallows, and dip, then rise; dip, rise. The breeze chills my skin into goose bumps before I return to shore to dress in the shade of overhanging oak trees. I have a late riverside lunch of soup with noodles heated over a Sterno can, then continue a few more miles and pitch my tent at Huckleberry Hill Campground near mile marker 62.

At 8:25 p.m. the fireflies resume their upward brushstrokes of light. I hear the songs of thrushes and the drone of distant jets: 33 dBA, bumping to the low 50s with the upward spiraling song of a thrush. At last light, just before I slip into my tent, two Jet Skis rocket across the mirrored Potomac, turning the calm riverbank into a lapping ruckus.

Early morning is misty and almost completely quiet except for a distant train horn (37 dBA) that bounces back and forth across the forested Potomac River. Then follows a series of owl hoots (33 to 45 dBA), fish splashes, and a mix of songbirds (wrens, cardinals, thrushes, warblers, vireos, and more) and the morning rush hour higher in the sky: 53 dBA.

The river is so wide at this point that it is more like a mist-covered field than a river, with many islands rising, house-like, out of the mist. Unripe pawpaws are everywhere. The sun has just begun to penetrate the mist, speckling the towpath with fussy leaf shadows. A series of shallow rapids (54 dBA) begins in the river and masks nearly all other sounds except the thunderous jets. Breakfast consists of handfuls of mulberries and wild strawberries.

As I approach the outskirts of Harper's Ferry, West Virginia, near mile marker 61, I discover that the Park Service's water pump has been disabled, the pump arm removed. I assume this is done when the water is unsuitable for drinking. So I walk the railroad bridge across the Potomac to John Brown's Coffee and Tea, replenish my water supply, and make my way back to the towpath, where I'd stashed my gear at the base of the bridge.

Milepost 59. The river is 20 feet below the canal and towpath. The birdsong is vibrant at 50 dBA. Then comes the deafening rumble of a passing train at 82 dBA. The only towpath traffic is a man in his rubber-tired electric wheelchair, breathing through a tube extending out of his neck, iPod in place, bumping along at a cruising speed of about 5 miles per hour.

Beyond Milepost 55, looking for a swimming hole to cool off from the noon heat, I come across the Brunswick Nutrient Removal Project, a waste water treatment facility under construction: 73 dBA. Just downriver some inviting campsites appear in the cooling shade where the bank slopes gently, offering easy river access. The river looks clean enough, and I see no visible evidence of sewage, so I assume safe entry. Floating in the water, I begin to feel a gentle nibbling at my skin and see numerous two- to three-inch-long fish clustered around my body, some just watching, others nipping at my skin, feasting on salt. I have been groomed by fish before, in pools below tropical waterfalls in Asia, but I didn't know similar manservant fish lived here. I enjoy being the center of attention until one gentle nibbler turns biter and I jump out faster than I went in.

Milepost 53. *Boom-boom-kaboom*. I recognize the sounds of a freight train, as a thundering diesel locomotive strains to take up the slack and pull its load. It took me by such surprise that I missed the decibel reading, but I have recorded this sound on numerous occasions in American train yards and know it's easily in excess of 90 dBA, if not 100 dBA, even at a distance of more than 50 feet. Europe's freight yards sound entirely different, nearly silent by comparison. Instead of large knuckles that rattle together, European freight trains use massive springs inside two large, piston-shaped chambers that act as shock absorbers. In almost every respect, from locomotive to rail engineering, ballast design to buffers, European trains are far quieter than American trains.

Milepost 51. A train horn, close to the towpath, blares above 90 dBA. A bit farther down the towpath, sawn-up pieces of discarded railroad ties litter the canal: creosote-soaked Pick-Up Stix that no one has picked up contaminating a national park.

At Milepost 48 I reach the Calico Rocks Campground and discover I've walked 14 miles through nature to camp next to the railroad tracks! But with no other choice, I pitch my tent and measure passing trains clanging above 90 dBA, a disconcerting approach to triple digits that mirrors today's overbearing temperatures.

As I take my evening bath the river's surface at eye level provides a mesmerizing pattern of slow swirls dappled by insects and fish. The moon is two-thirds full. Spade toads call from both shores in long, sheep-like bleats. Then a loud rumbling whine announces the arrival of a Boeing 737 on approach to Dulles International. To escape the din, I submerge my head completely underwater and let the cool Potomac flow over me as long as my breath holds out.

This morning's alarm clock was not the cheery song of a bird or the hoot of an owl, but the screeching, ear-tearing sound of metal on metal of a 5:55 a.m. freight train. This final assault capped a sleep-starved night of waking to one disquieting train after another. These awakenings averaged 75 dBA and peaked well above 90 dBA. As I break camp, yet another train pushes my sound-level meter even higher, to 96 dBA, making intelligible speech into my voice recorder impossible until it moves on down the line.

Milepost 47. The Potomac is glassy, mirroring the verdant banks on both sides. The towpath is lined with giant trees. Those with diameters of several feet surely offered shade to nineteenth- and twentieth-century mule drivers and canal boat workers, as they do for me now. But already the day is hot and muggy, as I predicted from last night's orange moon.

Woodpeckers have joined the morning chorus of songbirds, producing resounding echoes that fill the hardwood forest. Each tree has a different tone and each woodpecker a slightly different rhythm. Some birders can correctly identify the species of woodpecker just by listening to the drumming pattern. Others can deduce even more by close listening to bird sounds. In the Amazon, recording the daily cycle of sounds, I met locals who could listen to my day's "catch" and correctly tell the time within five minutes of when I made the recording. I knew because my digital recorder dates and time-stamps each recording. Those were some serious listeners of their land. That's my mission here, of course: to listen to our land and sort out my thinking while I do so. Walking, too, lubricates the mind.

Milepost 44. Distant prop plane and soft murmur of creek. The canal again has water. It's 10 a.m. and getting hot. Pretty soon I'll start looking for a swimming hole. I'm covering a mile every 20 minutes. My footsteps (43 dBA) have quieted over this trip. But will my mind?

By Milepost 43 the railroad tracks have veered off from the canal. I hear the *Nin-nin-nin nin-nin-nin* of nuthatches and the *Hummmm* of insects (31 dBA). A fox trots across the towpath carrying a squirrel in its mouth, probably headed back to its den to feed its kits. I pass into open sunlight, fully exposed to the heat of the day. I'm craving a swimming hole.

Milepost 41. Industrial sounds (52 dBA) reach my ears. There's an odd, ozone smell. The sounds get louder, then there it is: a PEPCO plant, belonging to the Potomac Electric Power Company. Downriver a bit, just short of noon, I slip into the river, desperate to cool off.

The water is warm. Even warmer as far as 50 feet from shore in deeper, flowing waters. I'm swimming in the power plant effluent. Farther along the towpath I come upon a cooler running stream and soak my swollen, blistered feet, charmed by an iridescent blue-bodied dragonfly hovering near the water's edge that quickly, silently darts to an eddy. If only nature's creatures ruled the skies over the towpath.

But they don't. The Potomac River and the C&O Canal serve as a visual approach to Reagan National Airport, and the closer I get to Washington, the more jets I hear along Justice Douglas's "long stretch of quiet and peace at the Capital's back door." In the late afternoon, as the air begins to still, they're audible every couple of minutes: 62 dBA, 71 dBA.

I arrive at the Turtle Run Camp around dinnertime, hot and exhausted. I've barely lowered my backpack when a nearby camper from a large group of bicyclists walks over. "We brought way too much food," he announces. "If you'd like, we have burgers with all the trimmings." He doesn't have to ask twice.

Another bath in the river before bedtime. Tonight it's not trains, but planes I'm counting instead of sheep: 55 dBA, 64 dBA. And they're still coming at 11:15 p.m. The plan was to contemplate the trip across the country here on the towpath and marshal my thoughts for the upcoming meetings with federal officials, but it's so noisy I can't concentrate my thinking long enough: 82 dBA for yet another overflight. Last night I walked 14 miles to camp next to railroad tracks. Today I hiked 18 miles to sleep underneath the approach to Reagan National. William O. Douglas must be holding his ears in his grave.

My eyes open the next morning to the sound of another jet: 65 dBA, about twice as much acoustic energy as normal human speech. I grab my camera and snap a couple of shots between the forest branches of aircraft flying overhead. Not that I'm likely to forget them, but I suspect that without a photograph, nobody will believe how low they are.

My gear is quickly packed and I'm back on the path. My blisters have popped and my feet remain tender. My psyche's a bit tender, too. With the help of my Treo, I've stayed in e-mail contact with Grossmann, who's kept me up to date on the meetings he's lined up for us in D.C. We're to sit down with former National Park Service deputy director Denis Galvin; Director of the National Park Service Mary Bomar; Secretary of the Interior Dirk Kempthorne; and a 30-year veteran in the noise regulation office of the EPA. Then I'll get to meet one of my senators, Maria Cantwell, at her regular Thursday morning coffee with constituents, and finally get to talk with officials at the FAA about my hope of rerouting planes around One Square Inch. But how am I going to prepare for these meetings without quiet to gather my thoughts? I will just have to trust my instincts, listen, and speak from my heart, a strength that the Hoh Valley has nurtured.

Milepost 30. My brow is already beaded with sweat, and this is the coolest part of the day. As temperatures climb and the humidity increases, sound (or noise) travels faster and is absorbed less by the atmosphere, meaning it actually remains intelligible longer than, say, on a mountaintop with a rarified atmosphere, cold temperature, and low humidity. In the air here along the canal, sound travels approximately 760 miles per hour, or 1,125 feet per second. In fresh water, like the Potomac River, sound travels 4,898 feet per second, more than four times as fast.

Like many, I'd always thought that the denser the medium, the more efficiently sound travels. But that's not actually the case, as I recently learned. The speed of sound depends on how efficiently the sound-wave energy is transferred. It is more a matter of stiffness than of density. A quick snap of the wrist on the end of a tight string will do a much better job of transmitting the wave energy down the line than a loose string. In the world of physics, water is stiffer than air and metal even stiffer. In a metal wire, sound travels at approximately 16,000 feet per second.

At Milepost 26 another water pump is nonfunctional and I'm out of water. But perhaps not out of luck. Coming toward me is a white Chevy pickup truck. When it draws alongside, I see the insignia of the park's caretaker: U.S. Department of Interior, National Park Service. I tell the two Park Service representatives about the broken pump, and they get out and take a look. They determine it's missing a handle part and will have to return and fix it later. But instead of a mere apology, the older of the two men reaches behind the driver's seat and drags out a big thermos, saying his wife packs him iced well water every morning. He nearly fills my empty jug. It proves to be some of the coldest, cleanest water that I have ever tasted.

The mile markers continue the countdown to D.C., and the noise continues to rise. Helicopter pass at 72 dBA. I sense my sensory balance shifting, my auditory attention giving way to visual attention. Why tune in to unpleasant sounds when there is something nice to look at? I can understand why we deem ourselves a visually dominated species, and perhaps we are, but we weren't always. Our natural way of being is through sensory balance, each sense providing essential and irreplaceable understandings that aid in our survival, no different from other higher vertebrates. Without this balance, how can mankind, de facto stewards of the planet, possibly hope to keep the Earth in balance?

Approaching Milepost 22, I'm nearing what the NPS map calls Violettes Lock, which I know as Violet's Lock, as dear a childhood spot as I can remember. This is where I dangled my first fishing rod over the canal, given to me by my father on my ninth birthday in 1962. I remember the muddy water of rainy Aprils and the crisp rustle of fallen oak leaves as I shuffled through them to reach the red sandstone wall of the lock, where I'd sit and lower my worm. I marveled at the softball-size Osage oranges that dropped from the overhanging branches with a tremendous *Ka-splash* that scared the fish. When a blue gill stole my last worm, I flipped rocks to retrieve more bait. Violet's Lock was my first natural haven.

A career Coast Guard officer before he worked for the FCC, my dad had hoped that I would follow in his footsteps and make my life at sea. Although he never said this, I always knew it. I did love the water as a child and spent years rocking to its rhythms as a young adult, first as a Z-card carrying seaman in the merchant marine on the Great Lakes, and then as a deckhand aboard the *Discoverer* for the National Oceanic and Atmospheric Administration in the Gulf of Alaska and Bering Sea. But a life at sea was not for me. I always longed for the smell of soil after a summer rain and the simple, reassuring touch of the earth beneath my feet.

Years later, after I'd heard my calling and set out to become a nature sound recordist (a profession that did not even exist at that time), my dad came to see me in Seattle. I took the day off from Bucky Messenger Service so we could take the ferry to the Olympic Peninsula and drive to my favorite beach, Rialto. It was a beautiful late afternoon. The sun was sliding toward the horizon. Pacific rollers streamed white manes before crashing near shore. The air smelled of a briny mix of salted fish and seaweed. I wanted my father to see me at work. I positioned my recording system at the water's edge and listened to and recorded the wave sweeps across the pebbled beach: a deep operatic breath, then a brief pause before roaring applause. What a wonderful concert, captured on my then state-of-the-art Nagra reel-to-reel recorder that cost me 7,500 bicycle deliveries.

I turned to my dad, expecting a look of affirmation, but saw instead perplexity and pain. "Hemp, why would anyone buy a recording when they could come here?" he asked. Unable to imagine any customers for my recordings, he feared his son was doomed to fail. I packed up in awkward silence and we talked very little on the four-hour drive back to Seattle. I knew he loved me and cared for me, but I was hurt and troubled. Why hadn't he listened? Not to me, but to nature's seaside concert. Not until 10 years later, when my work earned me an Emmy, did he recognize some value to nature recording.

Milepost 17. My right ankle is swelling. I pop a couple of ibuprofen. A helicopter passes over me: 72 dBA. The western horizon appears nearly black when it should still be light: severe thunderstorm approaching. I pitch my tent at Swains Lock Campground as the first gusts of the storm hit the forest canopy. I'm too tired to cook soup, so I dine on a few handfuls of granola while the storm passes, at least here, as a nonevent. The thunder rolls a few times, like wooden barrels down a long set of stairs, and the rain falls only moderately hard until I drift off to sleep.

At daybreak, my tent still drips last night's rain. I make tea and down a Balance bar. No need to break camp. I'll leave most of my gear here inside the tent so it will dry out, lightening my load. I'll tote my water jug and a side bag packed with lunch and my sound meter, recorder, and camera. I'll return later to retrieve my gear, after my dad picks me up in Washington. Time to make the final push.

The towpath has swelled to a wide gravel road nearly 10 feet across. The canal is three times as wide, calm and serene, overhung by large forest trees that accentuate a visual sense of privacy and peace. But my ears tell me otherwise. I hear the distant roar of traffic, both on the ground and in the sky. The soundscape now has a distinct urban feel, and the traffic on the towpath has increased as well. I've seen more joggers, walkers, and bicyclists in the first two hours this morning than I did on the first 50 miles of this walk. And most of today's path mates have a different demeanor. Smiles are rare, and hellos and hand waves, so common in rural areas, are as absent here as on any urban street. People seem turned inward, someplace else, perhaps thinking or feeling through business or personal matters or perhaps seeking an inner quiet.

Milepost 15 marks the approach to Great Falls, the southern limit of my high school days of wandering the forests, river, and canal à la Huckleberry Finn. A light rain starts to fall and feels good, promising to keep the day cooler. Another sign reminds bicyclists to use bells or horns to warn pedestrians. The distant roar of the falls and general urban din measures 44 dBA. A Boeing 737 on approach to Reagan measures 65 dBA. Another jet passes every few minutes.

After Milepost 10 the C&O Canal passes beneath the Beltway, Interstate 495, which circles the District. From beneath the overpass I hear the swirling roar of traffic: 70 dBA. Then, on the other side, a military helicopter passes: 78 dBA. I feel as though the towpath is now taking me into a sonic blender. My footsteps are inaudible.

The next mile marker has a very old post, like a headstone in a cemetery: "9 MILES TO W. C." The D has eroded.

On the side of a historical building in need of restoration and maintenance is posted a sign that reads "Save America's Treasures."

At Milepost 3, with McArthur Boulevard traffic a major part of the soundscape, my meter registers peak levels as high as 75 dBA and jumps to 86 and then 98 dBA with jet passes.

Milepost 2. The canal wall on the opposite bank is now concrete and pockmarked with drainage pipes. The predominant smell is raw sewage.

Milepost 1. The sounds? Road traffic. Snarling Weed Eater. Passing jet. Farther on, two mules pull a canal boat loaded with tourists. The boat is quiet. So are the mules. But my sound-level meter reads 75 dBA.

Milepost 0 is absent. The towpath morphs to 1000 Potomac Street in Georgetown. Sixty days after leaving home, I've made it to Washington, D.C. My walk is done, but not my journey across America. I've still got more ground to cover, most of it now indoors, inside federal office buildings. And I have one important picture to snap: the OSI stone in front of the Washington Monument.

Click.

Held up against the towering white limestone obelisk, my reddish angular stone, polished smooth by handling, seems so small and insignificant. Can I possibly transform it into something monumental, a symbol of national awakening and listening and concern that parts the planes around the Hoh Valley and helps preserve one of America's last great quiet places? Maybe the next few days will tell.

"Pioneer stock." That's something my dad always told me. "Hemp, you don't have a thing to worry about, because when it gets tough, remember, you're from pioneer stock." On my mom's side I have James Marion Wilson, a signer of the U.S. Constitution and the Declaration of Independence. On my dad's side, I trace back to a settler who made it to Nevada in a covered wagon. I saw this pioneer fortitude in both my parents the day I sat at one side of my mother's bed, across from where my father stood. With less than two days to live, my mother, unable to talk, removed her wedding ring and handed it to her husband, expressing her unspoken wish that he find love in this life again.

My dad and his second wife, Mary, pull up in the handicapped-access parking lot near the Washington Monument. Both greet me warmly. My gear at Swains Lock can wait for tomorrow. It feels good to be off my feet, traveling by car. And it feels good to be coming home.

I-270 takes us into Montgomery County, Maryland, where I spent most of my childhood. The burden of convenience is not to be underestimated inside zip code 20854, Potomac, Maryland, population 48,822. With a median house value of nearly $1 million and a household size of only three, the average adjusted gross income in Montgomery County has reached a staggering $230,000. It's no coincidence that I live 3,000 miles away. Joyce is my kind of place.

There it is, the lawn I hated to mow (thankfully, looking like it has been freshly shorn). And the American flag over the front door. I dump my satchel and empty water jug in my old bedroom, then come back downstairs, pass Mary in the kitchen, cooking up something wonderful, and join my dad in the family room with the overstuffed chairs, mail scattered about, and two prominent remote controls.

"Can we mute the TV?"

"Sure. This big TV is the best purchase I ever made. Demo from Radio Shack, twelve hundred dollars about fifteen years ago. Too big to get into a repair shop. Insured for about seven years. They fix it right here. We have so many TVs that we have to put an amplifier on the signal."

"How many TVs?"

"Five."

This is Montgomery County, all right. Before he collected TVs, Dad collected clocks, bringing back several from Switzerland, where he led the U.S. delegations at telecommunication conferences. Over the years, the house filled up with chimes. In the basement he kept a ship clock that would sound the bridge watch. Upstairs he displayed all the others, ornate ones that fit with the fine furniture he also loved to collect. I could always tell how Dad was getting along in life by how closely his clocks chimed together. But the clocks, I notice now, are nearly all gone.

"How's Oogie?" he asks, joking, "I don't know anything bad about him."

I don't either, so I offer this about my son: "Well, I've heard him swear a bit."

"That's good. You shouldn't swear a lot, but when I was skipper of a ship, if there was an emergency, I'd swear. And that really got people's attention, because I didn't swear. It got them to act a little faster."

"He occasionally uses the f-word too."

"Well, that's not good," Dad says, adding, "Let me fix us a martini."

"What time is it?"

"It's three-fifteen—like fifteen minutes late. All these time zones, can't we pretend we're in France or somewhere?"

"Merci," I say when Dad hands me a martini. Dinner is wonderful.

Later that night, lying in my old bed with the window open, I'm facing the closet door, which no longer features the big peace symbol and American flag I painted on it. My thoughts shift from the past. Tomorrow I want to stop in at the little-known museum at the Department of the Interior and see some of Washington's most celebrated memorials, all managed by the National Park Service.

Still awake at 1:30 a.m. It's absolutely quiet, except for the distant traffic from I-270. The measured sound level is an incredible 24 dBA, quiet enough that I can distinguish between passenger cars and 18-wheelers.

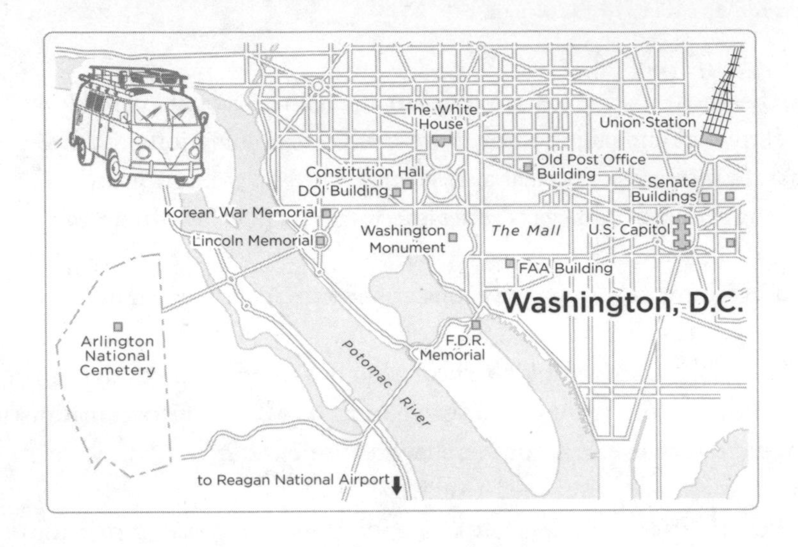

12 Washington, D.C.

> Nothing has changed the nature of man so much as the loss of silence. The invention of printing, technics, compulsory education—nothing has so altered man as this lack of relationship to silence, this fact that silence is no longer taken for granted, as something as natural as the sky above or the air we breathe. Man who has lost silence has not merely lost one human quality, but his whole structure has been changed thereby.
>
> —Max Picard, 1948, from *The World of Silence*

On Sunday morning at my parents' place, a bright, clear day, I lie in bed and listen to the mix of birdsong coming in through the window. Outside, the songsters have raised their voices to 40 dBA, as measured to the height of the tall oaks at 75 feet, and for once the birdsong masks the din of distant traffic coming from I-270.

I still need to prepare for the upcoming meetings with federal officials, since my strategy of collecting my thoughts while hiking the C&O Canal didn't work out as planned. Who would have guessed that a national park would become the visual approach to Reagan National Airport? So I've asked my dad and stepmother if we can go to the National Cathedral to focus my thoughts. There, I also want to study the interior architectural lines and to listen and remember the living cathedrals of the hardwood forest along the C&O Canal and its dawn chorus. Such once-quiet arching structures were our first natural places of worship. Perhaps there I can prepare myself to ask the right questions.

Dad has memorized every shortcut from decades of dodging traffic bottlenecks during his commutes, and in no time we're parking at 3101 Wisconsin Avenue NW, arriving after the service so that I can walk about freely.

Construction on the National Cathedral began in 1907, and it is fair to say it will never be finished. The Cathedral is a remarkable ongoing architectural achievement that stands more than 300 feet high at the central tower and spreads over 83,000 square feet. Entering the main space I immediately hear the familiar acoustics of a hardwood forest, an echoing blending of voices.

Tourists mill about, which is good news for me because I want to listen to the acoustics, the *behavior* of sound. Looking up I can see the grand, nine-foot-diameter fluted columns, arched at the top, that look to be about 20 feet apart, reminiscent of an ancient forest. A smaller set of arches represents the understory of the forest, at least to my eyes. I mention the forest-like quality of this majestic house of worship to a passing tour guide and she tells me that early cathedral builders were shipwrights. In the background an organ begins to play (64 dBA), then stops (55 dBA).

I feel here as I do in an arboreal forest, as I did near mile marker 95 on the towpath. Whether a church organ is playing or the birds are singing, I'm encouraged to let my intuition rise to the surface and balance my strictly rational thoughts. Because it's more easily communicated, rational thinking is overrated. By contrast, inner wisdom whispers. True listening is worship. The sacred feeling of silence is in all of us. This is my time to seek it again through silent prayer.

The following day, Grossmann and I meet near Dupont Circle and start our 10-block walk to the Department of the Interior Building on C Street.

Before we get there we face a fast-darkening sky. We step quickly, trying to beat the rain, but lose. Nature unzips a downpour strong enough to send curb-topping torrents of water down the streets, forcing a dash for shelter between the columns of a building. I know this building! It's the Daughters of the American Revolution Constitution Hall. When I was in high school, I came here Friday nights with my mother, who had bought us season tickets to the National Geographic lecture series. Now, with the rain angling in windblown sheets, thunder rocks the capital's downtown limestone canyons at 95 dBA, then 97 dBA, then a spine chiller at 112 dBA—not enough to turn heads at the Indianapolis Speedway but, except for glacier calving, earthquakes, and avalanches, one of the loudest sounds in nature.

Security at the DOI building is pretty intense. Metal detectors, X-ray, briefcase search; not only must my laptop be removed but its make, model, and serial number are recorded in the security log. Then come red name tags. I'm instructed not to take pictures. I guess that's one advantage of audio: nobody thinks to say, "Don't record sounds." A photo of Interior Secretary Kempthorne hangs below one of President Bush. Our appointment with Kempthorne is two days from now, and tomorrow we see National Park Service Director Mary Bomar.

"Is there an exhibit called America's Beautiful National Parks?" I ask a ranger stationed at the entrance to the DOI Museum. Years ago, David McKinney, the museum director, asked me to provide audio for an exhibit scheduled for 2005. I spent weeks preparing the soundtrack for the exhibit, which is why I'm eager to finally see it now.

"There was about a year and a half ago," he says matter-of-factly, "and some of that material is on the second floor and some is in other offices that you can't get into."

"Can we get to the second floor?"

"I can take you up there."

Before taking him up on his offer, I decide to cruise the current displays, called Reinventing Tradition and American Indian Design in Contemporary Clothing. There's an assortment of images by Edward S. Curtis, some incredible Indian beadwork re-creating the American Flag, a section devoted to the Bureau of Land Management. We have the entire museum to ourselves. Ventilation noise is the only audio (42 dBA), until we come to an exhibit devoted to another interior department, the U.S. Fish and Wildlife Service. *A Place for Nature* plays on a flat-screen monitor with a soundtrack of words, entirely without nature sounds. A wall placard asks:

> Why Should We Save Endangered Species?
> The voice of congress written in the Endangered Species Act reminds us:
>
> > Fish, wildlife and plants in the United States are of esthetic, ecological, educational, historical, recreational and scientific value to the nation and its people. The voice of a small child adds one more reason, "Because we can."

Natural quiet has become an endangered species, too. The child's voice, "Because we can," weighs in heavily, because the future success of preserving quiet places rests entirely on human choice.

Time for those second-floor photos. Ranger Kirk Deitz leads the way, telling us in the elevator that six of the photos from America's Beautiful National Parks now hang in the secretary's office, which is a restricted area. The rest line the walls in the second-floor corridor.

"These are four different views of Glacier National Park," says Ranger Deitz.

We look, and walk on.

"Everglades, Acadia," says Deitz, pointing as we walk by large photo portraits without reading their titles, having memorized them from his visits there. "My first ten years, I bounced all over the West," he states quietly. "I've been to all the big ones."

We are interested as much in Deitz as in the exhibit. Next month he'll backpack deep into bear country in British Columbia because he likes to "get away." The last town that Deitz lived in before here, on the San Juan River in the Four Corners area that borders the Navajo Nation, had only 120 people. His apartment at the time was once the kitchen of a bar called the Silver Dollar Saloon. A place where John Wayne used to drink 3.2 beer while making movies in Monument Valley.

"Redwood, Glacier, Grand Teton . . ." our tour continues.

The names are all familiar to me. To prepare the accompanying soundtrack to these photos I combed my huge sound archive. There was no budget for audio, only a long list of parks, most of which I had already recorded more than a decade earlier. I pushed aside all other work, believing that the formal setting of the museum would invite visitors to listen closely, and then, when visiting the real places, listen even more closely.

"Zion, Rocky Mountain, Great Basin . . ."

Galen Rowell, Marc and David Muench, Pat O'Hara, and Laurence Parent contributed a total of 38 large-scale photographs. Curated by Amy Lamb, this photo exhibit toured the nation. But without sound. At the last minute, the audio portion of the exhibit was cut to accommodate the wealth of photographs. Another case of the visual trumping the aural and a great opportunity missed: educating the public about our sonically alive natural treasures. This belated brush with the exhibit is bittersweet. The photographs are indeed stunning, but the display could have been so much more. I hope that One Square Inch will counteract our tendency to make natural sounds and natural quiet second-class citizens compared to natural scenery.

We turn in our red security badges and sign out, bound now for the Lincoln Memorial.

The ambience on the steps of the Lincoln Memorial is a mix of the shuffling footsteps of soft-soled shoes and a gentle babble, primarily foreign words (58 dBA). About four dozen people are inside. Several signs placed prominently on pedestals at the base of Lincoln's towering seated figure announce "Quiet, Respect Please." No one speaks loudly. Even children know not to shout. There is something innately right about being quiet here. And something inherently comforting, calming, and deeply deferential in silence, which is why we honor national heroes and lost loved ones with a moment of silence.

We stop at a nearby information booth operated by the National Park Service to ask for a map to help steer us to the FDR Memorial, built in 1997, long after I'd left the D.C. area. A jet rumbles loudly overhead. Then another, before we've even received our map and directions from the ranger seated inside. Reagan National is just across the Potomac River. It's late afternoon, the start of evening rush hour in the skies. We ask how much this jet noise affects visitors' experiences at the memorials.

"It's worse at FDR," she says. "I have to stop about three times when I give a talk. You can't be heard." The ranger's mild southern accent is from Arkansas or maybe Missouri, I'm thinking, until the 80 dBA roar of a third jet makes it impossible to hear her words clearly. Worse than this?

On the way to the FDR Memorial we stop at the Korean War Memorial, a striking set of GIs cast in bronze on patrol through a landscape of trees and short shrubs; each soldier has a battle-weary stance that suggests the personal trauma of war. Now the overhead noise is from a military helicopter: 78 dBA. One can only wonder about the effect on veterans standing here, their service memorialized, while the air above them beats with an aggressive helicopter *whack-whack-whack-whack-whack* capable of resurrecting ghastly war flashbacks.

The map steers us to the FDR Memorial, which covers seven and a half acres of a secluded, well-treed stretch of parkland between the Tidal Basin and the Potomac River. This horizontal, flowing series of four outdoor "rooms" was designed by Lawrence Halprin, the same landscape architect who designed Seattle's Freeway Park. I'm wondering if here, as over the noisy interstate in downtown Seattle, Halprin employed running water to mask urban din. At the visitors center we're told that a guided walk takes about a half-hour. A light sprinkle (the sky shaking off this afternoon's earlier deluge?) has begun, so our tour guide ducks back inside for a clear plastic cover for his Park Service hat. Then we're off.

"The design," the ranger says, "is supposed to make you feel that you're not in the middle of a major city."

Halprin describes his architectural aims in a diagram- and photo-filled book, *The Franklin Delano Roosevelt Memorial*. The word *quiet* appears twice, once referring to the huge blocks of Carnelian granite he employed throughout ("With its darker dapples and sparkling chips of mica, [the blocks] are striking yet subdued, and the overall impression is of quiet power and dignity"), and again in describing his intentions in linking the period rooms of Roosevelt's 1933–1945 presidency: "As visitors prepare to leave the second-term room, they enter a passageway which leads them toward room three. This passageway provides the same kind of quiet, contemplative space as did the first passageway."

A jet passes. My sound-level meter registers 78 dBA, similar to the noise in the urban classrooms beside the elevated tracks.

"I usually stop talking when they come this close," the ranger says, telling us that Reagan National is only about a mile down the river. I can understand him, but not easily. His interpretive tour tells us about FDR's life and presidency, and about the memorial itself. "It was dedicated by President

Clinton," he says, "but there was a little bit of controversy because nowhere in the memorial proper was FDR portrayed in his wheelchair. Of 125,000 pictures taken of him, only two showed him in a wheelchair." The sculpture he points to, of Roosevelt in a wheelchair, came later.

That dedication day apparently featured another, unspoken controversy. Eager to silence the passing jets during the dedication ceremony, the Clinton White House turned to the FAA for assistance, but their pleas fell on deaf ears. The FAA would not reroute, and so the planes flew as usual, drowning out the words of the speakers, including a none-too-pleased Clinton, who had to stop his speech and wait for the jet roar to subside.

So how well does Halprin's design succeed? I'd say it does a wonderful job of visual isolation, each room revealed in sequence, as if not one place but a walk-through landscape typical of modern landscape architecture. He does, in fact, employ water in many locales. In room two there's the TVA Fountain, which looks like a spillway or dam. From 35 feet away, the sound of water registers 75 dBA. Moments later, when a jet flies by, the reading jumps to 81 dBA. The ranger's voice is no match for waterfall and jet. So the flowing water does mask the sounds of the city to some extent, but the aural effect is not relaxing. When you have to strain your voice to be understood and lean closely to hear, the intimacy is forced and unnatural, and an innate sense of suspicion creeps in.

The water element in the funeral court, we're told, "is flat, calm, placid," meant to evoke the stillness of death. The ranger's comment is punctuated by another jet, this one at 91 dBA. The end of the memorial features a granite wall bearing FDR's famous four freedoms: Freedom of Speech, Freedom of Worship, Freedom from Want, and Freedom from Fear. After all the tour-interrupting jet intrusions, Grossmann can't help but ask: "Do you have a chisel? So we can add 'freedom from noise.'"

The next morning I wake up thinking about my high school years, then saunter downstairs. Breakfast is coffee, orange juice, soft-boiled eggs, and English muffins, which pop up from the same shiny chrome, two-slice toaster with a wide-braided cloth cord I remember from boyhood. How Dad keeps it alive, I'll never know. I see his prescription medicines occupying the entire side of the table, more vials than I can count in a glance.

"What are your plans today, anything special?" he asks playfully. He doesn't get to see me wearing a coat and tie often, and I know he's enjoying it. Far cry from my ponytail days.

"John and I will meet this morning with a former deputy director of the Park Service, someone who really knows the ropes. After lunch, we've got an appointment with Mary Bomar, head of the Service. We finish with a stop at the EPA to learn why its Office of Noise Abatement got axed."

"Well, I wish you luck, Hemp."

Dad reminds me to give him a call when coming home on the Metro and he'll pick me up. "Just remember to get off at Whiteflint Station."

The Metro's Red Line is fast and convenient, a far cry from the stand-still traffic jams of the 1960s and 1970s, before it was completed. On the commute I check e-mail and phone messages on my Treo. Our hoped-for interview with Interior Secretary Kempthorne tomorrow is in limbo. We're told he's sick and may not make it in.

I meet Grossmann at Dupont Circle, and we walk to 1300 19th Street NW and ride the elevator to the third floor. The doors open to the National Parks Conservation Association and its watchwords in raised letters on the wall: "Protecting Our National Parks for Future Generations." Isn't that the job of the National Park Service? It should be, of course, but the NPCA knows only too well the realities of government politics and thus works hard behind the scenes as watchdogs and preservationists. The Association is thankful for the experience, contacts, and vision of a long list of former high-ranking Park Service officials, who, like longtime deputy director Denis Galvin, are greeted on sight in the reception area. Even decades out of office, these dedicated individuals still proclaim and prove their love of America's unspoiled wilderness treasures by staying abreast of issues affecting the national parks and speaking up when they see the need. Galvin, National Park Service deputy director from 1985 to 1989 and again from 1998 to 2002, was one of 11 former high-ranking Interior Department officials to sign a March 26, 2007 letter addressed to Secretary of the Interior Dirk Kempthorne (see appendix C), a letter that said in part:

> We must express our alarm over a proposal in Yellowstone National Park that would radically contravene both the spirit and letter of the 2006 Management Policies. The proposal is to *escalate* snowmobile use as much as

> three-fold over current average numbers even though scientific studies have demonstrated conclusively that a two-thirds *reduction* in average snowmobile numbers during the past four winters is primarily responsible for significantly improving the health of the park for visitors, employees, and wildlife.
>
> The latest National Park Service study illuminates in detail that allowing Yellowstone's current average of 250 snowmobiles per day to increase—to as many as 720 snowmobiles—would undercut the park's resurgent natural conditions. Specifically, the study reveals that snowmobile noise would return to areas of the park where visitors are currently able to enjoy natural sounds and quiet.

The letter goes on to call the proposed increase in snowmobile use at odds with the Park Service mission to "preserve, to the greatest extent possible, the natural soundscapes of parks."

Years ago I lost my naive notions of the National Park Service as able stewards of America's wilderness after reading a book by Richard Sellars called *Preserving Nature in the National Parks,* published by Yale University in 1997. Sellars wrote the book while employed as a historian by the NPS and had access to information never before available. He told of policies to hunt and kill cougars to better enable grazing animals such as elk and deer to groom landscapes with scenic value. He wrote of the introduction of nonnative species, such as some varieties of trout, to improve fishing. Sellars put his finger on a key problem, one that persists to this day: "The central dilemma of national park service management has long been the question of exactly what in a park should be preserved," he wrote. ". . . Scenery has provided the primary inspiration for national parks and, through tourism, their primary justification. Thus, a kind of 'façade' management became the accepted practice in parks. . . . Façade management has long held more appeal for the public, for Congress, and for the National Park Service than has the concept of exacting scientific management."

Galvin steps out of the elevator just a few moments after we do. He's tall and thin, looking fit enough for some serious backpacking, though he's wearing a sport coat over a white polo shirt that has a pen clipped sideways in the space below the fastened button. A Lands' End canvas briefcase is slung over his right shoulder. Bryan Faehner, a legislative representative for the NPCA, who worked as an interpretive ranger with the Park Service at

North Cascades National Park, introduces us to Galvin and points us all to a sunny corner conference room.

I explain about my cross-country trip and One Square Inch and its zero tolerance for human noise intrusions. I hand them each my one-page fact sheet and tell them we'll be meeting with Director Bomar this afternoon. "I'm not expecting Mary Bomar to say, 'What a great idea, let's get this done.' But I do want to take every step I can, and the responses of those—whether FAA or EPA or National Parks—will be in this book. We hope that the dialogues will have a force in shaping public opinion."

Galvin, who served in the Park Service for 38 years and is now a trustee of the NPCA, provides some background on the Park Service's management policies with respect to natural quiet. "In 1970 there were no references to natural quiet," he says. "In 1988 there were two." He was deeply involved in writing the 1988 management policies and the 2001 update. In George W. Bush's first term, he reminds us, a big fight broke out over Park Service policies. "It started as an administration draft that essentially gutted the policies written in 2001. Ultimately, because of the ruckus [including protests lodged by organizations such as the NPCA, the Wilderness Society, and the Greater Yellowstone Coalition], the administration backed down, and the 2006 policies are substantially as they were in 2001. The commitment to protection of natural quiet is at least as strong as in 2001."

At least on paper.

"The principal difference between 1988 and 2006," Galvin continues, "is that natural quiet becomes desirable in more than the human sphere; it is desirable if you are trying to preserve all the living things in a natural environment, that quiet is important for them. And the reverse is inimical to them—that lots of noise, as is the lack of night skies, can be harmful. We know now that there are biological implications to that. Salamanders' breeding habits are interrupted in places like Santa Monica because of the absence of night skies. Turtle migration patterns, too. They hatch and head to the water. On a dark night, the water is brighter than the land. Well, now you've got Tallahassee behind the Gulf Islands, so when they hatch, those turtles head the wrong way. This is still an area with very little literature, not a lot of research, but it is accumulating."

Galvin's firsthand experience with noise in the national parks started in the late 1960s at Grand Canyon National Park, when he trained new rangers. He'd hike down to Phantom Ranch, spend the night, and hike out the next day. He remembers writing a memo to the regional director that went something like this: "Hiking in the Grand Canyon is like enduring the Battle of Britain. You hear aircraft all the time." This was before the onset of helicopter tours, before all the congressional attempts to curb the noise.

I mention the simple approach of One Square Inch. My stone on the log, my call for quiet, and, if obeyed, soundscape management over perhaps as much as 1,000 square miles of Olympic National Park. The world's first quiet sanctuary.

Galvin finds the concept both intriguing and reasonable. So I ask advice. "This is about trying to preserve a resource, which is already protected under management policy. How do I get that recognized?"

"There are always three officials important in any management mode," he says. "The park superintendent. An unsympathetic superintendent can pretty much torpedo what the director wants to do. There are 390 units out there and there's always somebody who 'didn't get the word,' as JFK said during the Cuban Missile Crisis. As director, which I acted as many times, and as deputy director, I had access to all the correspondence the director signed. I don't know how many times a month I'd pick up a clip and see a superintendent saying something and think, that's ridiculous, that's not the policy of the National Park System. And two other people, the director of NPS, the policy-making office; and the regional director can also be important."

I tell him of my hike to One Square Inch with Olympic National Park Superintendent Bill Laitner in 2006. "I think he's sympathetic to natural quiet but doesn't really want to challenge the FAA." (Superintendent Laitner would retire in early 2008.)

"An important part of this is the FAA," Galvin says. "And my experience with the FAA is they simply have no interest—and I'm not exaggerating—they have no interest in regulating aircraft with respect to protection of natural resource values or human amenities, apart from the White House."

"And yet by law," Grossmann points out, "they're supposed to be interested."

Galvin agrees, but admits, "My experience was 20 years ago, but it doesn't sound like they've changed very much. They're always talking about noise, they're never talking about natural quiet.

"There are certain analogies to this issue and the air quality issue," he says, again speaking from his long experience in the Park Service. "The Clean Air Act gave the Park Service a foot in the door on the clean air issue by saying that all national parks above a certain size would be Class One areas—that is, the air quality couldn't be degraded there. As a result of that, in the seventies, the Park Service started setting up an air quality office and monitoring air quality in the parks. Over a thirty-year period, the Park Service found out a lot about rural air quality, issues like regional transport—that some of the air pollution at Grand Canyon was coming from Los Angeles. It's hard to believe we didn't know that, but we didn't. It really surprised us, the extent to which these pollutants traveled. The Park Service became de facto a very important voice in this rural air quality issue on regional transport and regional haze. It seems to me that the issue of natural quiet is like where the air quality issue was twenty-five years ago."

He's made a key point: rather than look at its mandate to protect natural quiet in the parks as a burden, the Park Service should consider it an opportunity—an opportunity to lead the charge against noise pollution by supporting the idea of the world's first quiet sanctuary; by banishing snowmobiles (what possible reason allows them in the parks except the political power of local interests?); by cutting back air tours; and by using pack animals and foot travel instead of helicopters for much of its research activities.

Galvin counsels that Congress is the ultimate policy maker for land use in the United States.

"So," I ask, "if Congress was to pass a bill . . ."

Both Galvin and Faehner agree that might be the best approach.

"And your strategy is a good one," Galvin says. "To concentrate on one square inch of one park is a lot easier than taking on the system as a whole, with all its vagaries and variances and different political outlooks."

Grossmann and I thank Galvin for his time and ride the elevator down to the main floor, buoyed by his insights and optimism, in search of a spot for lunch. We find a place where the food is good, but the floor is hard, the ceiling is low, and the busy midday crowd pushes the noise levels to 75 dBA, making conversation difficult. Short on time, we hail a cab for the Department of the Interior Building, this time ready for the security blitz.

Bing. Today, the elevator doors open at the third floor, and as we step out, I can't resist. I take a sound-level reading here in the hallway: 58 dBA. Fan blades rotate overhead. Voices travel down the corridor. A single plant seems the lone bit of greenery. I feel a long way from the Hoh Valley. We announce our arrival to a receptionist and are promptly and warmly greeted by Director Mary Bomar, who personally comes out to usher us into her spacious office, suggesting we take seats on either side of her at a polished conference table. Joining us is Sharon Kliwinski, from the Park Service's Natural Resources Program, and the de rigueur representative from the department's Office of Public Affairs. Director Bomar hands each of us her card and a Park Service pin, like those worn by rangers, for a souvenir.

"I actually read about what you're doing. I saw it somewhere," she says.

"Thank you very much for making your time available," I say.

"It's quite all right, because it's very important to us," she says, positioning a couple of sheets of notes in front of her, then folding her hands in her lap. She's wearing a smart beige suit and gold earrings. Her warm smile is framed by an expensive haircut. A nearby wall is covered in photos, among them, President Bush, who appointed her to head the Park Service in 2006, and a shot of the director with First Lady Laura Bush.

Director Bomar asks if I've got the One Square Inch stone with me. So either she or a staffer has been to the OSI website and knows it's traveling with me. As she turns it in her hand I tell her where it came from and about the single-point soundscape management strategy that it symbolizes.

And then, with a glance at her notes, she begins to talk.

"I was raised in villages in England, but at the same time my family owned a manufacturing company in Leicester, eighth largest city in England—manufacturing hosiery. I'm used to the sounds of the city, but I travel a lot."

She lived in Philadelphia during an earlier stint with the Park Service, now lives in D.C., and before moving to the United States and becoming a U.S. citizen, lived in London.

"Last night, very quickly, when I got home at 7:30—I always check my e-mail before I go to bed, and I put some notes together, and these are some of the things I thought about. There's an excellent book called *Last*

Child in the Woods: Saving Our Children from Nature-Deficit Disorder. One of the centennial goals, when we went out on the forty listening sessions to the public—it was certainly a minute at the mike where they could get up and tell us how do we keep our parks vibrant for the next hundred years and what do you want to see happen in the national parks for the next generation and for your grandchildren. Soundscape. Nighttime skies. We heard the word 'impairment,' which I was quite surprised hearing from the American public, from people off the streets coming in to talk about impairment and keeping the parks natural. . . .

"But I really do think about the last child in the woods. We are concerned about losing children['s connection] to nature. We've got planned over the next seven to ten years seven pilot parks where we're bringing children back into the parks to do inventories of species—flora and fauna—and inventorying the natural resources of the park."

Her words have been spinning like a windmill obstacle on a miniature golf course. Finally I spot an opening. "Are any of these nine places in the Western United States?"

"They're across the United States. They started with Rock Creek."

"Point Reyes is one of them," says Kliwinski.

"I've visited nearly all of the national park units in the United States that have wilderness areas," I say. "There's nothing east of the Mississippi where I've found true quiet during daylight hours. Even in the West it's very scarce. And I've discovered that not only do the kids need the education, but at Olympic Park, even the rangers need the education. They couldn't tell me a dozen acoustic features of the park, and Olympic Park is truly—and I hope this is the one message that becomes apparent in One Square Inch—that Olympic National Park is the listener's Yosemite. That's why I moved to Olympic."

"So you live in that area?" the director asks.

"Yes, I live near Port Angeles," I say, telling her that I've hiked to One Square Inch with Bill Laitner, the park superintendent.

"I knew Bill Laitner many years ago. There are lots of golden nuggets being the director of the National Park Service. The gold nugget for me is that I'm a career employee who came up through the ranks so I've known many of these superintendents through the years. Peter Jennings said to me—I was in Philadelphia as superintendent at Independence—

'Boy, Mary, superintendents are really kind of a special group.' And they really are."

I glance across the table. Grossmann's eyes make a slow half-revolution. I know what he's thinking. We're penciled in for 45 minutes and about half of that has already ticked off.

"The National Park Service is made up of seven regions—I bet you know this already," Director Bomar continues. "We have seven regional directors, and the national capital region here in D.C. When we hold four meetings a year for the national leadership council we try to make at least two of them or one, if we can, out in the field . . ."

Grossmann shoots at the windmill. "What's your most recent backcountry listening experience, where you would have experienced profound natural quiet?"

Director Bomar thinks a moment. "A couple months ago, I would have said at Zion, because they've got such great park transportation in place at Zion. People can get on and off. You don't have the vehicle noise in Zion. So hearing the water, the stream, just hearing water, to me, I'm very in tune with hearing the trees, the bristling of the leaves in the trees. Hearing birds. I was pleasantly surprised sitting out in one of the little log cabins there in the lodge, in the evening."

"Did you hear, yourself, the air tours, while you were in the backcountry?" I ask.

"No I did not. And I would have noticed it, believe me, especially living on Slater's Lane here in Washington and with Reagan National Airport being so close."

I will never begrudge a person his or her experience, especially a serene one, but the usual reality at Zion is altogether different. Dick Hingson, noise/aviation specialist for the Sierra Club, has kept me abreast of his battles at Zion and Grand Canyon. He has informed me that Zion backcountry impacts by aviation are up to 15 aviation noise intrusions per hour (one every four minutes), each lasting two to four minutes.

I mention my affection for John Muir, the father of our national parks, and my close reading of his journal descriptions of the music of nature, especially in Yosemite National Park. I tell Director Bomar that the last time I tried to record there, aircraft noise was audible more than 50 percent of the time.

"I know the Park Service is embroiled in controversy—Grand Canyon,

Yellowstone, Hawaii Volcanoes, the list goes on and on," I say. "Those are long-standing, deep battles I don't think are going to be resolved soon. I do want to say that Olympic is an entirely different situation. In the back-country there, noise intrusions are rare. Noise-free intervals sometimes last for hours. That's unheard of anywhere else that I've been to in the park system."

"Now that surprises me," says Director Bomar. "So tell me more about that, because I sat in a canoe on Lake Crescent. I took out the five regional directors, and I often use this story—getting back out to the parks, getting back out to nature. I think I had just been confirmed. My national leadership council meeting, I actually delayed it, because it was very important to me, the message that I sent to the national leadership council, made up of the seven regional directors and six folks here in this office that are the associate directors, and some senior-level staff. I got in the canoe and it's a huge canoe. We all rowed in sync. We had two great young interns there as guides telling us the story of Lake Crescent. So I would row, and then I stopped because it was so beautiful in the middle of that lake. I heard no traffic, but somebody said to me . . ."

"What time of day was this?" I simply have to ask because this lake is intimately familiar to me, located only six miles from my house in Joyce.

"It was about two o'clock in the afternoon," she says, continuing. "So we rowed. I noticed when I stopped, the rest of the team stopped. I've used that story to say: I'm in here as a leader. I've been brought into this position because I have strong leadership skills and I'm not afraid to takes risks, and in my time in life have nothing to lose but to do the right thing. But I noticed when I stopped rowing, so did the rest. I said to them, sitting on the lake, 'Just stop and listen, because we don't get the opportunity. We're so caught up with our computers today and sitting in an office.'"

I'm stunned by her anecdote, and try to tell her so. "I have to say that a lot of Americans feel that it is still quiet in our national parks. My immediate assumption is, they've either had a rare opportunity or they've had an undiagnosed hearing impairment. Now Lake Crescent is bordered by Highway 101 and at two o'clock, the noise-free interval, if it exists at all, certainly has to be less than a minute. This isn't an exaggeration. I live six miles from there. I have both the concern that you have a hearing loss and also the concern that maybe you may not be able to appreciate the transformative quality of quiet and what it offers. I would like to extend a personal invitation to you to visit One Square Inch, because even though you think that you've experienced quiet at Zion or Lake Crescent, the Hoh is the place to go."

We continue to talk—me about One Square Inch, the director about quiet experiences she remembers as a girl in England, during which she interjects, "And I have good hearing, by the way." She mentions "the wind blowing through the grasses and leaves, the sounds of nature, of birds. Yeah, I think I can describe that. In my mind I have a very clear vision."

"And do you miss that?" Grossmann asks.

"Oh yes, there's a stress-free . . . calmness that's brought to you. It's a reason, I believe, that people go to national parks. It's enjoyable and it's a wonderful environment to be in. Sound is important to me—and I will piggyback on that and tell you that my husband was raised in the boot heel of Missouri and he's sixty-seven. I find I repeat myself now; my husband's hearing is not what it used to be, and as we get older, none of our hearing is what it used to be, and I think that sight and some of the notes that I wrote down about our senses—it's very important to have the luxury of good hearing and good sight and every day I'm very blessed that I get up and I can see and I can hear. And I thank God for that."

Director Bomar graciously allows the conversation to continue past our 45-minute allotment. A few minutes later, Grossmann pulls some papers from his briefcase, unfolding and holding up a poster that shows a deserted dirt road winding through a hardwood forest aflame in fall color. At the bottom appear the words *Quiet: A National Resource.*

"I love that," Kliwinski says softly.

Grossmann next holds up the October 1979 issue of the *EPA Journal,* an issue entirely devoted to noise and the environment. A smaller version of the poster, he shows all at the table, appears on the back inside cover of the magazine. "But what I really want to share with you," he says, thumbing a few pages back to a paragraph marked by a Post-it, "is a passage from the essay on 'Quiet as a National Resource' written by David Hales, who was then deputy assistant secretary for fish and wildlife and parks in the Department of Interior. Hales wrote this back in 1979:

> A most appropriate, in fact, necessary role of the National Park Service in years to come will be the preservation of some special places which are not

> polluted by sound, just as we would not allow them to be polluted by dirty air or water. In these places, the artificial and unnecessary introduction of sound into a natural environment is more than just an irritation caused by what you can hear. It is, in essence, an act of robbery, a theft of those sounds which naturally belong in these environments, and which are part and parcel of the natural and cultural heritage of this Nation.

"I couldn't agree more," says Director Bomar. "When I see that picture—I lived as a child beside a wood—that's the kind of picture that comes to my mind as genuine quiet in a natural resource. Some of the things I wrote last night: we cannot ban all outside sounds without banning people. We were talking about impairment—there are impacts that are going to happen—reducing traffic with new transportation is very important. We're talking about snow coaches yesterday, as an alternative to snowmobiles—there's got to be—I really do applaud what you're doing. Do we have all the answers? Absolutely not. I look behind you and see a video, *Planet Earth* from the BBC. I'm dying to see that video which was just sent to me. . . . Trying to balance visitor services and protect natural resources has always been a challenge and will continue to be a challenge to us and I will never shy away from any conversation, as I'm sitting with you here today, because it's very important to me."

Grossmann has one other document to reference, a May 2007 policy letter from Secretary of the Interior Dirk Kempthorne to President Bush, setting forth the priorities and goals of the national parks in advance of the Park Service's upcoming 100th anniversary. "In this Future of America's National Parks letter, Kempthorne says: 'The Park Service needs bold goals, clear objectives, and specific strategies for the future.' One Square Inch fits that to a T. It's a bold idea. It's very clear in its objective. And it's very specific in its strategy. We really hope this will be strongly considered as a way to meet the management goals and preserve for the next generations of Americans what's there. And we'll lose it if we don't pay attention."

"It's very important," says Director Bomar. "You're right."

We talk a bit more, then she signals that the meeting is over. "I'm going to leave you with one special thought," she says. "I always say I'm an American by choice, but I do strongly feel that there are special places that unite us all as Americans and I think national parks do that—and I always end my meetings when I'm with new folks—the one thing I can say—the sound of America is the song of my soul and is to many Americans. I often think, when we talk about soundscape, what did the original people hear and what did the explorers hear when they first came to this great country? I was down at the four hundredth [anniversary] at Jamestown recently and had the privilege of meeting with the queen just a few days later during her visit here and she questioned me about when did I come to this country. We have such great opportunities, and I hope that through the centennial that we are really able to be bold in many of our goals that will help keep our parks vibrant. So stay tuned, as they say."

As we're leaving, she adds, "I really look at us as being on the same side." And she asks if I have a website.

In the hallway, walking to the elevator, Kliwinski laughs, seemingly for no reason. But she's not smiling. I ask her why she laughed.

"Well, it's all relative," she says. "You're talking about the wilderness, but I live next to Reagan National Airport, and after nine-eleven, what did I hear? I heard footsteps for the first time. I heard my neighbors in their homes. I heard my community, and it went on for three weeks. And when the planes resumed flying and the quiet stopped, I cried."

Our next stop traces back to a scrap of paper handed to me 15 years ago by a now-retired acoustic consultant named Buzz Towne, who did a lot of work for the federal government. "Look this man up if you are ever in Washington," he said. On that piece of paper Towne had written the name Ken Feith and the letters EPA. Feith, who once worked in the agency's Office of Noise Abatement, still works for the Environmental Protection Agency, I discovered, when I called the number and introduced myself. I'm hoping he can tell me about the demise of a once-vibrant program that championed the nation's need for quiet. "Sure, come on over," Feith says, when I introduce myself on the phone. "I'm in the Old Postmaster General's Building."

Built by the Works Progress Administration, the building looks as though it belongs in an Indiana Jones movie. Marble floors, waxed to a fine shine, line the long, empty corridors and reflect the gleam from the ceiling globes that light the way every few doors. We find the EPA Office of Air and Radiation, announce our arrival, and are led to a tiny conference room barely large enough for a table and several chairs.

A few minutes later, in strides a big man with hair as white as his dress shirt, a colorful tie that might have come from the MOMA gift shop, and a welcoming, avuncular air. Feith greets us warmly, seeming much more like the kind of man who'd be right at home in a Santa suit than a savvy Washington survivor, the last man standing of the long-eviscerated federal program that once took dead aim at noise pollution. He's got a story to tell, and it's clear he's in no hurry to tell it.

"In the beginning, Moses created . . ." Feith says, jokingly cuing up his biographical overture. He explains that he got into the field of acoustics ("which I didn't know how to spell at the time") after applying for a research assistant's job at the Illinois Institute of Technology in the late 1950s. During the cold war he worked on projects for the Navy on antisubmarine warfare and "various clandestine agencies" on acoustic monitoring projects. "Are you football fans?" he asks. "Have you seen the device for diagramming the play on the TV screen? I'm the cofounder of the company that invented the Telestrator."

Feith came to Washington as an entrepreneur in 1969, three years before Congress passed the Noise Control Act of 1972. He began consulting with the Environmental Protection Agency in 1975, working in the Office of Air, Noise and Radiation for a program known by the acronym ONAC, Office of Noise Abatement and Control.

"There was a lot of good work getting done," he says. "We had ten regional centers of excellence on noise established at universities. We provided this ECHO program, Each Community Helping Others, with instrumentation and all sorts of materials with the understanding that they would reach out and you'd get this tree effect." ONAC published all manner of brochures: *Noise: A Health Problem; Noise around Our Homes; Think Quietly about Noise;* not to mention a thick *Noise Effects Handbook: A Desk Reference to Health and Welfare Effects of Noise*. These were also the days of the poster we showed Director Bomar earlier today titled *Quiet: A National Resource*.

And then, he explains, President Reagan took office. Soon thereafter, an axe-wielding agency administrator was appointed, a woman who looked high and low for offices to close. "Well, you don't close down water," says

Feith. "At that time we were up to our ears in radioactive waste. And you can't close the air programs when people are dying of respiratory illness. She got down the list, came to noise."

Feith interrupts that story for another. "The noise program at EPA was not initially formed for environmental protection. It was formed to facilitate interstate commerce. The first area was railroading. Back in the fifties, sixties, and seventies, communities were establishing their own standards and really curtailing traffic." He explains that the railroad associations went to Congress looking for a single, nationwide commerce law. When the trucking industry caught wind of this, they asked to be included, too. Ditto the airline industry. "So the federal government is responsible for major noise sources in commerce, when noisy products cross state lines," Feith continues, explaining the 1970s federal push into broader, societal noise issues was summarily aborted when somebody brought to the attention of the new EPA administrator "that it says in the Noise Control Act that state and local governments have the primary responsibility for protecting their citizens with respect to noise."

The axe fell. All but one of the regional noise centers was shut down, as was the D.C.-based Office of Noise Abatement and Control, where Feith was director of the noise standards and regulations division. A $10-million-a-year office employing 60 people essentially disappeared. "I'm all that remains," Feith says. "I'm a shadow noise program. The institutional memory. They've dropped noise from the logo. This is the Office of Air and Radiation."

"Did the states pick up the ball?" asks Grossmann.

"No. State and local governments said, 'The feds don't think this is very important anymore. We can do better things with our money.' Most noise abatement programs disappeared. Are we writing any new regulations? No. We haven't done that since 1982. Do we have a budget? We don't have a line-item budget, no."

"So what do you do?"

"Very little," he laughs. "No, my principal role is as a trade negotiator for the EPA. I just got back from two weeks in Europe. Under the World Trade Organization agreements, no country shall impose requirements on import products that are more stringent than they apply to their own products. But countries have a tendency to protect their industries. I deal in environmental regulations as they affect the U.S. transportation arena: automobiles, trucks, buses, motorcycles. Noise is part of my portfolio. I wear two hats. One: I'm playing the role of a diplomat representing the U.S. EPA within the United Nations organization that is developing 'globally harmonized' regulations for the environmental performance of all-wheeled vehicles. Because of my background in noise, I put on my engineering/physics hat and move into the group of experts that are developing noise regulations for automobiles, trucks, and motorcycles. I'm there to ensure whatever regulations come out of these U.N. groups are not going to be deleterious to us in the States.

"The Noise Control Act does not give EPA authority to address community noise. There's nothing legally we can do. What we do have the legal authorization to do is regulate the noise of products: lawn mowers, leaf blowers. Are we doing that? No. Because there are no funds."

I tell Feith about our ear splitting visit to the Indy 500 racetrack. That leads to talk of loud motorcycles and another admission of how the agency's reach comes up short in regulating noise. "We restrict the noise emissions for newly manufactured motorcycles," he says, "except for competition, where they can be as noisy as you want them to be." And that, he adds, cracks the door for more decibels on the city streets. "As just one example, Harley Davidson's Screaming [Pro] Eagle exhaust system is intended only for competition and should not be used on the street," he says. "But then it goes to the [independent] dealer—and he ignores that. He's got a three-hundred, four-hundred-dollar exhaust system he can sell and he's got this young guy here."

Agency regulations require that an ID stamp on exhaust systems match a corresponding number on the bike, and also forbid tampering with exhaust systems (a common method is to ram a solid metal rod up the exhaust pipe to break off the noise-damping interior baffles). Dealers who install illegal exhaust systems are subject to fines of up to $10,000. Operators of excessively loud bikes can also be fined. Feith admits that the EPA has no way to enforce its regulations. It's up to state and local jurisdictions. But few accept that responsibility.

Back in the 1970s, he tells us, ONAC had a "buy quiet" campaign. "We did a lot of research and put out literature—here's what to look for—a whole range of products," he says, praising the dishwasher in his house as so quiet "you can hardly tell when it's on" and mentioning that a prime consideration in choosing the heat pump for his home was that it, too, was the quietest on the market. Indeed, noise-level labeling for products has come a long way.

"We had a memorandum of understanding with the lawn care industry," Feith continues. "We were going to regulate lawn mowers, and the lawn and garden care industry came to us and said, 'Please, don't hit us with regulations. We will launch a labeling program listing the noise level of our products and we will take steps to reduce the noise.' We entered into this memo of understanding. Within a short period of time, Sears products all had hang tags on them. I went to a local store and asked the salesman what the number on one of those tags meant. He said, 'That's the power rating. The higher the number, the more power you get.'"

Feith shakes his head. "It is the mentality of the population that really governs the importance of noise or the dismissal of it," he says, speaking from long experience. "Noise tends to be transient. The guy in the house next to you gets up at seven a.m. Sunday morning and cuts his grass and you get upset about it. By eight-thirty he's finished. You cool down and have your second cup of coffee. Remember what happened with leaf blowers? Communities got upset and outlawed them. But they still exist, and the guy walking around with a leaf blower on his back is losing his hearing, because it's about a hundred ten decibels. Especially here in the States, we're very fickle. We respond to what's happening right now, but we don't think long term. That's the problem with noise: no bodies in the street. We can't show a direct link to something like cancer. People simply do not understand the significant adverse health effects of noise. They just don't.

"And it depends on who's in the catbird seat," he continues. "You know, we had an environmental vice president for eight years and could do nothing. The day after his inauguration he came to EPA and we had an all-hands meeting. He got to the podium and talked environment to us—and we never saw him after that."

At heart, Feith is an educator, which explains why he has been looking to seed two inner-city pilot programs to teach young schoolchildren about the harmful effects of noise. He mentions a model program in Portland, Oregon, called "The Dangerous Decibel" that could serve as the basis for an educational unit he'd like to see taught in, say, 10 training centers around the country, and then widely introduced into the schools. "The most successful program we had [back in the days of ONAC] was our program in the elementary schools. Young children take this stuff home and they are great communicators."

I ask if he thinks One Square Inch has the possibility of gaining traction, helping to reintroduce the notion of quiet as a national resource, and an endangered one at that.

"You know what you want?" he suggests. "Something like that commercial from years ago where the Indian had the tears running down his face. Maybe now he's got a splitting headache from the motorcycles. I don't know how you do it. I'm well past the age when most people retire. I've been in this business too many years, and as I said at a conference recently, my biggest complaint is that we experts are talking to the experts. We've redefined the problem so many times, and it doesn't change. We know almost everything there is to know about noise. We know how to measure it. We know how to quantify it. We don't know how to stop it.

"If people want quiet, they will get quiet," Feith says. "But you have to get enough people to want it. I think it can happen. EPA, with our little outreach program and the few bucks that we've got, we're going to try. And maybe it will grow."

Maybe it will, I think, after we thank Feith and say goodbye, heading for the Metro. Many people are concerned about noise and hunger for quiet, each in his or her own self-quieting way. Basel Jurdy and Mark Reddington, by designing quieter indoor places. Bill Worf and Tina-Marie Ekker, by reminding even environmentalists that scenic value alone is not enough. Jay Salter, by writing quiet-inspired poetry. Karen Trevino, by leading the NPS Nature Sounds Program in spite of bureaucratic disconnects. Elliott Berger, by helping design products and educating to protect against hearing loss. And nowadays, empowered by the Internet, plenty of grassroots organizations focus on noise pollution and quiet preservation, both in America and abroad. To name but a few: the Noise Pollution Clearinghouse; Noisefree America; the Noise Abatement Society; Right to Quiet, based in Vancouver; Noise Mapping England; No Boom Cars, a Louisville, Kentucky, effort seeking enforcement of their local noise ordinance; Lower the Boom, another anti–boom car effort; the Quiet Use Coalition, which argues against gas-powered engines in national forests; and the Alaska Quiet Rights Coalition, proving that noise is a problem even in such a big and remote and sparsely populated state.

First up today, a chance to meet one of my senators at her weekly coffee with constituents. We reserved a spot by calling ahead a few weeks ago. This won't be a private audience, we'll be part of a group. We have no idea how this coffee works, but it offers easy access to a Capitol Hill policy maker.

I emerge from the Metro tunnel on an escalator crowded with commuters at Capitol Plaza. The day is absolutely clear, and the absence of vehicular traffic (for security reasons) in the vicinity of the Senate Office Buildings near the Capitol dome is almost refreshing. At the Dirksen Senate Office Building Grossmann and I breeze through security and find our way to the suite of offices belonging to the junior senator from Washington State, Maria Cantwell, a second-term Democrat with a strong environmental record who, fortuitously, sits on both the Transportation Committee, which oversees the FAA, and the Energy and Natural Resources Committee, which includes the National Park Service under its umbrella of interests.

The immaculate reception area is stately: beige walls with eggshell-white trim, an American flag in a corner joined by the distinctively green Washington State flag. I spot a familiar Northwest Indian art print, a series of embedded faces suggesting that all are of one being, and a framed array of former Washington State senators, among them Scoop Jackson, who guided such landmark environmental legislation as the Wilderness Act of 1964 and the 1969 Environmental Policy Act (NEPA) that required environmental impact statements.

We sign in, and I also sign up for a photo with the senator. Here in the reception area, as the 8:30 start time draws nearer, we and several others are indeed offered coffee (Starbucks, of course) and are ushered to a conference room with a large table. I take the seat closest to the door and nearest the head of the table. Grossmann slides in beside me. Joining us are two couples and a single man. Not one of us is under 50.

Delayed just a few minutes by legislative concerns, Senator Cantwell walks in squarely, apologizes for being late, and remains standing at my immediate left. She's wearing a white business suit and a cameo necklace over a black blouse. She smiles warmly as she turns and looks straight at me. "So, tell me, why are you here today?"

I'm first up, apparently, of a round-the-table sharing of names and reasons for wanting a few words, face-to-face, with a member of the U.S. Senate. This, I can do.

"I am interested in Olympic National Park and protecting the quiet."

"What's the matter," she jokes, "are the animals misbehaving?"

Everybody laughs, including myself. Then I explain as quickly and concisely as possible what I've said again and again over the past couple of days: that I have traveled the world collecting natural sounds and year by year find that increasingly difficult as quiet locations, even deep within our national parks, are overrun with man-made noise. "Olympic is the listener's Yosemite, naturally quiet for long stretches," I say, "until air traffic breaks the natural soundscape. I have been woken up at four o'clock in the morning deep in the Hoh Valley by a jet in-bound for Sea-Tac."

"That's not right," says Senator Cantwell.

I talk a bit more, then she looks to my right and the introductions continue. Grossmann apologizes for being from New Jersey, scores an easy laugh, and explains his ties to One Square Inch. A rabbi from Mercer Island expresses his concerns about U.S. foreign policy and human rights violations. A state senator named Rod presses his small business concerns. His wife speaks of student loan repayment issues. The other woman at the table is worried about ever-rising transportation costs.

"You guys brought up some great issues," says Senator Cantwell, mentioning her support of the environment and talking a bit about each of the other issues raised, then switching to a topic obviously at the top of her mind: Iraq and America's foreign affairs policies. She goes on for nearly 20 minutes, offering candid insights and concerns, shooting right past the scheduled 9 a.m. end to the session. When she's done, she thanks everyone for coming, and we adjourn to the outer reception room, where I'm soon steered beside her for my flag-filled photo. I hold out the OSI stone.

"What's this?" she asks.

"This is the stone that marks One Square Inch," I say as I hand it to her. "By saving just one square inch and maintaining it entirely free from any human noise disturbance, I believe one thousand square miles of national park quiet can be managed."

Click, the photo is taken.

Then I give her my one-pager on One Square Inch, which she reads immediately.

A Quiet Place

Whereas, the "National Park Service will preserve the natural resources, processes, systems, and values of units of the national park system in an unimpaired condition, to perpetuate their inherent integrity and to provide present and future generations with the opportunity to enjoy them."—Sec. 4.0, NPS Management Policies, 2001;

And whereas, the "National Park Service will preserve, to the greatest extent possible, the natural soundscapes of parks"; and whereas the natural soundscape is defined to "exist in the absence of human-caused sound."—Sec. 4.9, NPS Management Policies, 2001;

And whereas, Olympic National Park has the greatest diversity of natural soundscapes with the longest noise-free intervals (natural quiet) found at any national park.

It is requested that the Hoh Valley at Olympic National Park be specially designated a *Quiet Place* by NPS Director's Order—a natural place set aside as a sanctuary of silence for present and future generations to enjoy unimpaired by noise pollution.

It is asked that this newly designated *Quiet Place* be protected and managed by a simple but effective soundscape management tool called One Square Inch—for the following reasons:

- A large area (potentially more than 1,000 square miles) can be managed by protecting a single square inch of backcountry wilderness.
- Only individuals or businesses that actually produce a noise impact are asked to change noise-producing behavior. Those individuals and businesses that do not produce an actual impact are unaffected.
- The problem is addressed by a simple and friendly solution. A noise intrusion at One Square Inch is either heard or not. The offender is asked to remove audible noise and not repeat it. This is a voluntary action.
- The corrective action is inexpensive: soundscape management using One Square Inch requires only one part time, non-technical person. The current independent research program at Olympic National Park costs about $2,000/year and is entirely supported through donations.
- This new approach provides immediate results without the need for long term baseline studies and will not interfere with long term natural soundscape management plans that may specifically address bioacoustics.
- Visit www.onesquareinch.org for more information.

As she reads, she bounces the One Square Inch stone lightly in the palm of her hand.

"I want to write this up," she says softly to herself.

Then she looks up at me and says, "I want to write a bill." She searches the room for one of her aides. "Joel, I want to write a bill. Will you take Gordon and John into the other room and begin the process?"

I shake Senator Cantwell's hand and thank her, my heart racing, my spirits soaring. In my elated state, I think Chief Joseph of the Nez Percé Indians was right: "It does not require many words to speak the truth."

Back in the conference room, we talk with legislative aide Joel Merkel Jr. and senior adviser Amit Ronen, who specializes in energy and natural resources, for more than an hour. They ask a number of pointed questions about One Square Inch, among them, how much airspace would need to become a no-fly zone. I tell them I imagine about 20 miles on each side of the valley. They explain that any bill would go to the Energy and Natural Resources Committee, because it deals with Park Service matters. They ask if I've spoken with my congressman, the long-serving Norm Dicks, known as the godfather of Olympic National Park, who also happens to be chairman of the House Interior Appropriations Committee. They say they'll reach out to his office. And they suggest stressing ecotourism as a potential benefit for this economically depressed area of the state. Any bill, they counsel, requires a broad base of support. I take one clear message from their words: This meeting is but Step Two of another long journey.

Back outside, in the crisp air, we agree: that was some coffee. *Click*. I take a photo of the OSI stone backdropped by the Capitol dome. Should a proposed bill actually make it into the Federal Register, I would need to testify. Maybe one day I'll hold up the stone in a congressional hearing. Boy, the mind sure races when fueled by unexpected good news. But back to the present. I check my messages. Secretary Kempthorne did call in sick. We're told we might be able to reschedule something on another visit to Washington. So just one meeting left, with the FAA, which we've heard one Park Service official call "the eight-hundred-pound gorilla."

On our walk to Independence Avenue SW Grossmann and I cross the National Mall on the sidewalk of a street that has been barricaded. We see dozens of police Harley Davidsons weaving through a maze of orange traffic cones set up as an obstacle course. It turns out that this is the eve of a national competition among motorcycle police. Today is practice day. The skill level is impressive, the smooth swaying of the policemen negotiating tight turns that cause many a bike's footrest to lightly scrape the asphalt—one after another, with not a single cone tip. But while all appear to be riding the same model motorcycle, not all sound the same. On the short sprints between sets of obstacles, some of the motorcycles tip my meter above 100 dBA, as loud as the spine-tingling thunderclaps of three days ago. I suspect that some of these Harleys have been modified and are illegal. I can only come up with one reason a policeman would want a louder bike: to enhance his sense of power. And it is true: each 3-dBA increase doubles the energy of sound. The louder bikes, legal or not, intimidate me. But wouldn't quieter bikes be more appropriate for peace officers? Wouldn't they make for more peaceful neighborhoods and also, possibly, a stealthier approach when apprehending criminals? And wouldn't quieter bikes make it unnecessary to ask this question: How can you expect noise ordinances to be meaningful if the police themselves ignore them?

The FAA building is easy to spot. It commands the entire block across from the Hirshhorn Museum. Outside ambience: 57 dBA. After we pass through airport-like security we see an actual airplane, albeit small, and probably of historical value, hanging from the ceiling. I'd like to ask about it while we wait for our escort to appear, but a more informative exhibit grabs my attention: Government and Industry in Partnership for Civil Aviation. This reminds me that the head of the FAA was a former lobbyist for the aviation industry. The nine photographs and seven paragraphs include something about a Southwest Airlines Boeing 737 making its first flight on April 9, 1967. That's the same model plane that I measured at 81 dBA inside the passenger cabin. *Click,* I take a photograph. My camera's flash brings a security guard stepping briskly my way. He tells me no photographs are allowed of these *public* displays. I stow my camera.

The FAA public affairs official who set up today's meeting, Tammy

Jones, meets us in the lobby and escorts us upstairs for our 1 p.m. roundtable. We exchange greetings and business cards with Nancy Kalinowski, director of Systems Operations Airspace and Aeronautical Information Management; Lynne Pickard, deputy director of environment and energy; and Tina Gatewood, an air traffic control and environmental specialist. An additional member of the FAA's public affairs staff, Henry Price, will also be sitting in. I ask for and am granted permission to record the meeting.

A question about the title Sound Tracker on my business card enables me to describe my professional need for and love of naturally quiet places free from human noise and my belief that the Hoh Valley is one of the very few such spots left in America. I hand across the table an audio CD that offers the natural soundscape of Olympic National Park and my one-page fact sheet explaining my campaign for One Square Inch of Silence.

"My aim," I explain, "is to establish the Hoh Valley as the world's first quiet place. This would be essentially a no-flight zone. I'm interested in gaining information from those around the table about how that could be done." I mention that we learned yesterday at the EPA that the FAA is in charge of its own noise levels and ask about FAA advisory circulars that caution pilots about noise-sensitive areas.

Jones interjects now with a procedural ground rule: "If there are any quotes that you'd like to use in your book, we like to ask you to run them through us. Pretty much everything we say should be public information, but just so we can review the quotes."

With perhaps no way of proceeding without complying, we agree to show the FAA any quotes we decide to use from this meeting.

"What," asks Grossmann, "is the noise standard that comes into play over national parks?"

"We're working to develop a standard," replies Pickard. "Can I give you a little introduction to what we do on aircraft noise? As you noted, FAA is responsible for dealing with aviation noise—and this gives you a little quick picture of what we consider a pretty good success story."

She hands each of us a sheet of paper bearing a colorful graph titled "U.S. Aviation Growth and Noise Exposure at 65 DNL: Actual/Predicted Noise Exposure and Enplanement Trends for the U.S., 1975–2005."

"This shows, if you look at the blue area here, that in 1975 there were approximately seven million Americans living around airports with significant noise exposure—very high levels. That number has fallen—2005 is the last year on this chart—to around five hundred thousand," Pickard says, confirming that significant noise is considered an average day-night exposure of 65 dBA. She stresses that the decrease has come even as the number of passengers flown—that is, the number of flights—has increased. Some families have been relocated. Some homes have been shielded against noise with the likes of triple-pane windows. But mostly, she says, federal regulations and newer model jets have helped lower noise-exposure levels.

"And this," she says, handing us a second sheet, "is our vision for the next twenty years. Between now and 2025 we're expecting aviation demand to double or triple in the United States. That is a huge growth." Even so, she adds, "our vision for noise is for absolute reduction of significant noise notwithstanding the growth in aviation."

Pickard talks about another environmental issue, the impact of high-flying jet emissions on global warming, but I'm back on the FAA's definition of significant noise impacts. "Is there any other definition of significance, other than sixty-five DNL, when we are talking about a noise-sensitive area?" I ask.

"If you mean for national parks," says Pickard, "we do not have the body of work that has been done for years. This is based on a body of work that was done over a few decades to arrive at what's the best way to measure and define significant aircraft noise and significant other transportation noise around airports. We are working right now to develop that body of work for special places like the national park system. That's my office, working with National Park Service, and we're investing some money in research.

"When you're talking about lower-level noise it's a whole different game to try to figure out than when you're talking high-level noise. It's harder to figure out for national parks: When is noise a problem?"

Time for another handout. This sheet bears both the FAA logo and that of the National Park Service and includes a graph of a "moment in time" showing decibel and frequency levels for several sounds audible in the Grand Canyon.

"Here, among other things on this chart, you can see aircraft noise," says Pickard. "This is low-level aircraft noise—it's between ten and twenty decibels. It's lower than birdcalls. Now you can hear both of them, if you're standing there. You can hear the birds, you can hear the insects. In this case you can hear the elk, which is much louder than the aircraft, and you can hear the aircraft, but you can't hear it a lot, and if you were walking and not paying attention or talking to somebody, you might not hear it at all, because it's a pretty low noise level.

"It's easy to figure out when you're creating very high noise levels around airports that you've got a problem. It's harder to figure out how much of a problem you have with aircraft noise at lower noise levels—and do you have enough of a problem to make a federal government decision that you should not fly over a certain area, that you should actually restrict it from flight, which is a big decision in this country. And Nancy can show you why it gets to be a big decision and a big problem."

"Just as a quick reference," I jump in. "This is Grand Canyon, but if this was the Hoh Valley—because I go there several times a month and do these same measurements using a sound-level meter which I have here today—the aircraft, instead of being under the twenty-decibel level, it's typically between forty-five and fifty-five dBA. That's actually louder than anything on this chart . . ."

"Sure, sure—and you can get different noise levels," she answers, reading from a different part of the sheet. "These are some examples of sound levels in national parks: sounds of leaves rustling [Canyonlands National Park], twenty dBA, crickets are pretty loud [Zion National Park], about forty decibels. This is the sound of a military jet [Yukon-Charley Rivers]—looks like a training flight, pretty low over the ground at about a hundred meters—at one hundred twenty decibels. Now everybody agrees, that's really loud. When you get down here in the lower noises levels, the question is, Is it loud enough that we should start taking action to reroute? Especially since it's not an easy thing to do. It poses problems."

"So what are the kinds of values that the FAA uses in making that kind of determination?" I ask.

"We're trying to get a better fix on how people in national parks react to noise. It's harder than you would think to even figure out whether people are hearing aircraft noise. We figured out fairly recently—actually, we've got a good model now, but just in the last two years—whether a person with normal hearing could hear an aircraft. We can calculate audibility using measurements and using computer models. It's based on a lot of work that the military did on detectability of noise with submarine work to detect an enemy sound."

She reaches for another document, this time a thick publication: *Report on Effects of Aircraft Overflights on the National Park System: Executive Summary Report to Congress.* In it, she says, are the results of surveys of national park visitors that "identify the phenomenon that I'm talking about, about how hard it is to get at some of the impacts of lower level noise, because they've got some statistics on visitors' responses about when they didn't hear noise when we know that it was measurable. That complicates it, when you're trying to figure out effects on visitors. We're also interested in effects on wildlife and borrowing studies from other wildlife studies—when is noise too much for a national park where it may not be too much for another environment."

"So," I ask, "if the FAA agreed that park visitors at the Hoh Valley or Olympic National Park felt that it should be a quiet place or quieter than it is now or that a jet intrusion they just heard shouldn't exist, and you felt like it was a fair representation of the visitor experience, does that mean the FAA would route flights around Olympic Park?"

"Not necessarily. Again, I'll draw from our analogy of most of our experience around airports. We know that we've still got five hundred thousand people living in significant noise exposure around airports, and we've got additional plans to try to make aircraft noise quieter than it is today. Our goal is to pull that sixty-five DNL inside the airport boundary. We've got plans for improvements in noise-abatement flight procedures and we've got plans for improvements in land-use planning around airports to try to reduce that number, but we've got those people out there right now. It doesn't mean we go out and restrict airport operations. We still have to balance out needs for the national air transportation system with environmental impact, so it doesn't necessarily mean you shut down an airport runway or, in your case, shut down an air traffic route. And I'll let my air traffic folks talk a little bit about some of the difficulties of doing that."

"We did a study, because the National Park Service asked us to consider, what if we created the Grand Canyon area as a flight-free zone," says Kalinowski. "They were concerned about high-altitude noise as well as the work we'd done for the last fifteen years with them in terms of air tour operations. You think about the West as a wide-open air space, but it really isn't. There's a significant amount of air space that's devoted to military activities. There's significant geological boundaries that make some flying in certain areas difficult, and we have a lot of air traffic that operates in that area. So rerouting that traffic created, number one, a significant safety problem, which was of the greatest concern to us. And it also created great inefficiencies and greater delays, which also puts greater emissions into the atmosphere. We basically could not do that at Grand Canyon."

I mention my recent rerouting around an Olympic Park size thunderstorm flying to Chicago.

"I'm not saying it's impossible to reroute, because every day we do reroute," Kalinowski replies. "We reroute for a major military exercise. We reroute for storms, but it always is a great strain on the system. People take tremendous delays. We hold them on the ground. We hold them in the air. Our main purpose and need is always safety, and then for the greatest efficiency and to reduce delays in the air traffic system, and we design the routes to try to be as efficient as possible. Now you know, since the beginning of aviation, routes have been designed based on a ground navigation system. And we are moving toward a satellite-based navigation system. Many of the airlines have that capability now, and we're trying to provide an airspace and an infrastructure that allows them to do as much satellite navigation as possible. It's easier, certainly, west of the Mississippi than east of the Mississippi.

"That means more of a point-to point rather than following a ground trajectory from VOR to VOR, beacon to beacon, where they might zigzag, currently, to some extent. They want to minimize as much fuel and minimize the time to fly."

"Wouldn't this new system make it easier, in the case of Olympic National Park, to route planes around it?" asks Grossmann, noting that Sea-Tac is not nearly as congested an airspace as McCarran, near the Grand Canyon.

Kalinowski says that most recent airspace redesigns have focused on such heavily populated areas as New York–New Jersey, Philadelphia, Chicago, and Atlanta. "Seattle has had some work in the air-space redesign arena. And in that, our purpose and need has been, number one, safety and then efficiency and reduction in delays, but we always do an environmental study with any of our changes either in the airport or in routing across the country. If it's above ten thousand [feet] there are different environmental standards. Certainly above eighteen thousand we're not required to do an environmental assessment, but we're taking into account any noise-sensitive areas we would like to avoid. In Philadelphia we've taken a hard look at the John Heinz Refuge Area for birds. [At] Grand Canyon we've certainly been very focused on the environmental aspects."

Pickard chimes in: "And we have already committed as part of our preparation for the next-generation system to see if we can find more opportunities, as we get away from ground-based navigational systems, to avoid certain particularly noise-sensitive areas like national parks. So that is specifically part of our future plan: to see what advanced capabilities it's going to give us that we don't have now."

"So something like Gordon's One Square Inch is on your radar?" Grossmann asks.

"It's on our radar. I can't make a commitment because, number one, I need to study the new capabilities, and two, I need to consider—there's no such thing as a free lunch. Just from an environmental standpoint, not even getting into safety and capacity, we'll be looking at trade-offs. How much noise reduction do you get, compared to how much emissions increase. If you're talking about routing aircraft in a way that gets them more directly to a place, because they don't have to follow the ground-based nav aid and we can avoid a certain area, like a national park with a more direct routing, that's a win-win. If we can't, if we've got trade-offs, what's the best trade-off? We'll be looking at that."

"The thing that's important to get across," offers Grossmann, "and I'll let Gordon speak to it since he's the listener, is how truly special a place the Hoh Valley is and why he's selected it. There are virtually no quiet places left in America."

"What makes you say that? Just taking my ears out around the country, I think there are a lot of quiet places left," says Pickard.

"Most people do," I say. "That's because most people live in urban environments, and when they're in a *quieter* place than where they have been, they experience it as quiet. If they remain in that quieter place, then things begin to get a little noisier as they open up—just like when you go into a darker room and your eyesight improves and the shadows emerge and all of a sudden you're even able to read when you thought it was a black room. The same thing happens with hearing.

"I've been to every one of our United States in search of quiet, and certainly this trip across America was also in search of quiet. I believe quite confidently that there is no natural quiet left east of the Mississippi River and that the noise-free interval—that time between noise events—is generally under a minute when you're west of the Mississippi River. Sometimes it's a few minutes, and if it exceeds fifteen minutes during daylight hours, that's truly exceptional. The Hoh Valley is the only place I've been to where the noise-free interval is measured in hours. And that is really worth protecting."

I see it's our turn for show-and-tell. Grossmann unfolds the EPA poster from the 1970s. "Quiet: A National Resource," he says. "There was a time when our government was very interested in quiet. And we've gotten away from this."

"I'd disagree," says Pickard. "We put a tremendous amount of effort and money into reducing the aircraft source noise, and we're continuing . . ."

I interrupt. "I would like to say that there's a subtle but significant distinction here. There's a difference between reducing noise, which is important, and we both agree on that, and protecting quiet. Just by reducing noise levels you do not protect quiet. You do not create a quiet experience."

"We are striving for quieter and quieter aircraft technology," says Pickard. "I don't know if you've talked to NASA. They do the longer term foundational research, and they're looking at the silent aircraft. We want to get there."

I'm all for that, I say. "At One Square Inch I have to hear a sound, it has to be audible for it to be significant. So I agree with you: if a plane can fly undetected by my ears at eighteen thousand feet, let it fly. But that is not our situation today."

What's needed, I say, is a shift of emphasis, particularly in our wilderness areas, from reducing noise, which is filled with all kinds of technical issues and formulas and modeling and expenses and time, to protecting quiet, which One Square Inch can do very simply.

"Shall we show them the map?" says Pickard. "You're talking right now about flight restrictions, because we don't have the source-noise reduction that you would like to see." A map is unrolled on the table (see appendix

D). It's labeled "United States Parks and Reservations Special Use Airspace and Airways" and is about the size of a wall poster. Yes! I see at a glance a glaring flaw in their intended ace-in-the-hole.

I concede that there is a problematic, spaghetti-like tangle of lines over much of the nation, including Grand Canyon National Park, and admit that planes probably can't be rerouted around most national parks. Then I point to the northwest tip of the map, at the nearly bare space over Olympic Park. There, I point out, the mess untangles. Just three stray noodles on one border of this big plate of spaghetti need to be moved off Olympic Park. That's all I'm asking to help bring quiet to the world's first quiet sanctuary.

"As I said, we have to do a balance," Pickard replies. "We are responsible for the national aviation system, we have a very robust environmental program, and we are spending quite a bit of resources right now looking particularly at national parks and what sort of quiet—I'll say quiet instead of noise—what sort of quiet is needed to protect the national park environment. We don't have that answer yet. When we have that answer we will see what our prospects are for addressing that."

I realize that our allotted hour is drawing to a close. I ask if an individual airline pilot, when he or she files a flight plan, could request, say, to fly around Olympic National Park.

"Basically," explains Kalinowski, "as a pilot, you're taking your flight plans from your dispatch office. And the dispatch office is making the decision based on what they prefer to fly, in other words, what the company has established as the correct routing for the destination you're going to based on the weather, the winds, and the least amount of fuel you're going to use that day."

"So it's an airline decision?" asks Grossmann.

"It's both an airline and an FAA decision. It's a process called collaborative decision making. In general, they tend to file the same flight plans every day that they know where they're going to fly. They know the routing. It's loaded into the computers onboard in the flight management system, the FMS. In general, unless there is a weather problem, they tend to fly the same routes."

"But if a passenger made a request?" asks Grossmann.

Tina Gatewood speaks up: "If it was an individual passenger who asked a pilot to deviate, then, I think, it would be up to the air traffic controller working that sector to determine whether he had enough leeway in the amount of traffic he was controlling to approve or disapprove that request."

"So those decisions can be made," I say.

"It depends on traffic," says Gatewood.

(Note: In its review of its answers to our questions, the FAA stated: "In reference to Gatewood's comment above: An individual passenger cannot make a request to an airline pilot." Pilots, however, have told us otherwise.)

"You're concerned over that area," says Price, no longer a mere observer from the FAA Public Affairs Office. "Have you spoken to people getting on the jets, saying, 'This is how much more you'll be paying per ticket to make this flight go around the park. Are you willing to pay that?'"

Price presses on. "Seven hundred million people flew last year. I don't know how many million people go to parks each year. Now, we've got to go to everybody in the nation and say, 'Are you willing to pay this much more per ticket to fly around a national park?' That's what you're up against, too."

"As for cost-benefit analysis," I say, "we have three million visitors to Olympic Park each year. We've had two timber mills close. I have seen the poverty in the town of Port Angeles. I live there at the park. To be designated the world's first quiet place and to develop quiet tourism in that area—let me tell you, I do a lot of traveling and it is so noisy. There is a tourist need for this quiet place. It would be a tremendous benefit."

Grossmann mentions that we've been told, by both a commercial pilot and a former FAA controller, that a plane would have to veer only a little bit to skirt the Hoh Valley.

"Veer a little bit off?" says Price. "You wouldn't believe the increase in fuel costs."

"It's more than you think," says Pickard.

"Do you know the cost per minute to fly a commercial jet?" asks Grossmann.

"Nancy?" says Price, tossing the ball.

Grossmann intercepts. "I was told by the Air Transport Association sixty-six dollars per minute, and it's only going to take a couple minutes to make that adjustment."

"And those poor people who need to fly to see Grandma are going to pay more for that ticket," says Price.

"We did a study," says Pickard. "Sometimes it is a misperception that it's just a few minutes, particularly with a ground-based nav aid system. That was certainly the sense of people at the Grand Canyon who wanted to have a flight rezoned over the east end of the Canyon, and you start seeing a domino effect on the national airspace system. So I wouldn't presume it's quite that simple, that cost-free. I don't know. I haven't studied it. But it tends to be more complex than you might think with the interactions of the air space system."

Price restates his question: "Do you think the average person would be willing to pay more?"

"I think a lot of people might be willing to pay something," says Pickard.

I think so, too. And with more public visibility for One Square Inch, it's not unfathomable to think the airlines, in seeking to very publicly do right by the planet and take a green stance by not violating a quiet sanctuary, would see this as positive public relations and, even in hard times, absorb some of the incremental costs. But there's no more time for discussion and debate. I conclude by reiterating that if my idea of preserving quiet in our national parks is possible anywhere, it's possible in the Hoh.

"I understand. That's your belief," says Pickard. "Good luck with your book."

We leave bearing the FAA handouts, including the poster.

Back at my parents' home that evening, I'm both elated and exhausted. When I tell my dad what Senator Cantwell said, he seems both proud and amused, judging by the quiet, wry smile on his face. I know what he's thinking: Fine for now, just wait till the lawyers get involved.

Over the years my father and I have had our differences on a lot of issues, but one fact remains. He is the kindest, most generous and big-hearted man I know, and not just to his family. He wants the world to be a better place for everybody. I consider it a privilege to call him my father. I know that he will be buried someday in Arlington National Cemetery, just across the Potomac River from the nation's capital. But we have never been there together, so as my last stop in Washington, I suggest we go. He says yes, a bit cautiously.

We arrive at the gate to the vast and well-groomed resting ground for tens of thousands of soldiers and dignitaries who have earned the right to be buried at our nation's most honored cemetery. I lower my driver's window and inquire about handicapped access. The woman interrupts: "Excuse me. I can't hear you right now, because there's a plane." My meter climbs. 75, 81, 85 dBA as a jet climbs steeply out of Reagan and flies right over the C&O Canal. I'm directed to a separate office nearby while my father waits in the Acura, air conditioning running, windows up, in a parking area filled with tourists, RVs, and tour buses. He's not as interested to be here as I thought he would be. Maybe it's a little too close for comfort.

"Take a left," Dad directs.

I read from a large stone archway: "On Fame's Eternal Camping Ground Their Silent Tents Are Spread and Glory Guards with Solemn Round the Bivouac of the Dead. Here to rest 15,585 of the 315,555 citizens who died."

"That's just in that section," adds Dad.

We continue our auto tour past the fields of headstones. Then pass a knoll of pilgrims headed to the Kennedy grave site. It is about 11:30 a.m. on a beautiful day.

"I was in all three wars," Dad murmurs. He means World War II, Korea, and Vietnam.

We make a right and stop near the Tomb of the Unknown Soldier. I will go alone. Eighty-five years old, with artificial knees and a bum back, my father can't manage that distance on foot.

Every minute or so, another jet obscures the sounds of footsteps and wind through the trees. At the Tomb of the Unknown Soldier sparrows are chirping, that bright-pitched call echoing off the surrounding marble architecture with stately white columns. I'm one of easily 100 visitors. Hardly anyone is speaking. The loudest sound is the 61 dBA heel snaps measured at 30 feet, made when the guard quickly pivots. The words on the Tomb proclaim: "Here rests in honored glory an American soldier known but to God."

As the noon hour approaches, the crowd swells for the famous Changing of the Guard ceremony. The bell tower chimes, then comes a series of deeper gongs: 66 dBA. After the fourth of 12 gongs, the bell is no longer the loudest sound. Yet another jet (75 dBA) intrudes upon this sacred ceremony.

"Ladies and gentlemen, please, may I have your attention," a soldier announces in a firm but respectful voice. "I am Staff Sergeant Dickmyer of the Third Infantry Regiment of the United States Army to hand over the relief of the Tomb of the Unknown Soldier. The ceremony that you are about to witness is the Changing of the Guard. In keeping with the dignity of this ceremony it is requested that everyone remain silent and standing."

Those not already standing rise, and an enormous swoosh of clothes cues me to how big the crowd has become.

"Thank you," says Staff Sergeant Dickmyer.

The crowd remains respectful and quiet during the ceremony. But not the sky overhead. About every minute another jet passes, shattering the respectful silence. Here at the Tomb of the Unknown Soldier. At the resting place of President Kennedy. And in the not too distant future, the same will occur at the Arlington Cemetery plot where my father will come to rest.

Will it ever stop? Forgive me if I want to shout this message to the sky. The constant noise is demoralizing. A thunderstorm, if it were to suddenly appear over Arlington, would cause the FAA to reroute air traffic immediately and without question. A different kind of thunderstorm needs to build, a thunderstorm of public concern and indignation. It is time for us to raise our voices in the name of silence, not just to restore endangered natural soundscapes in our national parks, but to restore a sense of balance in our lives—a balance that, sadly, many people don't even notice has gone but that may be essential for our survival.

I started this journey by asking for just one square inch of silence at Olympic National Park. Let me say this: FAA airspace over the United States would be reduced by only four one-hundredths of one percent (0.04 percent) if Olympic National Park were to become off-limits to aircraft. Now, at journey's end, I want more: I want my right to quiet restored at home, at work, and in my schools and community. And elsewhere. Near the Tomb of the Unknown Soldier is a circular cast-iron sign that says "Silence and Respect." It does not take many words to speak the truth. The FAA should, as a bare minimum, restrict aircraft into and out of Reagan National for the noon Changing of the Guard on two days a year: Memorial Day and Veterans Day. Enough said.

We need silence to honor and remember those who died in service of this nation and to help inform us of who we are and who we want to become. After returning from One Square Inch, Kathleen Dean Moore, a philosophy professor at Oregon State University, wrote this:

> When wind plays across the maple leaves and sets them in motion, it's we who are most deeply moved. No one knows why music speaks so directly to the human spirit, but it's possible to imagine what it says—that we are not separate from the world, not dominant or different. Like stone, like water, we carry the shape of the world in our rustling. We are all music, we are all matter in motion, all of us, together sending our harmonies into a black and fibrillating sky.

Saving silence is not an inconvenient chore, but an awakening joy. When we listen to silence, we hear not absence, but presence.

I cannot imagine a future without quiet, nor do I care to try.

Epilogue

Echoes

The process of this book, as of now a year and a half in the making and still short of publication, has been personally challenging. I consider myself shy at heart and enjoy my solitude. Working as the Sound Tracker, hiking and recording nature sound portraits, visiting with friends, and bodysurfing, I was living my dream. Occasionally I would get a call to travel and visit other fascinating places. But One Square Inch of Silence changed all that. When I heard Olympic Park's aural solitude eroding and took it upon myself to establish the nation's first quiet sanctuary and protect it from noise intrusions, I set in motion a process that took me to Washington, D.C., and, frankly, places I would have preferred not to go, such as onto noisy roads and up elevators and into one conference room after another.

But one of those elevators took me to Senator Maria Cantwell's office. At last check, I learned that the senator remains interested and is evaluating the best way to move forward.

Let me provide a few more updates.

The sun set on Earth Day 2008 and the latest unmet deadline for the National Park Service and the FAA to agree on means for a substantial restoration of natural quiet in the Grand Canyon. That deadline had been set eight years earlier.

It's now 21 years and counting since the 1987 passage of the first congressional legislation to control air traffic over Grand Canyon National Park. It's clear the two agencies are communicating, because a couple of weeks prior to Earth Day, on April 9, 2008, the National Park Service gave "clarifying" notice in the Federal Register to remove all aircraft flying above 17,999 feet MSL (above mean sea level) from their near-term, long-overdue rule making for restoring natural quiet at Grand Canyon National Park. Despite a 2002 Federal Court of Appeals Decision (yes, there have been lawsuits slowing things down), which ruled in part that the Grand Canyon Overflight Act did apply to high-flying jets, the Park Service appears to be letting go of the rope on high-altitude commercial and private jets in its tug of war with the FAA, at least at this one battleground national park. Even with air tours someday curtailed, this would signal a death sentence for natural quiet. If ever applied nationwide, there would be no way to achieve quiet without futuristic, truly silent jets at Canyonlands, and One Square Inch of Silence in Olympic National Park would never live up to its name.

"Take people's words and reflect upon their actions," Karen Trevino told me in Colorado when I stopped in at the National Park Service Nature Sounds Program. I'm reflecting, all right.

Here's something that just arrived in my e-mail box from Skip Ambrose, whom I interviewed just outside of Canyonlands. He has attached a copy of a letter from Dan Elwell, assistant administrator of FAA Aviation Policy, Planning, and Environment, to David Verhey, assistant secretary for the Department of the Interior's Fish and Wildlife and Parks, dated March 6, 2007. Ambrose points me straight to this FAA statement to the Department of the Interior: "The inclusion of all aircraft within the purview of this Act [National Parks Overflights Act of 1987], following a 2002 U.S. Court of Appeals decision, leaves us in an untenable position." Ambrose emphasizes that this is the first he has seen the FAA acknowledge that high-altitude jets make noise en route. But what really catches my eye is the very next sentence: "It is impossible for the Secretary of the Interior to fulfill the Act's mandate to substantially restore natural quiet, using NPS definitions, without moving routes that would severely affect the safe and efficient management of airspace."

So there it is in black and white. The FAA admits that if we want natural quiet at the Grand Canyon, jets have to fly around, at least by current NPS definitions. But lower the bar, or in this case the ceiling, as the National Park Service just proposed in the Federal Register, and exempt aircraft flying at 18,000 feet or higher and, bingo, the "problem" goes away.

Should air traffic actually double or triple during the next several decades, the FAA's argument has merit, at least for most of the 390 national park units, many centered in or near urban areas. But I doubt that all national parks need be under flight paths to keep air traffic safe, especially in the upcoming switch to GPS navigation. Certainly not Olympic National Park, where only three jetways need to be moved or abandoned: J54, J523, and J589.

Closer to home, the long-awaited Final General Management Plan for Olympic National Park, a plan that will likely guide the park for more than a decade, was finally issued. The copy I hold says many wonderful things about the value of natural soundscapes and natural quiet, but provides hardly any examples and lists none of the unique acoustic features of this park. Nor does it mention that Olympic has the longest noise-free interval of any national park. More telling, this new management plan replaces "degradation due to human-caused noise" with "unacceptable impacts." Here, too, the National Park Service would lower the bar.

When I e-mailed Cat Hawkins Hoffman, the chief of the Natural Resources Division of Olympic National Park, a week ago, asking for "as little as fifteen minutes or as much time as you can offer" to discuss the management plan, she e-mailed back:

> I talked with other staff here regarding your interests, the future of natural quiet/soundscape plans for the park, etc. I know that all I can offer right now—speaking for my staff and our current obligations—is an unsatisfying response that there are numerous high priority issues we're barely keeping up with. A near-term need (in the planning realm) is beginning a wilderness management plan which will be a large effort. And although dam removal (Elwha, Glines Canyon) is still 3–4 years away, many tasks are underway in my division to prepare for it, such as several fisheries projects, plant propagation, exotic plant removal, documentation of baseline conditions, etc. We are chronically at a dead run.

Apparently, with little time or resources (not even a sound-level meter) or will, natural soundscape management in Olympic National Park, the listener's Yosemite, boils down to a stone and a jar. They work together—the stone marking ground zero in a call for the preservation of natural quiet and the jar collecting the thoughts, public opinion, on the significance of silence as a natural resource and endangered national treasure.

As you know, when I put the Jar of Quiet Thoughts at One Square Inch I pledged that the written messages of the quiet pilgrims would never be published, for they are meant to be read only at that one, very special spot. But let me lift the lid a little and share a few words: *soft, peaceful, solitude, peace, love, eternal, hope, wonder, golden, supreme, God, present, bliss, honor, space, growing, prayer, truth, depth, still, patience, gratitude, resonance, serene, innocent, dream, music, dance, brave.* How different are these words, written in silence, than, say, those of urban-dwelling graffiti artists who travel noisy urban streets and write on walls and signs and subway cars? The thoughts in the jar reassure me that I am not alone in my love and wonder of quiet. But now the jar, too, is endangered.

I learned last week that Olympic National Park intends to remove it. Barb Maynes, the park public information officer, told me in a phone call that the jar has to go because I do not have a permit for it. And if I applied for a permit, she guessed, I'd be denied. She revealed her agency's concern that the popularity of the site has worn a path and damage has occurred. I assured her that the path was more popular with the elk than with pilgrims, and that no damage has occurred. Just yesterday, in fact, I hiked to OSI to check on the jar and make a backup copy of its contents. I saw lots of elk prints in the muck, but no signs of human damage from the trail to the moss-covered log with the stone atop it.

When I returned, my e-mailed request for a permit application to Jerry Freilich, research coordinator at Olympic, brought a sympathetic but hardly encouraging reply. It said in part:

> I have heard about the one square inch and had seen your website before these latest e-mails, so I am generally aware of what you're doing. I am sure that all of us in the Park Service are supportive of the idea that natural soundscapes should be protected in the parks. But I do not understand how your project would qualify for a scientific permit. Scientific permits are issued for two main reasons: to protect park resources and to ensure that we gather the information and analyses that come from the research. Although not every project involves collecting something, in general, scientific permits are issued when specific data are being collected, hypotheses tested, or there is a formal study design. Each of our scientific permits is peer-reviewed and every investigator is required to submit an end-of-year report.
>
> Although there are gray areas that we consider individually on a case-by-case basis, in general we do not permit class projects for general collection. We do not issue permits for private individuals who are merely observing birds or flowers. And we do not issue permits for "installations" (that's the word used in the Wilderness Act) of equipment in wilderness unless no other site could be found in a non-wilderness area.

> I would say that your project does not fit the usual sense we have of scientific permits and your project does not fall within my usual activities. The park issues "special use permits" that might cover your situation but I am not sure of their requirements.

Freilich gave me the phone number and e-mail address of someone to contact about a special-use permit. So the dance continues. But it appears that the tacit support for One Square Inch that I had from former park superintendent Bill Laitner may not be shared by the new acting superintendent, Sue McGill.

The bigger surprise: I'm not discouraged. I'm actually optimistic. Anything worthwhile usually comes hard and has to be fought for. I was reminded of this, blessedly so, in perhaps the best of all possible ways.

When last seen in these pages, back in chapter 4, my daughter, Abby, had bid me goodbye after basically trashing One Square Inch as a stupid idea and breaking her promise to accompany me on the early stage of my journey. Hoping that her attitude might have changed, I asked her to come along when I replaced the stone after returning from Washington, D.C. I invited her brother, Oogie, too, and we all piled into the Vee-Dub, which, by the way, I had shipped back across the country. No sense pushing my luck any further. The three of us hiked up the Hoh Valley in horrible weather, sloshing through the mud but joking much of the way. Everyone's spirits seemed high. But after hiking 3.2 miles, when it was time to leave the trail for One Square Inch, Abby again turned mulish. She refused to walk the last 100 yards. On the hike back my frustration boiled over. I tried to get her to explain how she really feels about One Square Inch, questioning her sincerity for choosing it as the topic for her senior project, a requirement for graduation. She got sarcastic. I got angry. Farther down the trail I apologized. Abby accepted my apology, and we hugged. But my heart continued to ache.

Two weeks ago, Abby shared with me a copy of her senior project, called "Preserving Natural Quiet," which asks: "How can preserving one square inch of natural quiet at Olympic National Park change noise pollution worldwide?" I braced myself as I started to read, then felt a great burden lifting when I came to these words:

> I know that you can't redo the past, but if I could, I definitely would. . . . I have grown up a lot. . . . This project has inspired me to believe that one person can make a difference and I will make a difference. Who would have thought that preserving just one square inch of land from noise pollution would virtually affect an entire 1,000 square miles surrounding it? Just imagine the possibilities if all of those 390 National Parks execute the same plan and helped in the preservation of silence. . . . I think people hear the words, "No that's impossible" or "Well that sounds great, but that's asking a lot" excessively. I learned that in order to truly succeed people need to be more optimistic and ask themselves, "Now what can I do to make this possible?" In my opinion, that is the winner's way of looking at something. . . . From this experience, I have learned not to let anyone tell you no. Just do what you have to do to succeed and be as optimistic as possible.

One Square Inch has progressed from a "me" to a "we" organization. My board of directors now consists of Samara Kester and Elliott Berger. Our first annual meeting in February included a hike to One Square Inch. We hope to receive 501c3 tax-exempt status by the end of this year. Donations are being accepted at www.onesquareinch.org.

The profusion of sound level measurements I took crossing the country serves two enlightening purposes. First, grounded in dBA values of events familiar to us, we gain a basic acoustic fluency and will better understand when someone, a scientist or a politician or a government bureaucrat, speaks about noise levels. Second, there will be an equally educational, *déjà son* recognition of the sounds captured in the Sonic EKG of America I've included in appendix E. While whimsical, the graph is also grim, filled with noise spikes. Quiet is rare. Noise has indeed become a modern plague found nearly everywhere, and often at unsafe levels. Noise has become so prevalent it's taken for granted. So overlooked, so systematically unmonitored, that it is not among the 25 metrics (such as drinking water, indoor air pollution, trawling intensity, burned land area, industrial CO_2 emissions, and pesticide regulation) that constitute the annual rankings in the Environmental Performance Index annually issued by the Yale University Center for Environmental Law and Policy. "There's no data. We need data that's collected on a methodologically consistent basis over 150 plus countries," explains the Center's director, David Esty. "There is no such thing in the noise category." (This 10-year-old global Environmental Performance Index also fails to factor in other environmental issues, such as chemical exposures, wetlands protection, and recycling efforts—also for lack of consistent data.) Meanwhile, as our cities grow more toxic with noise, and we drift toward a nation of shouters, the sound of our footsteps has all but disappeared. Is it any wonder we've lost our way?

To find the path we need only return to the very legislation that created the national parks, the Organic Act of 1916, and heed seven words worth repeating, a thousand times if necessary: "unimpaired for the enjoyment of future generations." Back then, Congress could hardly have imagined today's helicopters over Haleakala and sightseeing planes over the Grand Canyon, the jets over Canyonlands and the snowmobiles roaring through Yellowstone. But they would recognize the still predominant, stunning quiet of the Hoh Valley.

"Unimpaired for the enjoyment of future generations."

It is time for us to prepare to celebrate the 100th anniversary of the national parks, not for what our parks have become, but for who *we* have become because of them: defenders of wilderness. As Edward Abbey said, "The idea of wilderness needs no defense, it only needs defenders."

One Square Inch is all it takes. Defend that, and we have the world's first quiet place.

Gordon Hempton
—Welcoming spring at Joyce, Washington, May 1, 2008

Appendix A

Correspondence with James Fallows

Sometime after my meeting with Skip Ambrose, described in chapter 6, I tracked down the essay that had so infuriated him. It appeared in the 2001–2002 edition of *Wilderness Magazine,* which is published by The Wilderness Society. Called "Loving the Land from Above," it was written by James Fallows, a respected national correspondent for *The Atlantic* and author of six books, most recently *Free Flight: From Airline Hell to a New Age of Travel*. I e-mailed Fallows after reading the essay; our ensuing electronic conversation is included here.

Loving the Land from Above

By James M. Fallows

Wilderness and machinery are usually a bad match. Machines are noisy. They grind things up. To the extent that we value wild forests, prairies, and canyons as specimens of life before the industrial age, the mere presence of machines diminishes the tranquility and sense of time-transport that can come from pristine vistas.

So I took it for granted that, through the first few decades of my life, my most memorable and enriching experiences in the wild would also be the least mechanized. Hiking the length of Santa Catalina Island on a week-long Boy Scout trip and camping out each evening. Cross-country skiing through the Methow Valley, east of the Cascades in Washington. Drifting on a raft down rivers in Colorado, Utah, Texas. Swimming with a snorkel mask among coral reefs.

But in the last few years I've come to appreciate wilderness in an additional way. This leads me to make the contrarian case for one particular noisy machine as a friend of wilderness. I have now seen much of the nation's wilderness from above while flying small airplanes, and I am convinced that if more people did the same, they would feel more deeply about the importance of preserving wild space.

Let me mention the obvious caveats. Precisely because airplanes are noisy, the benefits they provide their passengers and pilot often come at the cost of people on the ground. Fortunately, the effect is transient—unlike a paved road or a jeep's track across the desert, the marks of a plane's passage dissipate quickly. More important, it is controllable. Noise impact falls off dramatically the higher a plane goes, which is why flight charts are full of markings requiring pilots to stay 2,000 feet above sensitive bird-nesting areas, 4,000 feet above wilderness hiking or canoeing areas, 14,500 feet above the most scenic parts of the Grand Canyon—so high that pilots are required to use oxygen masks there. (So why is the Grand Canyon still noisy? Because commercial tour operators are allowed to fly lower.) And what about pollution? The small airplane I fly gets 25 miles to the gallon, better than many cars.

There is a less obvious drawback of the aerial view, affecting the people inside the airplane. Everything about scale and proportion is altered by seeing the world from a few thousand feet up. The fine-grained details of nature are blurred or invisible—no individual trees or glens or bends in a stream. The vertical perspective is foreshortened. Neither the natural splendor of Niagara Falls nor the man-made heft of the Grand Coulee Dam is as impressive from above as from the ground.

Moreover, the initial impact of the aerial view is subtly anti-wilderness, in that it makes the North American continent seem to be nothing but open space. You fly eastward from the crowded San Francisco Bay area, and in 30 minutes you're at the Sierra foothills—and then for what seems like an eternity, all you can see is the mountainous lunar desert of northern Nevada, with rarely a road or structure in view. Fly westward from New York or Washington, and within half an hour you're trying to make out hamlets amid the forest growth. If you took only one small-plane flight, you might think: What "wilderness problem"? There's empty land wherever you look.

But after taking several hundred such flights in recent years—up and down the two coasts, across the Rockies and the Great Plains, over most of the country except the Deep South—I have reached the opposite conclusion. Viewing the land from above is among the most effective ways to appreciate its wild areas.

The connected, knit-together nature of the landscape is the most lasting impression of the aerial view. Everything is tied into everything else. The city runs into the suburb, the suburb peters out into the woods. As you fly eastward from the 100th meridian toward the Atlantic Coast, with every mile you can see the land growing moister, the trees less sparse, the roads and property lines more regular as prairie gives way to farmland and then to industrial towns. You can see the exact frontier along which wilderness and settlement meet. Once, over the Midwest, I could see factory smokestacks in the distance ahead of me—and out the left window, small ponds that were waterfowl spawning areas, dotted with thousands of birds. The wild areas seemed the more important and the more vulnerable for being so directly connected to the towns. And you can judge more acutely the pressures on wild land. The "beauty strips" that shield clearcut areas from highway travelers do not fool anyone in an airplane.

I think I am beginning to understand why Charles Lindbergh spent his last decades as a passionate environmentalist; why Beryl Markham, the English writer who as a young woman learned to fly in Kenya, as an old woman became a Kenyan naturalist; and why, of course, the early astronauts knew they had transformed public consciousness of the natural environment with the first photo of the entire shimmering globe. They all depended on machines—in the case of the astronauts, the noisiest and most complex machines ever invented. But with their machines they helped us see what Antoine de St. Exupéry called "the true face of earth," including its wild areas.

To: James Fallows
Subject: "Loving the Land from Above"; invitation to OSI

Mr. Fallows,

I recently read your article, "Loving the Land from Above," in the 2001–2002 issue of The Wilderness Society's magazine and while I enjoyed your descriptions of the view out the window of your plane, I'm troubled by the very premise of your piece.

I am a professional nature-recording artist. I make my living recording in America's most pristine wilderness, an increasingly difficult job because of ever-increasing man-made noise intrusions. I've been told by a commercial pilot that his plane, flying at 36,000 feet over Yosemite, is unheard on the ground. I know that to be untrue, just as I know that similar flights over other national parks like Canyonlands and Olympic National Park, where I've established One Square Inch of Silence, also break the soothing spell of natural quiet on the ground.

You, too, do not seem to realize that your flights also disturb the natural silence enjoyed by backcountry hikers like myself, for you dismiss the noise as "transient." Yes, you fly at higher altitudes than the much more disruptive air tour helicopters and planes, and your aircraft's noise intrusion (on the order of 30 to 35 dBA) is quantitatively much lower than the 60 dBA of many scenic flights. Nevertheless, the noise of your plane, introduced into a natural ambience [whose decibel level is] in the low 20s, scars an existing peaceful moment, for it is 10 times as loud as the natural sounds—of birdsong, the wind through the trees, a burbling brook. Like other natural-quiet seekers, I may have hiked for days to reach a peaceful spot, hoping to escape all human noise—the very definition of natural quiet that our national parks are charged with providing. But your flight shatters my peace.

You take pride that your small airplane gets 25 miles to the gallon. On foot, you can do even better. And do something you cannot high above the ground. Listen to the landscape.

One of my favorite nature "sound recordists" is John Muir, who wrote about what he heard during a winter storm in 1874:

> I drifted on through the midst of this passionate music and motion, across many a glen, from ridge to ridge; often halting in the lee of a rock for shelter or to gaze and listen. Even when the grand anthem had swelled to its highest pitch, I could distinctly hear the varying tones of individual trees—spruce, and fir, and pine, and leafless oak. . . . Each was expressing itself in its own way—singing its own song, and making its own peculiar textures. . . . The profound bass of the naked branches and boles booming like waterfalls; the quick, tense vibrations of the pine-needles, now rising to a shrill, whistling hiss, now falling to a silky murmur; the rustling of laurel groves in the dells, and the keen metallic click of leaf on leaf—all this was heard in easy analysis when the attention was calmly bent . . .

You can read about my efforts to protect the natural soundscape at www.onesquareinch.org. Would you like to go for a hike to OSI? I think that we would have some interesting conversations and enjoy listening to the land too.

Sincerely,
Gordon

~~~~~

From: James Fallows
To: Gordon Hempton
Subject: RE: "Loving the Land from Above"; invitation to OSI

Dear Mr. Hempton:

I understand what you are saying, about my article from six years ago.

If it makes you feel any better, I have been living in China for the last nearly two years and have done virtually no flying of any kind. And if you want to see real noise pollution (and pollution of every other sort), I have places to show you that you would hardly believe.

J Fallows
~~~~~

From: Gordon Hempton
To: James Fallows
Subject: RE: "Loving the Land from Above"; invitation to OSI

Mr. Fallows,

I'm guessing that when you say "I understand what you are saying," I am not the first to write you about flightseeing.

I would like to know more about your impressions of China. Murray Schafer wrote the book *The Tuning of the World* and coined the term "soundscape." In that he suggests that a noise standard for cities should be as simple as the ability to hear footsteps—something that is not achieved in most U.S. cities.

In Venice, I have listened to leather soles tap on stone walkways, clearly, from over a block away; but there was no vehicle traffic in the older district where I stayed. Can you offer any examples of noise pollution in China or, better yet, the presence of quiet anywhere?

I'm very much interested in a quick comment about this. I have been around the world three times but never to China.

All the best,
Gordon

~~~~~

From: James Fallows
To: Gordon Hempton
Subject: RE: "Loving the Land from Above"; invitation to OSI

Dear Mr. Hempton:

Actually, you are the only person to have written in that regard.

The only places I have seen where it is quiet in China are where people are so poor they can afford no mechanized equipment of any kind—they plow with oxen or pull the plows themselves, they harvest wheat with scythes, they winnow the wheat by throwing it up in the air and letting the chaff blow away. This is indeed quiet, but it has some drawbacks. Plus there is the true wilderness/desert of Qinghai, Xinjiang.

So, I am sensitive to noise, especially whirring "white noise." If I could, I would outlaw leaf blowers in America, which I detest. One reason I don't like New York is how noisy it is, especially at night. I am glad this is your cause. But oddly it is a sign of America's good fortune that it has worries of this scale. (One thing about noise, compared with other forms of pollution, is that its half-life is extremely short. When it stops, it STOPS.)

As I write there is a pile driver going all night, across the street.

J. Fallows
~~~~~

Appendix B

Indianapolis Noise Profiles

The concert and speedway profiles were provided by Elliott Berger of Aearo Technologies. Each plot has two lines. The upper line is the Lmax value, the maximum-sound-pressure level, measured in decibels, for each minute during the event; and the lower line is the Leq, the equivalent average-sound-pressure level, also measured in decibels.

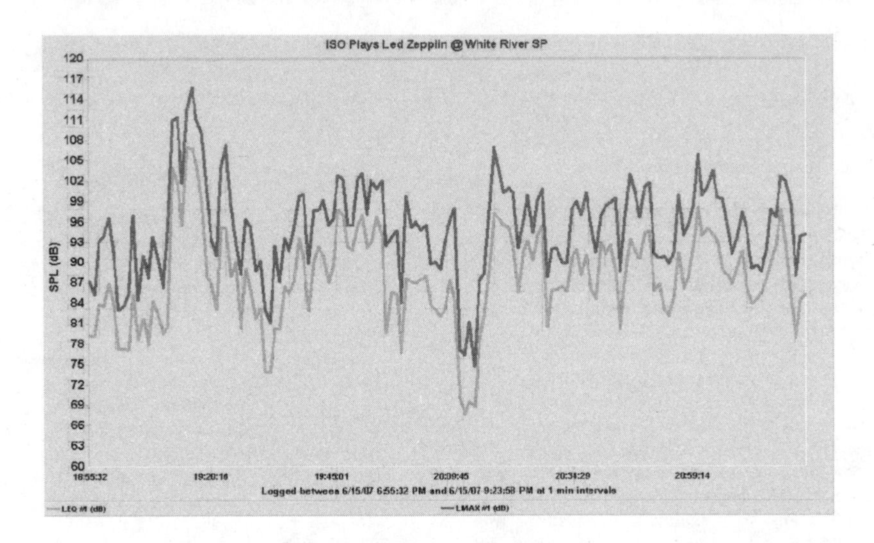

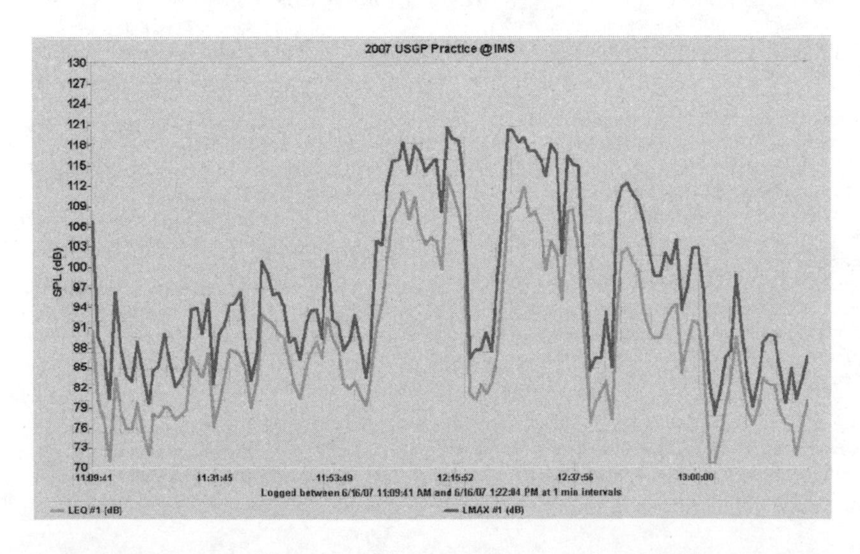

Appendix C

Kempthorne Letter

March 26, 2007

The Honorable Dirk Kempthorne
United States Department of the Interior
1849 C Street, N.W.
Washington DC 20240

Dear Secretary Kempthorne:

We write to thank you for your efforts on behalf of our National Park System. Your requested increase in general funding for the national parks is striking and urgently needed. We are hopeful that Congress will approve your request and thereby significantly bolster the National Park Service's ability to preserve and interpret our common heritage for the benefit and enjoyment of the American people.

We recognize that there are great strains on the federal budget. Your extraordinary support of a history-making operations increase for the national parks merits our enthusiastic support. You are absolutely right to emphasize that a pressing need exists for this significant increase in park operational funding. The eroding condition and health of irreplaceable natural, historic and cultural treasures is painful to witness, as is the decline in resource protection and visitor education programs. We admire your commitment to reverse these declines.

It was our privilege to be stewards of the National Park System. Collectively, our high-level management experience spans half a century. We were heartened when you reaffirmed the keystone of that stewardship, that the fundamental mission of the National Park Service is the conservation of park resources. Indeed, your strong declaration of support for the longstanding management policies that have governed the life of the parks reassured the American public and the Congress that you will insist upon the highest protection of park resources and values and will not allow uses and activities that conflict with this founding principle of the national parks.

Given this, we must express our alarm over a proposal in Yellowstone National Park that would radically contravene both the spirit and letter of the 2006 Management Policies. The proposal is to *escalate* snowmobile use as much as three-fold over current average numbers even though scientific studies have demonstrated conclusively that a two-thirds *reduction* in average snowmobile numbers during the past four winters is principally responsible for significantly improving the health of the park for visitors, employees and wildlife.

The latest National Park Service study illuminates in detail that allowing Yellowstone's current average of 250 snowmobiles per day to increase—to as many as 720 snowmobiles—would undercut the park's resurgent natural conditions. Specifically, the study reveals that snowmobile noise would return to areas of the park where visitors are currently able to enjoy natural sounds and quiet. It demonstrates that exhaust would increase in Yellowstone's air. It sidesteps a recent recommendation made by Park Service scientists: that in order to minimize disturbance of the park's wildlife, traffic should be kept at or below current levels, not expanded. The study also provides clear evidence that reducing snowmobile numbers still further—from 250 per day to zero—while expanding public access on modern snowcoaches, would further improve the park's health.

The development of four-stroke snowmobiles has brought reductions in air and noise emissions compared to traditional two-stroke snowmobiles. But emissions from the newer snowmobiles remain significantly greater than those of modern automobiles. Moreover, in the context of Yellowstone's winter season, impacts from four-stroke snowmobiles are frequently accentuated by inversions, lack of breeze, the park's intrinsic quiet, and the fact that wildlife in a weakened condition tend to concentrate where thermally influenced rivers and thinner snow cover provide more accessible food. These areas are precisely where Yellowstone's roads are located. Enabling every 100 visitors to move through these sensitive areas requires ten modern snowcoaches or, by contrast, 80, 90, even 100 individual snowmobiles.

In each of four separate studies since 1998, costing a cumulative $10 million, the National Park Service has verified conclusively that greater volumes of traffic required by an emphasis upon snowmobiling add dramatically to air and noise pollution and disturbance of Yellowstone's wildlife. On at least three occasions, the Environmental Protection Agency has independently corroborated that providing access by modern snowcoach and phasing out the use of snowmobiles will provide Yellowstone's visitors, employees and wildlife with dramatically healthier conditions. By 4-to-1 margins, the American public has said throughout these studies that it wants nothing less for Yellowstone than the best available protection.

We admire your support of the 2006 National Park Service Management Policies and your declaration that upholding these policies is fundamental to the nation's commitment to preserve its national parks. It is our profound hope that in our country's oldest national park you will insist that your commitment be upheld—that the traditional conservation emphasis of the national parks will be continued. The current proposal to accommodate increasing snowmobile use in Yellowstone is at odds with these policies:

- *"...the Service will seek to perpetuate the best possible air quality in parks..."*
- *"The National Park Service will preserve, to the greatest extent possible, the natural soundscapes of parks."*
- *"Where such use is necessary and appropriate, the least impacting equipment, vehicles, and transportation systems should be used."*
- *"NPS managers must always seek ways to avoid, or to minimize to the greatest degree practicable, adverse impacts on park resources and values."*

We note that during the protracted discussion over winter use in Yellowstone, visitors adventuring to Old Faithful and other destinations in the park have increasingly been choosing modern snowcoaches as their means of access. These "least impacting" vehicles, which minimize "adverse impacts on park resources and values," are also considerably more affordable for visitors than snowmobiles. Snowcoaches are more accommodating of older visitors and children than snowmobiles. And because they facilitate conversation between guides and visitors and among family members, they have given rise to a boom in visitor education. In all these respects, the growing popularity of snowcoaches has been enormously positive for Yellowstone and its visitors.

Mr. Secretary, we join as former stewards of the national parks in urging you to demonstrate in our country's oldest national park the wisdom and value of the 2006 Park Service Management Policies. You were right to call them the "lifeblood" of our country's commitment to its national parks. Ensuring that these policies are upheld in Yellowstone is one of the greatest contributions that you can make to the future of our National Park System.

Sincerely,

Nathaniel P. Reed
Assistant Secretary of the Interior
1971-1976

George B. Hartzog, Jr.
National Park Service Director
1964-1972

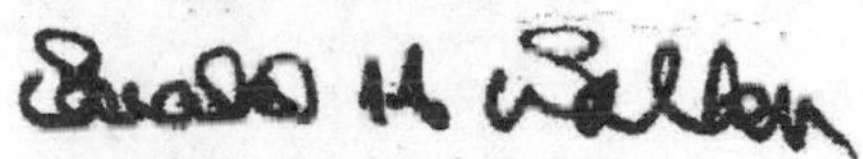

Ronald H. Walker
National Park Service Director
1973-1975

Gary Everhardt
National Park Service Director
1975-1977

Russell E. Dickenson
National Park Service Director
1980-1985

James M. Ridenour
National Park Service Director
1989-1993

Roger G. Kennedy
National Park Service Director
1993-1997

Robert Stanton
National Park Service Director
1997-2001

William J. Briggle
National Park Service Deputy Director
1975-1977

Denis P. Galvin
National Park Service Deputy Director
1985-1989 and 1998-2002

Michael V. Finley
Yellowstone National Park Superintendent
1994-2001

CC: Sen. Jeff Bingaman
Sen. Pete V. Domenici
Rep. Nick J. Rahall
Rep. Don Young
Lynn Scarlett, DOI
Brian Waidmann, DOI
Mary Bomar, NPS
Mike Snyder, NPS
Suzanne Lewis, NPS

Note: When this letter was sent, Fran Mainella, director of the National Park Service (2001–2006), was not among the signees. Less than a year had passed since she left her job, and federal policy governing the interactions of former presidential appointees with their former agencies prohibited her from adding her name to the list of supporters. In November 2007, Mainella did so, stressing, "When there is a conflict between conservation and use, conservation is predominant."

Appendix D

FAA Map of the Continental United States

United States Parks and Reservations Special Use Airspace and Airways (Air Traffic Airspace Laboratory, 5.19.2003, insert of Olympic Park added)

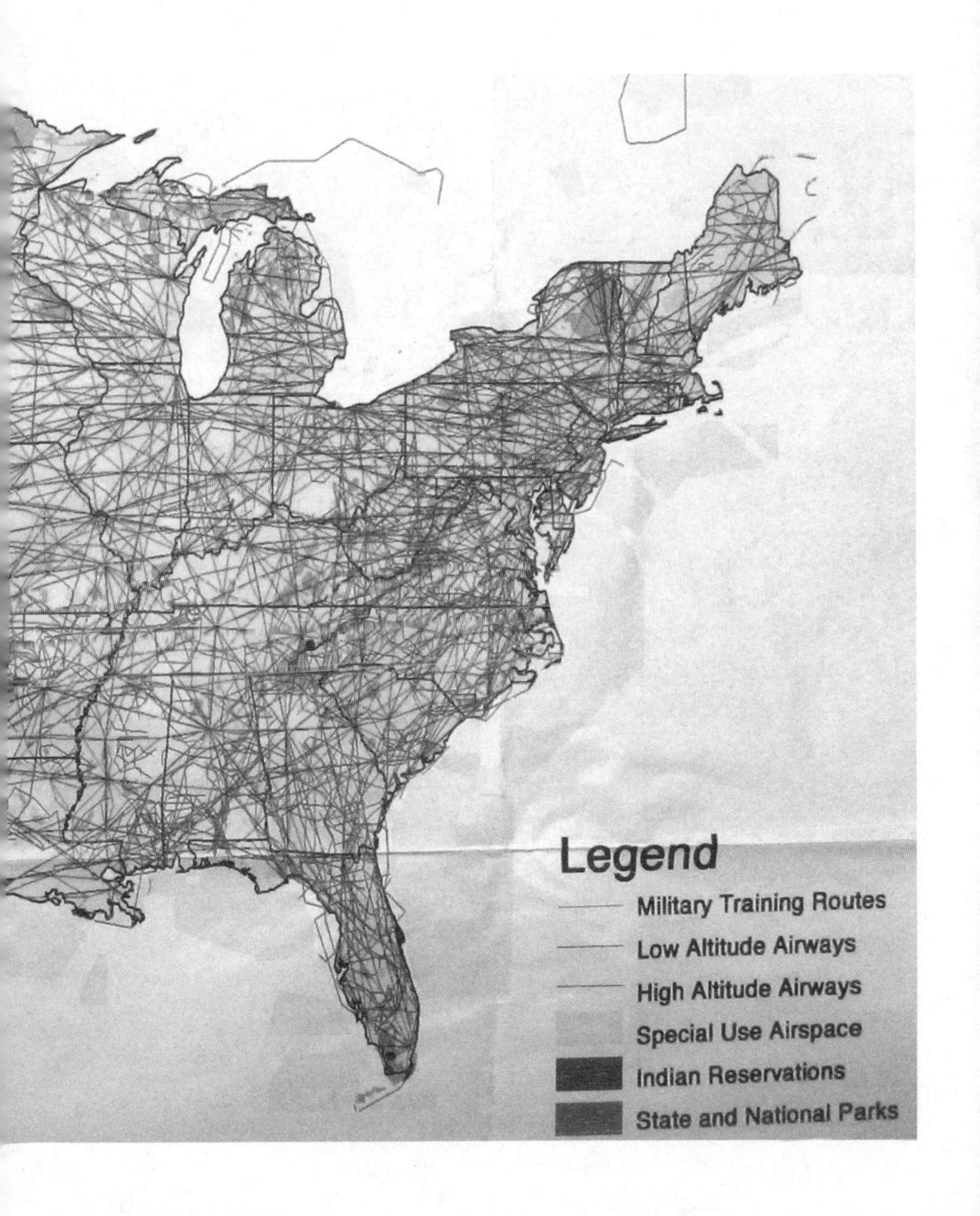

Appendix E

Sonic EKG of America

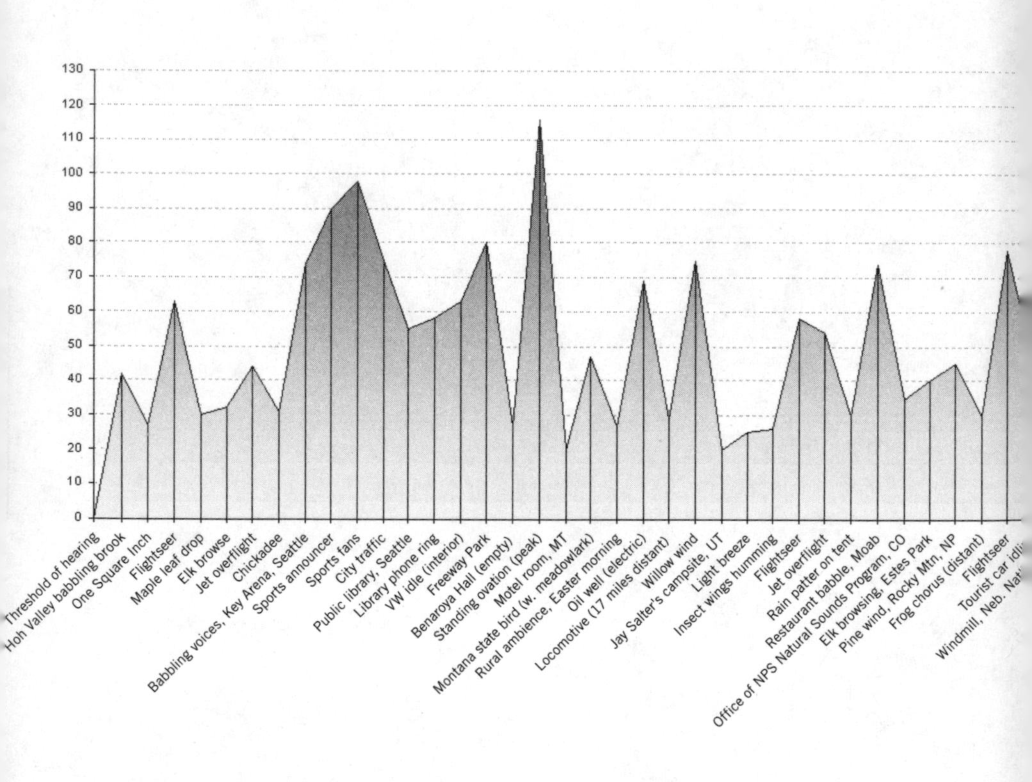

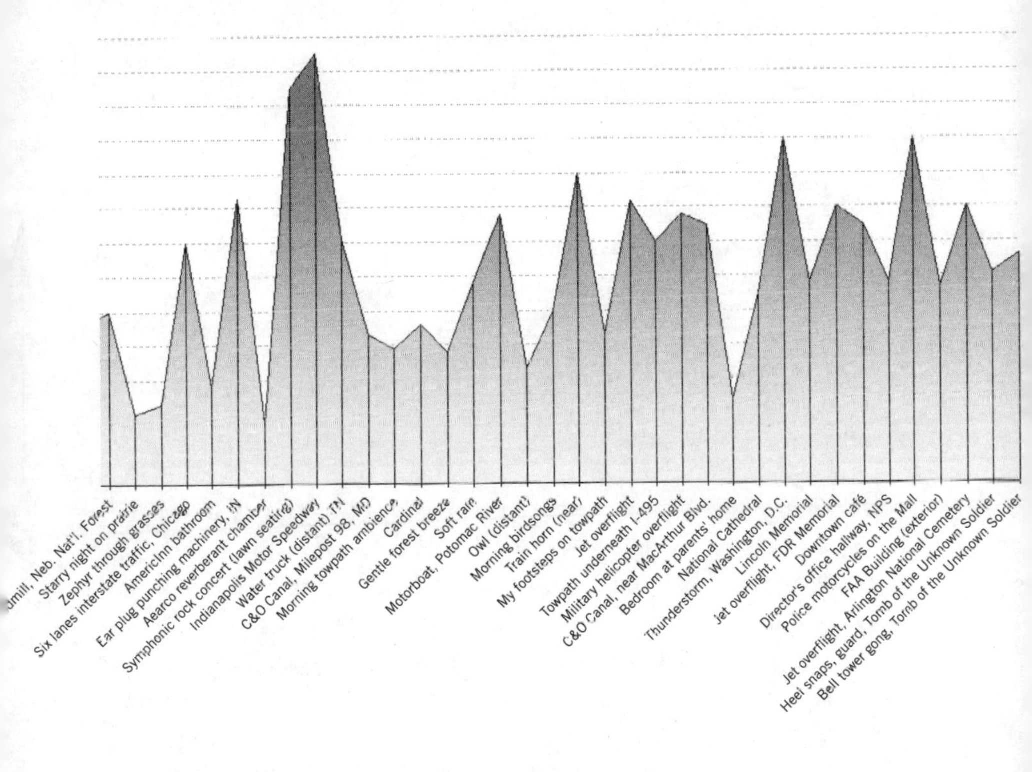
dmill, Neb. Nat'l. Forest
Starry night on prairie
Zephyr through grasses
Six lanes interstate traffic, Chicago
AmericInn bathroom
Ear plug punching machinery, IN
Aearco reverberant chamber
Symphonic rock concert (lawn seating)
Indianapolis Motor Speedway
Water truck (distant) TN
C&O Canal, Milepost 98, MD
Morning towpath ambience
Cardinal
Gentle forest breeze
Soft rain
Motorboat, Potomac River
Owl (distant)
Morning birdsongs
Train horn (near)
My footsteps on towpath
Jet overflight
Towpath underneath I-495
Military helicopter overflight
C&O Canal, near MacArthur Blvd.
Bedroom at parents' home
National Cathedral
Thunderstorm, Washington, D.C.
Lincoln Memorial
Jet overflight, FDR Memorial
Downtown café
Director's office hallway, NPS
Police motorcycles on the Mall
FAA Building (exterior)
Jet overflight, Arlington National Cemetery
Heel snaps, guard, Tomb of the Unknown Soldier
Bell tower gong, Tomb of the Unknown Soldier

Appendix F

Your Personal Quest for Quiet: A Mini-User's Manual

Top 5 Ways to Quiet Yourself in the Wilderness

Advice from author and grizzly tracker Doug Peacock, Livingston, Montana

I travel into the wilderness to experience wildlife, especially critters like grizzlies. I want to see these animals, not avoid them. When possible, I always travel into the wind and usually bushwhack off the trails. I sneak around making as little noise as possible. You have to depend on your ears and nose, and these rusty senses, often dulled by urban overload, are always better than you think. Here are five ways to quiet yourself in the wilderness:

1) Travel alone. Solitude is the deepest well I know, and the need for it can hit like a drowning man's gasp for air or simply tap you on the shoulder. You should take advantage of a solitude fix anytime you can. Going solo eliminates the need for talking. That's a good start.

2) Travel like an animal. Stop and listen every five minutes or so, more often in brushy country. Smell the air and lean your best ear into the wind for at least a minute. If you are bushwhacking off the trail in irregular terrain, glance at the next 15 feet of ground and memorize the sticks and stones. Then you can cover the distance quietly while your eyes sweep the tree line around you. Looking at your feet precipitates noisy imbalance, and lack of alertness could bring you nose to nose with a great big carnivore.

3) If you travel with others, especially off the trail in grizzly country, devise some simple hand and arm signals for alerting the people behind you (I usually walk point), indicating they should stop, take cover, move off the ridge, etc.

4) Once you enter the woods, speak in whispers. It's a good habit to develop—quietness.

5) If you're looking for animals in valleys and basins and the wind isn't blowing like hell, sit silently just below the ridges and listen. I do this in the early evening and at dusk, for hours sometimes, just as animals come out to feed. About half of all the grizzly bears I have seen I first heard moving through the timber as they emerged for evening feeding.

Adds Andrea Peacock:

Travel with respect for others' sense of solitude: Make sure your clothing blends in—silence is visual as well as auditory. Leave the technology at home: no cell phones, GPS gizmos, and for godsakes, no iPods.

Top 5 Ways to Quiet Your Neighborhood

Suggestions from Les Blomberg, founder and president of the Noise Pollution Clearinghouse in Montpelier, Vermont

Here are five ways you can reduce your acoustic footprint, assuming you don't drive an unmuffled Harley-Davidson (in which case, step one would be to put the muffler you took off back on) or drive a boom car (in which case, step one would be to turn down the bass).

1) Buy, use, and share electric lawn equipment. This is the best single way to quiet the suburban soundscape in the next 10 years. Electric lawn equipment tends to be 10 to 20 decibels quieter than gas-powered equipment. But if you want to experience a quiet neighborhood, it's your neighbors who need to use electric equipment, so share your equipment with them. For electric mowers, keep the blade sharp—you won't have extra horsepower to hack your lawn into submission; you'll actually have to cut the grass.

2) Buy quiet air conditioners and carefully choose where you put them. Listen to the urban soundscape on a summer evening, and you'll hear the drone of air conditioners (that is, if you're not listening to traffic). Check *Consumer Reports* before you buy a window air conditioner, and check the manufacturer's specifications for central air units—some are as quiet as 63 dBA at three feet from the unit. Also, don't put central air units under your bedroom window, or under your neighbor's, either.

3) Limit your noise to less than your neighbors'. For a neighborhood to get quieter, people have to start quieting it. If you make the same amount of noise as your neighbor, the neighborhood does not remain the same; it's actually 3 decibels louder. So, come fall, if your neighbor's clearing his lawn with a leaf blower, remind him of the quieter way. Reach for your rake.

4) Throw a party. Not a loud one, of course. And make sure to include all your neighbors. If you want a quiet neighborhood, you need a healthy community. People tend to noise-pollute anonymously, so don't be anonymous, and don't let your neighbors be anonymous, either. Get to know them, share tools, car-pool, invite them over. It is unlikely they'll wake you at 2 a.m. if they are going to be your guests at 6 p.m.

5) Don't book airline flights before 7 a.m. or after 10 p.m., and send overnight mail only in a true emergency. Nighttime flights wake millions each year—a nighttime takeoff or landing is the acoustical equivalent of driving through a neighborhood at night while laying on the horn—this is one of the anonymous ways we noise-pollute. If you're in a hurry, United States Postal Service Priority Mail is the quietest choice, since it usually flies on commercial aircraft during the day.

Top 5 Ways to Quiet Your Home or Office

Advice from Steven Orfield, president of Orfield Laboratories, Minneapolis, Minnesota, "The Quietest Place on Earth" according to Guinness World Records

Most of us spend the lion's share of our lives at home and work. Quieting those two places can dramatically improve our quality of life. Here are five ways to reduce the impact of noise in your home or your office.

1) Windows are the biggest sound leaks in most buildings. Double-pane or so-called sealed units can decrease incoming noise by 10 dBA or more. Installation of jamb liners can offer more attenuation by reducing the noise slipping through the window-to-jamb joints.

2) Many attics are very "live," meaning they amplify the sound that enters. And with little sound-absorbing material in the space, they, too, have many noise leaks, especially in one-story homes and near bedrooms. Install thick insulation in attic spaces and build simple noise traps below roof vents and side gable vents. These can be made from rigid fiberglass, folded and taped to make an L-shaped box, open at each end, that mounts on the interior of the roof vent.

3) Many rooms are also very "live." You can add carpet, rugs, wall hangings, and ceiling tiles (in basements and offices) to control the amount of

sound entering a room. Hallways tend to amplify sound. Here, too, carpeting and wall hangings will help. In addition, if you want to control amplified sounds from your TV, radio, and stereo, it's useful to know that bass sounds transmit through walls much more easily than high- and mid-frequency sound. Lowering the bass frequency level on your TV or radio (by adjusting the tone control or equalizer) will make the room quieter than merely lowering the volume.

4) Old appliances can also be very noisy, especially when compared with newer ones. Many new appliances now operate so quietly their output cannot even be measured in the manufacturer's acoustic lab. Many new dishwashers, washers and dryers, and air conditioners are half as loud as their predecessors. So if your appliances are interrupting your conversation, it may be time to select new ones. *Consumer Reports* is a good resource for appliance noise levels.

5) Don't look outside for major improvements in battling noise. Adding landscape berms, barriers, and large plantings to your yard to reduce highway or road noise actually accomplishes very little. Even the professionally installed highway barriers are not very effective unless you are very near to them (within a distance comparable to their height). They are installed mainly as a political solution to noise complaints. Look inside your home or office using the guidelines above.

Top 5 Ways to Protect Your Hearing

Guidance from acoustics and hearing expert Elliott Berger, senior scientist, Aearo Technologies, Indianapolis, Indiana

Hearing naturally declines due to the aging process, affecting mainly the ability to clearly hear high-pitched sounds such as birdsong, rustling leaves, critters scurrying through the brush, and children's voices. We can't stop that decline, but we can limit additional hearing loss from occupational and recreational exposures. Life can be loud. Be prepared:

1) Avoid or limit your exposure to hazardous sounds. Noise hazard depends on the level (sometimes called *intensity*) of the noise, and its duration, and how often the exposure occurs. A good rule of thumb: If you feel the need to shout in order to be heard three feet away, the noise levels are probably 85 dBA or more and hearing protection is recommended.

2) Listen to your ears. If, after the noise stops, you notice a ringing, buzzing, or whistling in your ears that wasn't there before, consider this a warning. Called *tinnitus,* this annoying internal noise is like a "sunburn" of the nerve cells of your inner ear, indicating that they have been irritated and overworked. Tinnitus is especially noticeable in a quiet place, such as when you are trying to go to sleep at night or listen to quiet natural sounds. If you don't protect your ears from noise, tinnitus can become a permanent, constant annoyance in your life.

Additionally, be aware of a second warning. An apparent muffling or softening of sounds after exposure to noise signals a temporary hearing loss, called a *threshold shift,* one that can worsen and become permanent with successive exposures.

3) Be prepared for noise. Just like you carry sunglasses to protect your eyes from bright sunlight, keep a pair of earplugs handy. (Foam earplugs are often your best choice because they combine high comfort and attenuation.) Should you forget them and suddenly get caught by a loud event, use your fingers. When pressed tightly into the ear canals they are the equivalent of a well-fitted pair of earplugs, though obviously not nearly as convenient to keep in place.

4) Learn how to properly insert earplugs. You should try different brands and types to find what is best for you. Be sure to carefully read the instructions and practice proper insertion. Two of the most common consumer complaints I receive about foam earplugs are that "they don't block enough sound" and "they don't stay in." Nine times out of ten the reason is incorrect fitting. The goal is a proper, very tight and crease-free roll down (thinner than a pencil), accompanied by pulling the ear upward and outward to open the ear canal and ease a full insertion. This takes practice. Without it you will still get protection, but the fit won't be as comfortable or secure, the noise attenuation not as great, and the boomy sound of your own voice (due to the occlusion effect) will be more annoying. You can learn more about fitting and using foam earplugs at: www.e-a-r.com/pdf/hearingcons/tipstools.pdf.

5) Be especially vigilant near explosive sounds, such as firing guns or fireworks. These pose significant dangers. One blast can cause tinnitus and hearing loss, and if you are unlucky, the loss can be permanent. That your ears recovered in the past is no guarantee that the next explosion won't be the one that inflicts a permanent acoustic nightmare. Therefore, *always* wear hearing protection when firing guns. Do likewise if you shoot off fireworks.

The Single Most Important Thing You Can Do to Save Silence

By Gordon Hempton

To truly appreciate the need to save natural silence, you must first experience it. Though now much harder to find, natural silence is no less powerful today than in John Muir's time. I invite you to make a silent pilgrimage to One Square Inch of Silence, but if you're unable to visit Olympic National Park, then I suggest you begin your own quest for quiet by seeking silence closer to home.

Begin by studying the image of Earth at Night at http://apod.nasa.gov/apod/ap001127.html. Light pollution is the evil cousin of noise pollution. Find an unilluminated place on earth, and you are more likely to find periods of natural quiet than simply studying a map. If you are choosing the United States, first check your dark spot against the FAA's "United States Parks and Reservations Special Use Airspace Airways" [appendix D of this book]. Once you find a promising location, check it against a local topographic map. Look for roads, power lines, gas lines, and other indicators of intrusive noise sources. You will likely spend more time planning your trip than actually taking it, so completely has the earth been touched by man. As a final predeparture step you can sound out someone who's been there.

Make an online post to one of various listening groups that have sprung up around the world. The largest and most active online group I know of is naturerecordists@yahoo.com, whose members painstakingly seek to avoid having noise intrusions ruin their valuable recordings. Other resources include The Nature Sounds Society, the Wildlife Sound Recording Society, and Cornell University's Laboratory of Ornithology. But be advised: Like anglers who guard their favorite fishing holes, many recordists prefer not to share their most pristine listening spots.

The obvious is still worth stating. Please be respectful of a quiet place by being exactly that—quiet. Many people have been critical of my effort to save silence by inviting people to One Square Inch and posting its very location on my website. There is no other choice. Either silence changes us or silence will go extinct. If one thousand people walk silently to OSI in a month, could we ask for any greater tribute? Surely the silence of the Hoh Valley would walk out with us. No telling what thoughts might occur, what resolutions might be made, and what actions might evolve to change the world.

Acknowledgments

There are two names on the cover of this book, but numerous people helped in many ways. Diane Bartoli, formerly of Artists Literary Group, gave birth to the book after reading coauthor John Grossmann's magazine article about One Square Inch of Silence. We are grateful to Diane for her vision of a much bigger stage for my quest for natural silence.

At Free Press/Simon & Schuster, we were equally fortunate to receive enthusiastic support and fine editorial guidance from our editor, Leslie Meredith, and diligent management, especially in crunch time, of incoming and outgoing e-mails and document attachments, by editorial assistant Donna Loffredo. We are thankful, too, for the vigilant eye of copyeditor Judith Hoover.

We wish to acknowledge the contributions of Elliott Berger, Les Blomberg, Doug Peacock, and Steven Orfield to the personal quiet manual in Appendix F—and to thank Elliott and Les for additional help in fact checking. Thanks, too, to Dick Hingson of the Sierra Club's national parks and monuments committee, for his dedication to the cause of preserving natural quiet and for keeping us constantly updated about legislative documents and meetings. And to Denis Galvin and Bryan Faehner of the National Parks Conservation Association for their guidance and insights into the ways of that other Washington, Washington, D. C.

I would like to thank my former wife, Julie, for her devotion and sacrifice during my early years as the Sound Tracker. At no small financial and emotional hardship, she allowed me to disappear for days on end to listen to the world and develop my thinking and my craft. Otherwise, there would be no One Square Inch.

I would like to thank my two children, Abby and Oogie, for simply being themselves and reminding me of the saying, "Life is something that happens while you're busy making plans." May you both find quiet in your lives.

I would especially like to thank my friends Nick Parry, Mathew Lee Johnston, Peter Comley, Kelley Guiney, and Jay Salter for cheering me onward through my long absences from home and, when back, long absences from them during periods of writing. Your encouragement really counted! John would like to single out his brother Bob Grossmann and friend Ed Adler for their interest, encouragement, and suggestions.

Finally, I would like to thank everyone I met on my journey to Washington, D.C. That simple cup of coffee you served me with a smile, the directions you shared to what you thought was a quieter place, the just-when-I-needed-it camaraderie from the buoyant brotherhood of fellow VW owners—all of that shared warmth made a difference and helped me reach my destination. My journey instilled in me a further appreciation of what it means to be an American.

Index